How the Ocean Works

How the Ocean Works

An Introduction to Oceanography

Mark Denny

Princeton University Press   Princeton and Oxford

Published by Princeton University Press, 41 William Street, Princeton, New Jersey 08540
In the United Kingdom: Princeton University Press, 6 Oxford Street, Woodstock, Oxfordshire
OX20 1TW

Library of Congress Cataloging-in-Publication Data

Denny, Mark, 1951–
How the ocean works : an introduction to oceanography / Mark Denny.
p. cm.
Includes index.
ISBN 978-0-691-12646-3 (hardcover : alk. paper)—ISBN 978-0-691-12647-0
(pbk. : alk. paper) 1. Oceanography—Popular works. I. Title.
GC21.D398 2008
551.46—dc22 2007034543

British Library Cataloging-in-Publication Data is available

This book has been composed in Sabon

Printed on acid-free paper. ∞

press.princeton.edu

Printed in Singapore by Imago

1 3 5 7 9 10 8 6 4 2

For Susan Rose

Contents

Illustrations

Tables

Preface

I have the privilege of an office that looks out on the Pacific Ocean. From my window I can watch fishing boats at work and sea otters asleep in the kelp. Gray whales, humpbacks, and once even a blue whale, have wandered by. The sounds and smells that waft in through my door are a constant reminder of the sea, a habitat that forms 71% of our planet's surface area and 99% of its livable volume. With its magnificent presence at my doorstep, writing an introductory oceanography text might seem a natural idea, but it took a trip to landlocked Falls Church, Virginia, to kick me into gear.

This book was conceived on the back porch of my friends' house one summer night in 2004. I was in town for a meeting, and John, Michelle, and I were reviewing the events of the day when the discussion turned to oceanography. A news report had appeared a few days before suggesting that melting ice in the Arctic—a result of global warming—might lead to a drastic reduction of temperatures in northern Europe. John wanted to know, "How is that possible? How can global warming produce colder winters in England?"

Good question. And even better, it was a question to which I knew the answer. Not one to turn down a teaching opportunity, I launched into a description of thermohaline circulation, a physical process in part responsible for controlling Europe's temperature. Thermohaline circulation might be drastically curtailed by melting sea ice. But before I could explain that, I had to back up and explain how seawater freezes in the first place, a strange story in which all the salt is squeezed out and the ice is fresh. Halfway into that explication I had to digress to explain why seawater is salty and where the salt comes from. The questions and answers entwined, the scope of the discussion expanded to include marine biology as well as marine physics, and we had a grand time learning how the ocean works. We didn't finish, of course, and as we were adjourning to our beds, John asked the fateful question: "Where can I read up on this stuff?"

I had no answer. The obvious place to start would be an introductory marine biology or oceanography text, but I was hesitant to suggest that solution. First, it was clear from the evening's conversation that any understanding of the ocean as a whole requires information about both marine biology and oceanography, but the two fields are seldom effectively combined in textbooks. Introductory texts on marine biology tend to gloss over the physics and chemistry essential to oceanography, and introductory oceanography texts tend to downplay the biology.

Second, the current crop of introductory marine biology and oceanography texts adheres to a disturbing publishing trend in which authors compete to see who can pack the largest number of glitzy photos, color-coded diagrams, and new vocabulary words into the least number of pages. Reading one of these texts is analogous to watching a cable news show. While you are trying to pay attention to the sound bite offered up by the commentator, your attention is diverted by the box scores crawling across the bottom of the screen and the ads for the next great reality show blinking in the upper right corner. As fact-filled and colorful as these texts are, I could not in good conscience recommend them as an effective means to begin learning about the sea.

Instead, what John needed was a book that presented a coherent story about how the ocean works. It need not be a complete story—to hew to the story line, some facts would have to be left out—but it would need to provide a solid basis on which someone could build. It would need to teach one how to think about the ocean, leaving the details for dessert.

Over the next few days, I searched for a book that met those requirements and came up empty. In growing frustration as I wandered through libraries and bookstores, I kept remembering how much fun it had been sitting on that porch, telling the ocean's story. Eventually, I decided that if I couldn't find the right book, I'd just have to write something myself.

A Road Map

How the Ocean Works is an ambitious (even audacious) title. It has been a challenge to live up to it, and there is no guarantee that I have succeeded. But in the attempt, I have taken an approach suggested to me by Dick Barber of Duke University, using as a central theme the fact that the thermocline separates the ocean into two parts: a well-lit, warm, and nutrient-poor surface layer and a dark, cold, and nutrient-rich deep layer. This theme debuts in chapter 6, and the biological and physical consequences of the two-layered ocean are developed in chapters 7 to 9. Appreciating the nuances of the two-layered ocean requires some background, however. Chapters 2 to 5 provide the reader with necessary facts concerning the shape and size of ocean basins, the physical properties of seawater, and the relevant characteristics of photosynthesis and consumption, marine plants and marine animals.

This core of the text (chapters 2 through 9) is placed in perspective by three surrounding chapters. Chapter 1 provides a brief history of humankind's interaction with earth's oceans. Chapters 10 and 11 put our understanding of the ocean to work, first to explain the anomalous "high-nutrient, low-chlorophyll" areas of the sea and their potential role in global climate change (chapter 10), and lastly to examine the evolving problem of extractive fisheries (chapter 11).

In telling the ocean's story, I have been selective in the facts and concepts presented. In some cases, I have left out vast areas of current knowledge often covered in introductory texts. For example, I deal primarily with pelagic organisms—the plants and animals that swim or float in the water column—largely ignoring benthic organisms such as corals, sea grasses, and coastal seaweeds. In other cases, I include more detail than is traditional for introductory texts. For instance, a whole chapter is devoted to explaining the Coriolis effect and geostrophic flow. These nonintuitive concepts are integral to understanding the global patterns of winds and currents, and therefore to understanding how the ocean works. But the abbreviated explanations given in most introductory texts are, at best, incomplete and likely to confuse anyone who tries to apply them. I felt compelled to provide a more thorough discussion. Other details that are too interesting to leave out, but too peripheral to include in the flow of the story—such as how long it would take to melt the Greenland ice sheet, how to measure great circle distances, and the physics behind convection—are relegated to footnotes and appendices.

After each chapter, I provide a list of articles and books to which readers may turn for further instruction. Again, I have been selective. These sources are the articles and books I have found most interesting and informative, but they are only the next step toward a full understanding of the ocean. The bibliographies contained in these sources provide access to the primary literature.

Acknowledgments

I owe a great debt to Dick Barber. Not only did he suggest the central theme of this text, but he also provided a basic lesson in telling the story. When I was an undergraduate at Duke University, I took a course in biological oceanography from Dick. As an exercise, he had the class read three papers, each attempting to model temporal fluctuations in ocean primary productivity. One model had three variables and explained about 80% of the variation seen in nature. The second paper had seven or eight variables and explained 90% of the pattern of fluctuation. The third paper had fifteen variables and explained a whopping 95% of what happens in the ocean. Dick then asked us which paper represented the best science. We all picked paper number 3, of course. How could you argue with a model that accounted for 95% of reality? But Dick was incensed. Clearly the three-variable model was the most valuable! Those three variables captured the essence of what was happening in the real world, and the twelve extra variables of our favorite model were just window dressing. I have tried to apply that lesson to the selection of topics in this text. If I have succeeded in streamlining the ocean's story to its essentials, it is largely to Dick's credit. If I have failed, the fault is mine for not having absorbed his message completely.

A host of other people contributed to the development of this story by providing pertinent facts and much-needed feedback: Michael Boller, Caren Braby, Brandon Cortez, Rob Dunbar, Pat Harr, Nick Holland, Luke Hunt, George Leonard, Patrick Martone, Luke Miller, Michael O'Donnell, Danna Shulman, George Somero, Freya Sommer, Dale Stokes, Emma Timmins-Schiffman, and Sam Umry. Early drafts of the text were imposed on students enrolled in "Principles of Oceanic Biology" at Stanford's Hopkins Marine Station, and I thank them for their forbearance and constructive comments. Katie Mach read the entire manuscript several times, correcting my grammar and pinpointing flaws.

It has been a pleasure to work with Princeton University Press on this project. Robert Kirk saw the value of a different approach to introducing oceanography, and he put the Press's professionals at my disposal: Terri O'Prey, who deftly shepherded the manuscript through production; Dimitri Karetnikov, who, with the help of Erin Suydam, Shane Kelly, Sandy Rivkin, and Anne Karetnikov, orchestrated the construction of figures; and Jodi Beder, who polished the prose and vetted the logic. I thank them all.

And, as always, I thank my family—Sue, Katie, and Jim—for their unwavering support of a husband and father whose mind was all too often off at sea.

How the Ocean Works

CHAPTER 1

Discovering the Oceans

Throughout history, men and women have been drawn to the sea. In ancient times, people felt the same desire you and I feel today, the urge to stand and gaze upon the ocean. Then, as now, the ocean meant different things to different people. For the hungry, the sea was a source of abundant food: fish and lobsters, seals and seaweeds, whales and shrimp. For the audacious, the ocean was a route to faraway lands, an avenue of commerce and conquest. For the bereaved, the ocean epitomized grief: the oblivion into which their loved ones sailed, never to return, or the source of waves that consumed their villages. For the lucky, the sea was a source of pleasure and recreation.

These historical perspectives persist, but if you go down to the shore today and stare out to sea, your perspective is likely tempered by a modern point of view. With our recently acquired ability to travel into space, we now see the earth and its oceans as a whole (figure 1.1). Using satellite cameras we can locate storms and guide ships out of harm's way. We can count the phytoplankton that fuel the fisheries and direct fishermen to fertile grounds. Our sensors detect climatic anomalies as they begin, allowing us to predict their consequences. These are exciting times for humankind's interactions with the sea.

But this new perspective has two dangers. First, our capacity to see the entire earth can foster a sense of arrogance. It is easy to forget that the ability to observe is only the first step toward the ability to understand and control. As we learn how the ocean works, we must be careful to note the limits of our knowledge. Second, the grand view from space has the tendency to diminish our sense of personal contact. Awe-inspiring as it is, figure 1.1 cannot convey the infinite expanse of the ocean at night when viewed from the deck of a small ship, the tang of salt on an ocean breeze, the crash of a breaking wave, or the shiver of water against your skin. To arrive eventually at a full appreciation of earth's seas, we first need to anchor our perspective in human experience. To that end, we retrace the steps of our ancestors and explore the arduous path by which human society discovered the oceans. In this chapter, we have time only for an outline of the full journey—a mad dash through history. So, brace yourself as we begin with legends.

Figure 1.1. Earth from space. A photograph of the western hemisphere from the Visible Earth project at NASA (visibleearth.nasa.gov/view_detail.php?id=2429).

Ancient Myths

In many cultures, myths of the sea tell mostly of destruction. From the islands of the Pacific to the coasts of Central America, India, and the Middle East, when the ancient gods were displeased with men or women, they often chose the ocean rather than fire or wind as the instrument of their wrath. In the Bible, for instance, forty days and forty nights of rain caused the sea to rise and wipe the earth clean, sparing only Noah and those on his ark. Similar legends of earth-cleansing floods are common among civilizations in the Middle East. The Babylonians, for example, had a flood myth similar to that of the Bible, with a

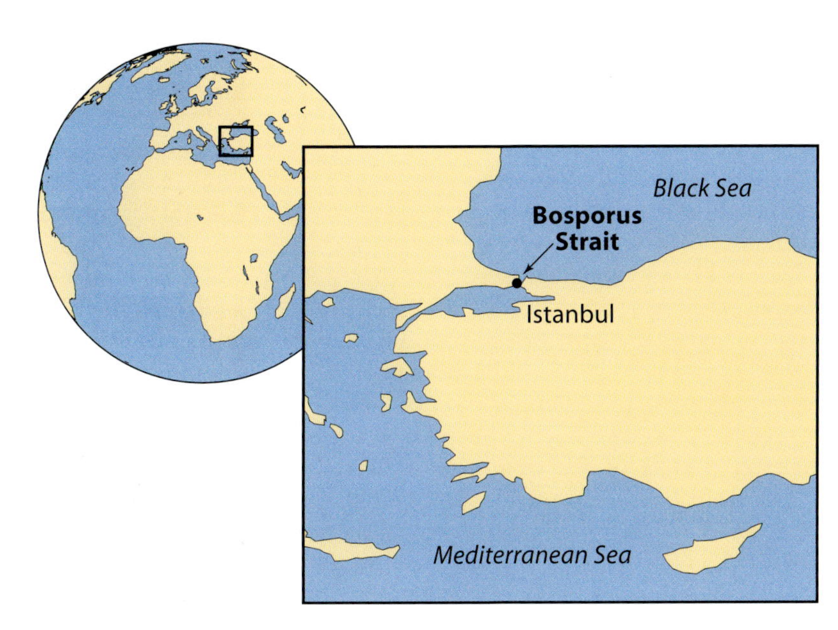

Figure 1.2. The Bosporus Strait, in what is now Turkey, separates the Black Sea from the Mediterranean Sea. Before the strait opened, the Black Sea was a freshwater lake.

man named Utnapishtim playing the role of Noah. Likewise, the Sumerians had King Ziusudra.

Given this ubiquitous mythical reference to floods, some historians have suggested that there are historical bases for these tales. For example, legends of a great flood among coastal Indians of the Pacific Northwest may refer to tsunamis (tidal waves) caused by earthquakes, and the flood myths of Pacific islanders may describe waves resulting from volcanic eruptions.

A potential, although controversial, source of the Middle Eastern flood legends concerns the Black Sea. Today, the Black Sea is connected to the Mediterranean Sea by the Bosporus Strait, a narrow passage adjacent to Istanbul (figure 1.2). However, about ten thousand years ago at the end of the last Ice Age, a sill (essentially an earthen dam) closed the Bosporus, and the Black Sea was a freshwater lake. As the glaciers receded and earth's climate warmed, the level of the Black Sea dropped due to evaporation, while the level of the Mediterranean rose as the world's oceans absorbed the water from melting ice. Eventually, the Mediterranean broke through the sill separating it from the Black Sea, and the consequent flooding would have been catastrophic to the villages along the Black Sea's coast—an event worthy of a legend.

Regardless of their precise origin, the flood myths convey the mystery and fear that tinged ancient encounters with the sea.

Commerce and Expansion

For all its destructive potential, the ocean has always tempted humans to risk its dangers in search of food and rapid transport. Boats built of wood and reeds traveled the waters of the Nile River in Egypt as long ago as 4000 BC, and many ancient civilizations of the Middle East used boats for fishing and coastal commerce. In particular, the Phoenicians were adventuresome merchants and adept sailors. In the first and second millennium BC, they developed a complex web of trade routes around the Mediterranean, and sailed as far north as the British Isles, trading for tin to use in making bronze. Similarly, in the Far East, coastal commerce using sampans and junks flourished in China. In the Arctic, the Inuit developed sea-going kayaks and umiaks, capable vessels from which they could hunt walrus and whales.[1] The indigenous people of Chile, Peru, and Ecuador used small reed boats and large rafts for fishing and commerce.

In fact, some anthropologists suggest that the expansion of humans into North and South America occurred most rapidly not by land, but rather by sea. Experts agree that humans spread from Asia into North America after the last Ice Age, and have long assumed that humans spread throughout North America before subsequently expanding to Central and South America. It has recently been proposed, however, that the leading edge of this expansion was not through the middle of the continent, but rather along its western shore as groups used small boats to travel south. For example, a village at Monte Verde on the coast of Chile probably dates back to at least 15,000 BC, a time when humans were only beginning to populate central North America. Clearly, from very early times, our fear of the ocean has competed with—and often lost to—our urge to go exploring in boats.

Perhaps the best example of this wanderlust is the spread of humanity to islands in the Pacific. Starting in the Philippines about five thousand years ago, modern humans rapidly expanded their range in the open expanse of the Pacific Ocean. By 1600 BC they had reached New Guinea and the nearby Bismarck Archipelago. Forced onward by population growth, Polynesians next sailed to Samoa around 1200 BC, and by 500 AD, they had traveled all the way to Hawai'i and Easter Island (figure 1.3).

It is clear that this expansion was not the accidental result of a few fisher folk being blown offshore and ending up on other islands. Hawai'i, for example, was settled by sailors traveling from the Marquesas across 3700 kilometers (2300 miles) of open ocean. That monumental leap required ocean-going outrigger canoes large enough to carry not only people but also the plants and animals of their culture. And finding Hawai'i required exceptional working knowledge of navigation and the sea. Each step in the Polynesian expansion involved skill, planning, and a group effort.

Even for the Polynesians, however, there were limitations. They sailed from one island to another within a circumscribed area of the Pacific, but this expansion was constrained by the availability of uninhabited islands. As well planned and skillfully executed as they were, Polynesian expeditions were incapable of invading previously occupied mainland territory. For instance, if prehistoric Polynesians reached South America—the next step east from Easter Island—their numbers were far too small to gain a foothold among the indigenous people who had arrived millennia before. Thus, despite their sailing prowess, the Polynesians could not move beyond the Pacific, and as a result, they had no knowledge of the entirety of the world ocean.

[1] Kayaks are small one- or two-person boats constructed from skins stretched over a wooden frame. The skins form a deck, so that when the occupants are settled into their cockpits, the boat is watertight both top and bottom. Kayaks are propelled by two-bladed paddles. In contrast, umiaks are larger skin-covered boats with no upper deck. They are capable of holding twenty people or more and much gear, and are propelled by oars, paddles, or sails.

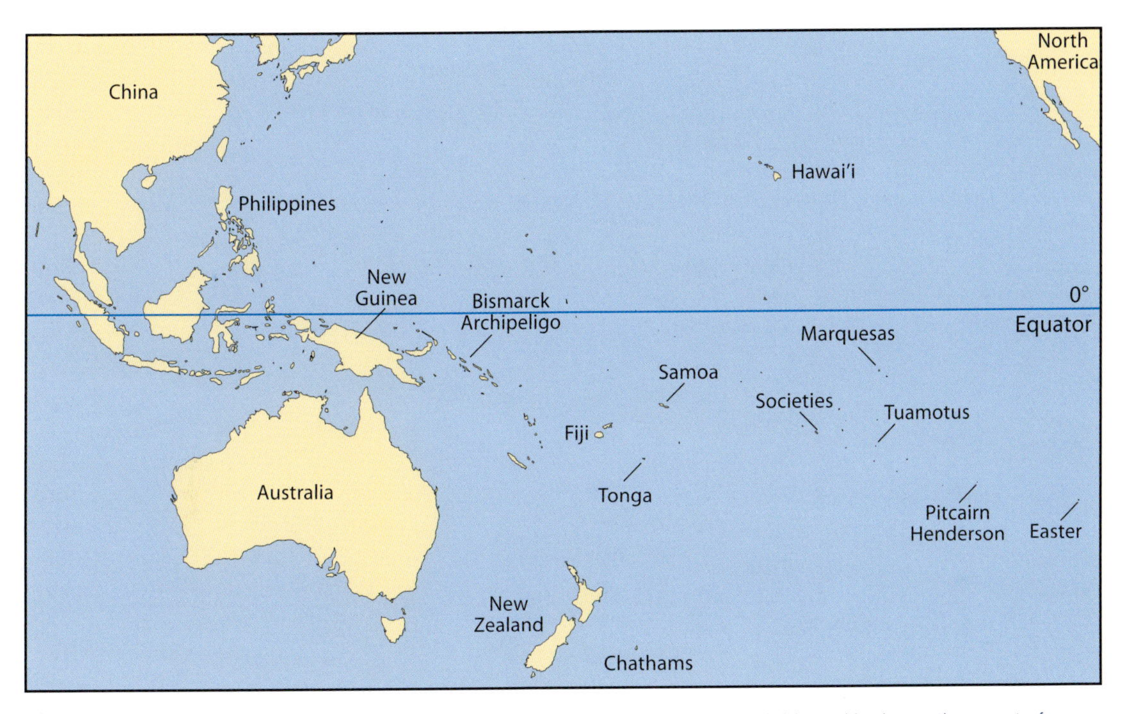

Figure 1.3. Islands of the Pacific, the realm into which the Polynesians expanded. Monte Verde, on the coast of Chile—site of a village dating back nearly 15,000 years—is 3600 kilometers southeast of Easter Island.

The course of Polynesian exploration in the Pacific was mirrored by that of the Vikings in the northern Atlantic. In the first few centuries AD, the climate in Scandinavia was unusually warm, allowing crops to flourish and populations to grow. As the countryside filled up, Norwegians and Danes looked for opportunities elsewhere. Utilizing their superior shipbuilding technology, the Vikings sent raiding parties eastward along the rivers of Russia and southward along the coasts of Europe, at times ranging as far as Constantinople (now Istanbul). These raids significantly impacted European history. For example, William the Conqueror defeated the English at the Battle of Hastings in 1066—the sole successful invasion of the British Isles—in large part because the English troops were exhausted from doing battle a month before with Viking invaders.[2]

In addition to raiding established societies, the Vikings undertook an island-hopping expansion westward into uninhabited territory (figure 1.4), first to the Orkney, Shetland, and Faeroe Islands (around 800 AD), then to Iceland (by 874 AD) and southern Greenland (in about 980 AD). Ships plying the trade routes between Iceland and Greenland were occasionally blown off course, and one of these (captained by Bjarni Herjolfsson in 986 AD) sighted a forested coast to the west of Greenland. Spurred on by the lure of abundant timber, which had become scarce in Greenland, Leif Ericsson bought Herjolfsson's boat and formed an expedition to explore this new land. Ericsson and his crews, accompanied by women and livestock, traveled along the coast of Baffin Island and Labrador and built a small outpost at L'Anse aux Meadows in Newfoundland. However, they skirmished with the local Indians, and in light of the prospect of

[2] William himself was the descendant of Vikings who settled in Normandy.

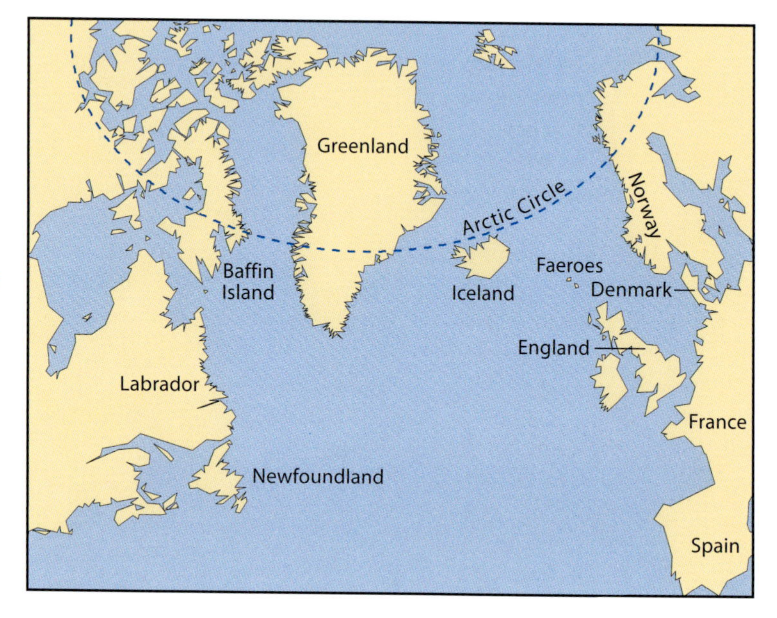

Figure 1.4. The realm of the Vikings.

continued conflict, the Vikings abandoned their attempt to colonize the New World.

Like the Polynesians, the Vikings explored the ocean not out of a thirst for knowledge, but rather in response to the demands of population growth and commerce. And, like the Polynesians, they were stymied in their expansion by the overwhelming size of indigenous populations. Viking travels were thereby limited, and they never developed an understanding of the scope of the world ocean.

In summary, as a result of its expanding population, by 1000 AD, humankind had dipped its toes into each of earth's seas. Many cultures used coastal waters for fishing and commerce, and the Polynesians and Vikings traveled a few open-ocean routes. But our knowledge of the sea was piecemeal: we hadn't truly discovered the ocean. That achievement resulted from a different journey, one that began with the Greeks.

Science and the Greeks

Growth of Greek civilization marked the beginning of abstract scientific thought. Whereas the Chinese, Egyptians, and Babylonians were masters of technology and astronomy, the Greeks were masters of concept. This distinction is evident in the way the different civilizations viewed the world. Early Chinese maps, for instance, give detailed information regarding the disposition of armies and the location of rivers and cities in the Middle Kingdom, but they show nothing beyond China's boundaries. In contrast, by 2500 years ago, the Greeks were actively speculating on the shape of the entire earth.

In fact, their speculations were highly logical and specific. For example, Aristotle (384–322 BC), the famous Greek scientist and philosopher, deduced that earth was a sphere, an assumption he based on four observations. First, there was the appearance of ships as they sailed into the distance. If the earth were flat, a ship sailing toward the horizon would appear smaller and smaller until it vanished from view. Instead, ships descended into the horizon, their hulls disappearing first, then their masts, suggesting that the surface of the water (and thus, of the earth) was curved.

The same logic applies from the perspective of the ship: as it travels away from shore, the coastline descends into the horizon. This perspective led Aristotle to note that stars appear and disappear on the horizon as one travels north or south, further evidence of earth's spherical shape.

Lunar eclipses provided a third clue. It was apparent from simple observations that the moon is a sphere. For instance, as the moon goes through its phases, the line separating dark from light changes shape as it travels across the moon's face. It is curved when the moon is a crescent, and straight when the moon is half full: what one would expect for a sphere lit from different angles. Shadows similarly reveal the shape of the earth. During a lunar eclipse, the shadow of the earth falling on the moon is always curved, as it should be if earth is a sphere.

And finally, there was the matter of aesthetics. To Greek mathematicians, the sphere was the perfect shape, the only object uniform about its center. The sun was clearly spherical, as was the moon. Why should the earth be any less perfect?

Acceptance of the earth as a sphere immediately led Greek scientists to two important questions: how big is the sphere, and where on its surface is Greece? In an extraordinary example of cultural genius, they devised answers to both.

In the fourth century BC, Alexander the Great (a pupil of Aristotle's) conquered much of the known world, sending many of the treasures he obtained to his museum in Alexandria, near the mouth of the Nile. After Alexander's death, the museum was supported by the royal rulers of Egypt and continued to acquire and catalog the fruits of civilization. To that end, it established a "think tank" (Euclid and Archimedes worked there, among others) and a vast library. The second Librarian of Alexandria, a man named Eratosthenes (ca. 276–196 BC), took it upon himself not only to accumulate the written knowledge of the world (as any good librarian would), but also to synthesize that knowledge. One of the things he contemplated was the size of the earth.

It came to Eratosthenes' attention that each year a notable event occurred in the city of Syene, due south of Alexandria. At noon on the summer solstice—the day of the year when the sun is highest in the sky—sunlight shone directly down into the wells at Syene, indicating that the sun was precisely overhead. Eratosthenes found this curious. In Alexandria there were vertical stone pillars called gnomons. If the sun were also directly overhead in Alexandria, the gnomons would not cast a shadow, but at noon on the summer solstice, shadows persisted. Clearly the angle at which sunlight approached the ground differed between Alexandria and Syene.

Enter the notion of a spherical earth (figure 1.5). Eratosthenes deduced that the difference in the angle of sunlight resulted from the different locations of Alexandria and Syene. At noon on the summer solstice, Syene (near the present-day city of Aswan) was at a point on earth directly under the sun, whereas Alexandria, north of Syene, was at a point on earth's curve where a vertical gnomon pointed off at an angle. Eratosthenes' genius was to use this realization to measure the size of the earth.

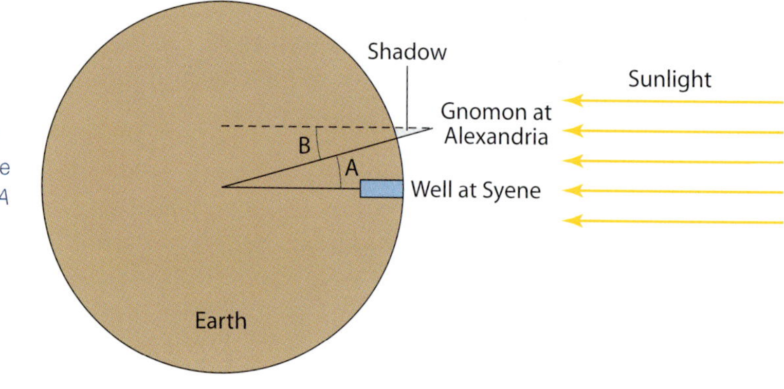

Figure 1.5. Eratosthenes' scheme for calculating the size of the earth. Angles *A* and *B* are equal.

To do so, he needed two measurements. First, he had to measure the angle between Alexandria and Syene (angle *A* in figure 1.5). This was easily done. Noting that angle *A* equals angle *B*, Eratosthenes realized he could use the shadow cast by a gnomon to measure the angle between Alexandria and Syene. At noon on the solstice, he wandered over to the nearest gnomon, measured its height and the length of its shadow, and used geometry to calculate first angle *B*, and then angle *A*. It turned out to be 7.12°, almost exactly 1/50 of an entire circle. This angle in turn meant that the distance from Syene to Alexandria was 1/50 the entire distance around the earth. Thus, if Eratosthenes could measure the distance from Syene to Alexandria, he could measure the earth's circumference.

Surprisingly, Eratosthenes didn't even need to leave home to measure the distance from Syene to Alexandria. As Librarian, he was privy to all sorts of information, including the fact that caravans of camels typically required 50 days to travel from Syene to Alexandria. It was common knowledge in those times that camels can travel 100 stadia per day, so he calculated that the distance from Syene to Alexandria is approximately $50 \times 100 = 5000$ stadia.

Now, a *stadium* is a measure of length used by the Greeks—a common foot race was one stadium in length, which set the size of a standard running track and eventually led to the current use of the term "stadium." Current archeological estimates suggest that one stadium is roughly 0.185 kilometers. Thus, in present-day units, Eratosthenes calculated that the distance from Syene to Alexandria was 925 kilometers. Multiplying by 50, he then estimated that the earth's circumference was approximately 46,250 kilometers.

This calculation is extraordinary. The current estimate of earth's circumference is 40,008 kilometers, only about 13% smaller than the distance Eratosthenes calculated 2300 years ago. One Greek, scratching his head in an ancient library, collecting a few facts from camel drivers and gnomons and making a simple calculation, measured the earth with astounding precision.

The Greeks' understanding of the earth did not stop there. In Eratosthenes' time, Greeks used a primitive version of latitude and longitude to locate points on earth's surface. In the beginning, the basis for this coordinate system was a line drawn east to west through the island of Rhodes (just south of Greece), essentially a line placed across the middle of the Greek empire. Parallel lines could then be drawn to the north and south, each expressing a line of more-or-less constant climate. These parallels—called *climata* or *klimata*—served as the basis for present-day lines of latitude.

Originally, the climata were not equally spaced; as indices of climate, unequally spaced lines sufficed. But as civilization became more complex and more cities needed to be located on maps, the informal system of climata became unworkable, and a new, more uniform system gradually evolved. Hipparchus (ca. 190–120 BC) placed the climata at uniform distances and added lines running north-south to form a grid. To standardize the angular distances between points on earth, he borrowed a tradition from Babylonian astronomers and divided the circle into 360 degrees. Thus, Hipparchus invented the modern coordinate system of geography, complete with units, and of equal importance, his system was popularized by the Greek author, Claudius Ptolemy (ca. 90–168 AD).

Ptolemy excelled not at creating new knowledge, but rather at sorting and chronicling existing knowledge, which he did superbly. In two books, the *Almagest* ("The Greatest") and *Geography*, he recorded his best estimate of all knowledge of mathematics, science, and geography, and introduced a few innovations, such as dividing each degree of latitude or longitude into 60 minutes, and each minute, in turn, into 60 seconds. As their titles suggest, Ptolemy's books were a tour de force, and they became the standard references of his day.

Although Ptolemy successfully recorded much of the Greeks' knowledge, he was not infallible, and one of his mistakes had historical consequences. For reasons that are unclear, Ptolemy did not accept Eratosthenes' estimate for the size of the earth and instead chose a value calculated by another Greek scientist, Poseidonius. Unfortunately, Poseidonius had proposed that the circumference of the earth was 30,000 kilometers, 25% smaller than the actual value. By accepting Poseidonius's figure, Ptolemy assumed the earth was considerably smaller than it actually is.

Compounding this mistake, Ptolemy also thought that Asia extended to the east substantially farther than it really does. With an undersized earth and an oversized Asia, Ptolemy's globe had no room for features such as North and South America or a large Pacific Ocean. In fact, by Ptolemy's estimate, Asia extended so far east that it came close to wrapping around the earth and touching Europe. The mistaken idea that the east coast of Asia was but a short distance west of Europe eventually led to Christopher Columbus's voyage to the New World, but that lucky miscalculation was far in the future. In the interim, the Dark Ages descended on Europe.

The Dark Ages

The notion of a spherical earth survived the Roman absorption of Greek culture, but it was nearly lost to Western civilization with the fall of the Roman Empire. In 391, for instance, Christian mobs overran the library at Alexandria and burned its invaluable contents. Copies of a few books, Ptolemy's *Almagest* and *Geography* among them, had been translated into Arabic and disseminated, and thereby survived, but it would be nearly a thousand years before the Greeks' insights resurfaced in the West.

During the Middle Ages, knowledge of the earth in Western civilization was governed more by Christian doctrine than by scientific inquiry. Maps were drawn not as an expression of reality, but rather to match the dictates of the Bible. For example, the earth was deemed flat, with Jerusalem at its center. The Mediterranean

Sea derived its name from its assumed location: the "middle of the earth." Surrounding the known lands of the earth was a mysterious ocean, unknown and—to a medieval culture closed in on itself—largely unknowable. Thus, for nearly 700 years after the sacking of Alexandria, knowledge of the ocean stagnated.

The Mongol Empire and the Role of Commerce

Although Christianity was largely responsible for the suppression of science in the Middle Ages, it also facilitated events that eventually brought the science of geography back to life. Beginning in 1096 AD with the First Crusade, Christian European society periodically waged war on the Muslim Middle East. This and subsequent crusades primarily sought recapture of the Holy Land for Christianity, but ancillary consequences abounded. For example, in traveling to and from the Holy Land and in governing the areas they conquered, the crusaders came into contact with Middle Eastern culture. As so often happens, the culture of the conquerors had less impact on the culture of the conquered than vice versa. Christian soldiers acquired a taste for intriguing riches available in the region—silk cloth, perfumes, and exotic spices—and soon active trade passed between the two cultures. Silks and spices from Asia arrived overland in the Middle East, where local traders acted as middlemen, selling to the highest bidders from the West.

This commerce had two major effects. First, the desire for silks, perfumes, and spices in Western culture gradually spiraled into an addiction. In an ancient analogy to the current commerce in oil, aspects of European society became dependent on commercial ties to the Middle East. Second, commerce opened Western eyes to the existence of Asian cultures. Although Europeans bought goods from the Arabs, they realized those goods came from powerful and intriguing civilizations farther to the east.

Events in China soon strengthened the interaction between Europe and Asia. In 1214 AD, Genghis Khan and his army captured Beijing, solidifying their hold on the Mongol Empire. For the next 154 years, the Mongols ruled all of the Far East.

The Mongol Empire impacted European history largely through commerce. In their administration of the Empire, the Mongols maintained the overland trade routes between China and the Middle East. Trade flourished, and the ready flow of goods heightened Europe's dependence on Asian goods.

The existence of safe routes of trade also made it possible for Europeans to travel to the Far East. Perhaps the most famous and influential of these travelers was Marco Polo (1254–1324 AD), a merchant from Venice. In the company of his father and brother, Polo traveled the Silk Road across the high Hindu Kush to China and spent 24 years at the right hand of Kublai Khan, Genghis's son. He acted as a military adviser to the Mongols during the siege of Saianfu, served as a court diplomat in Yunnan, and returned from the East as the escort for a princess betrothed to a Persian king. In his version of these events, Polo sailed home in a fleet of fourteen ships, accompanied by 600 courtiers and sailors, and was paid in jewels. He was subsequently captured in a sea battle between Venice and Genoa, and spent several months in prison, where a cellmate named Rustichello recorded his stories. Published in 1299, the *Travels of Marco Polo* was an instant success, and the book sparked Europe's interest in the Orient.

No empire lasts, however, and the burgeoning interaction between Europe and the Mongol Empire collapsed in 1368 AD when peasants in China overthrew the Mongols and established the Ming dynasty. In the wake of this revolution, overland trade routes with the West deteriorated, and the flow of silks, perfumes, spices, and cultural information declined. Flow ceased altogether when the Turks captured Constantinople in 1453.

European Exploration

Our story now splits in two as China and Europe follow separate paths. We begin with Europe.

The collapse of overland trade between China and Europe occurred at an auspicious time. The late 1300s saw the flowering of the Renaissance; Europe opened its eyes to scientific inquiry, and bits and pieces of Greek knowledge resurfaced, among them the works of Ptolemy, now translated into Latin from the Arabic in which they had rested for nearly a thousand years. With Ptolemy's vision of the earth at hand, European sailors were emboldened to extend their reach, searching for a sea route to the Far East to replace the defunct paths over land.

Two routes beckoned. First, because of Ptolemy's underestimation of earth's size, sailing west from Europe to reach Asia seemed possible—possible, but not easy, because this sort of transoceanic voyage could not be done in increments. Sailing halfway across the ocean and then turning back would be pointless; unless one sailed all the way across, there was no monetary profit in the venture and therefore little reason to go.

In contrast, the second route could be accomplished piece by piece. For centuries, Europeans had been trading with the coastal inhabitants of northern Africa. If these maritime trade routes could be extended south along Africa's west coast, perhaps someone would eventually find a southern tip of the continent and, from there, a route to the East. And, unlike a blind voyage across the ocean, this coastal route promised commercial returns for incremental journeys. Traveling an extra hundred miles down the African coast held the promise of perhaps finding the golden route to India. But even if a ship's captain didn't accomplish that goal, he could at least buy slaves or ivory to make the trip worthwhile.

What was needed, then, was a source of financial incentive and a mechanism to track the incremental progress. These two requirements were combined in one man: Prince Henry of Portugal, known as Henry the Navigator. Born in 1394, Henry was the third son of King John. He therefore had little hope of rising to the throne, and after a brief career in the army, he shifted his energies to exploration.

At that time, people assumed earth's equatorial region presented an impenetrable barrier. It was common knowledge that, as one sailed south along the coast of Africa, the climate steadily grew hotter. Surely, they concluded, at some tropical latitude the air must become too hot to breathe and man could go no farther. Conventional wisdom set this limit at Cape Bojador (the "bulging cape") in what is now Western Sahara, latitude 26° North (figure 1.6).

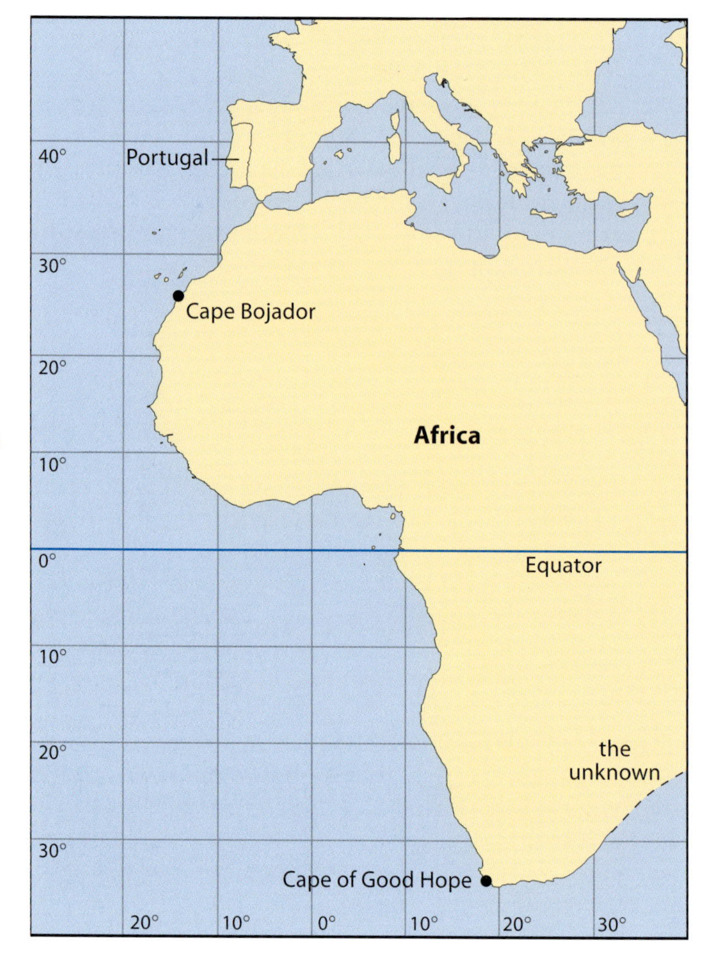

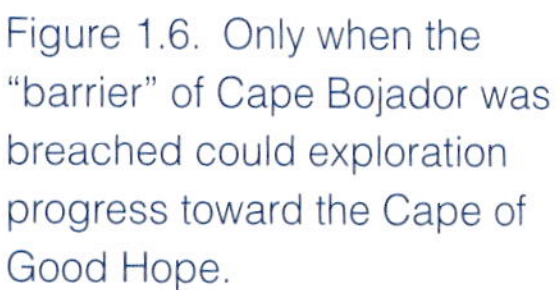
Figure 1.6. Only when the "barrier" of Cape Bojador was breached could exploration progress toward the Cape of Good Hope.

Henry was determined to break that barrier. Between 1424 and 1434, he funded fifteen unsuccessful expeditions to Africa, a record of failure that would have daunted most entrepreneurs. But Henry persisted, and finally found a captain willing to sail past Cape Bojador. When that captain, Gil Eannes, turned the Cape in 1434, the impenetrable barrier collapsed, and the rush to India was on. Bit by bit, with Henry's support, the Portuguese worked their way down the African continent, mapping the coast and buying slaves as they went.

Prince Henry died in 1460, but by then the mechanism of progress was clear, and the quest continued after his death. Lopo Gonçalves crossed the equator in 1473. Nearly three decades after Henry's death, in 1488, captain Bartolomeu Dias and his crew were blown off course during a journey to southern Africa and, fighting their way back to land, found that they had rounded the Cape of Good Hope (although at the time they called it the "cape of storms") and with it the southern tip of Africa (figure 1.6). Dias pushed on another 500 miles to the east, far enough to know that the route to the north and to India was indeed open. But facing mutiny from his crew, he was forced to return to Portugal. Ten years later, Vasco da Gama completed the route, arriving in India on May 22, 1498.

In opening a sea route to India, the Portuguese accomplished more than first meets the eye. Yes, the route to India was a huge commercial success. It assured trade between Europe and the Orient in the face of any turmoil that might close the routes through the Middle East, and the Dutch quickly followed the Portuguese in exploring this new realm of commerce. But even more importantly, Portuguese exploration expanded Europe's perspective on the world as a whole. Until Dias rounded the Cape of Good Hope, it was possible for geographers and sailors to think of the Atlantic and Indian Oceans as separate. Suddenly, in European minds, two small oceans became one. And this amalgamation was accomplished not by chance, but by concerted exploration. In essence, Portuguese exploration opened the world's oceans to discovery. And this new perspective on discovery, not the physical fact of rounding the Cape of Good Hope, fundamentally changed humankind's interaction with the ocean.

Chinese Exploration

Let's now pick up the second thread of our story, and return to China. We have already seen that Polynesians and Vikings, the greatest of ancient seafarers, never mapped the world despite their travels to far-flung islands. The story is similar for the Chinese. Shortly after the fall of the Mongol Empire and the rise of the Ming Dynasty in the late 1300s, a court eunuch named Zheng He initiated a series of expeditions to impress the world with the might and wealth of China. Known as the Admiral of the Triple Treasure, Zheng He (or Chêng Ho, depending on your preference in phonetic spelling) oversaw the formation of a fleet of ships such as the world had never seen. The largest vessel, the Treasure Ship, was 135 meters long, with 9 masts and a beam of 55 meters. Even the smallest ships in the fleet were large by the standards of the day—55 meters long, with 5 masts.

Every year between 1405 and 1433, Zheng's fleet sailed to islands south of China (Java and Sumatra) and on into the Indian Ocean. By 1421 (when Henry the Navigator was still a young man and 77 years before Vasco da Gama's famous voyage), the Chinese had reached India, Arabia, and even Zanzibar in Africa (figure 1.7).

Unlike the Portuguese, who arrived in India with thoughts of spices and the slave trade, the Chinese desired to demonstrate the superiority of China by handing out gifts. Thus, the Treasure Ship carried treasure *from* China *to* the rest of the world to dazzle natives with the majesty of the Middle Kingdom. And who would not be impressed? The largest of the treasure fleets consisted of 317 ships and more than 30,000 crew members. Even the most world-wise Arab trader would be awed by a fleet of more than 300 gigantic ships sailing over the horizon.

But, as with the travels of the Polynesians and Vikings, the voyages of the Chinese treasure fleets sought commerce rather than discovery. As exemplified by their trade in silks, perfumes, and spices, China had for centuries been involved in commerce with virtually all the known world. Thus, in sailing into the Indian Ocean, they simply used new routes to visit old trading partners in known places. The Chinese, too, lacked the impetus to sail into the unknown.

When Zheng's patron, the emperor Yung Lo, died in 1424, Zheng's power declined, and the treasure expeditions were doomed. Soon after the last

Figure 1.7. Zheng He's route from China to India and Africa.

voyage in 1433, China turned in upon itself. The Great Wall, begun piecemeal in earlier dynasties, was consolidated to keep foreigners out, and it became a crime for anyone in China to leave. The path to Chinese discovery was closed.

The New World and Beyond

We now return to European explorations and pick up that thread of our story with Christopher Columbus (1451?–1506). Born in Genoa, Italy, Columbus was a sailor with grand aspirations. He focused his energies on Ptolemy's suggestion that Asia might lie a short journey west of Europe, and in 1488, Columbus and his younger brother Bartolomeo proposed to the rulers of Portugal that an expedition should be funded to explore that route.

Unfortunately, the timing of their proposition coincided with the return of Bartolmeu Dias from his triumphant rounding of the Cape of Good Hope; in fact, Columbus was on the dock when Dias returned. With a southern route to India within their grasp, the Portuguese were understandably reluctant to fund a chancy voyage to the west.

Frustrated in Portugal, the Columbus brothers spent the next four years peddling their "Enterprise of the Indies" among the courts of Europe. Finally, in 1492, they acquired funding from Ferdinand and Isabella, rulers of Spain, and Christopher set forth with three ships (the *Niña*, the *Pinta*, and the *Santa María*) for a voyage into the unknown.

The plan was simple. The best maps of the day (based largely on Ptolemy's view of the earth) suggested that Europe and Asia were closest at the latitude of the Canary Islands, off the coast of North Africa (about 28° North). Thus, Columbus sailed south to the Canaries and then turned west. Thirty-three days later, on October 12, a lookout on the *Pinta* sighted the island of San Salvador. After a brief tour of the Caribbean islands, including Cuba and Hispaniola, Columbus sailed north to the latitude of the Azores (about 38° North) and then turned eastward and returned to Spain.

Following so quickly on the heels of Dias's rounding the Cape of Good Hope, Columbus's expedition electrified Europe. The dual discoveries, in 1488 of a route around Africa and in 1492 of a route to the New World, strongly justified the European strategy of discovery, and other voyages soon followed.

Columbus himself made three more voyages to the New World, but a flood of other expeditions eclipsed his efforts. The Spanish quickly established colonies on the islands of the Caribbean and at various points on the Caribbean coasts of Central, South, and North America. The Portuguese followed suit in present-day Brazil, and the English, Dutch, and French sent colonial expeditions to the Atlantic seaboard of North America.

Although the lure of commercial exploitation drove much of the European invasion of the New World, discovery continued to play an important role. For example, in 1513, stories told by Panamanian Indians suggested to Vasco Nuñez Balboa that a great body of water could be found across the Darien mountains. Intrigued by this lead, Balboa marched across the Isthmus of Panama, and with 190 Spaniards, several hundred native guides, and much effort, he reached the Pacific Ocean near what is now Panama City. It would take a few years for the Spanish to realize the true enormity of Balboa's discovery, but, 89 years after Prince Henry the Navigator sent out his first expedition, the last of the great oceans was finally on the map.

Balboa's discovery of the Pacific led to an enduring, although somewhat misleading, bit of terminology. The Isthmus of Panama runs roughly east to west, with the Caribbean Sea, Balboa's starting point, to its north (figure 1.8). As a result, when Balboa first sighted the Pacific, he was facing south, and he dubbed his discovery the "Southern Sea," a name still in use, as when we refer to places such as Tahiti and Samoa as "South Sea Islands."

In 1519, Ferdinand Magellan (1480?–1521) led the next great voyage of discovery, a five-ship expedition to circumnavigate the earth in search of spices. Mirroring the voyage of Dias in Africa 31 years earlier, Magellan (a native of Portugal, but sailing for Spain) sailed south along the east coast of South America searching for a passage to the west and to the spice islands. Time and again he and his crew probed rivers and bays with no luck. Finally, in an arduous journey of 334 miles and 38 days, Magellan worked his way through the narrow straits that now bear his name, and entered the Southern Sea (figure 1.9).

Whereas Dias had fallen back after rounding the tip of Africa, Magellan continued on. Taking advantage of a lucky absence of storms in the ocean he described as "pacific," Magellan reached the Philippines in 1521. His crew

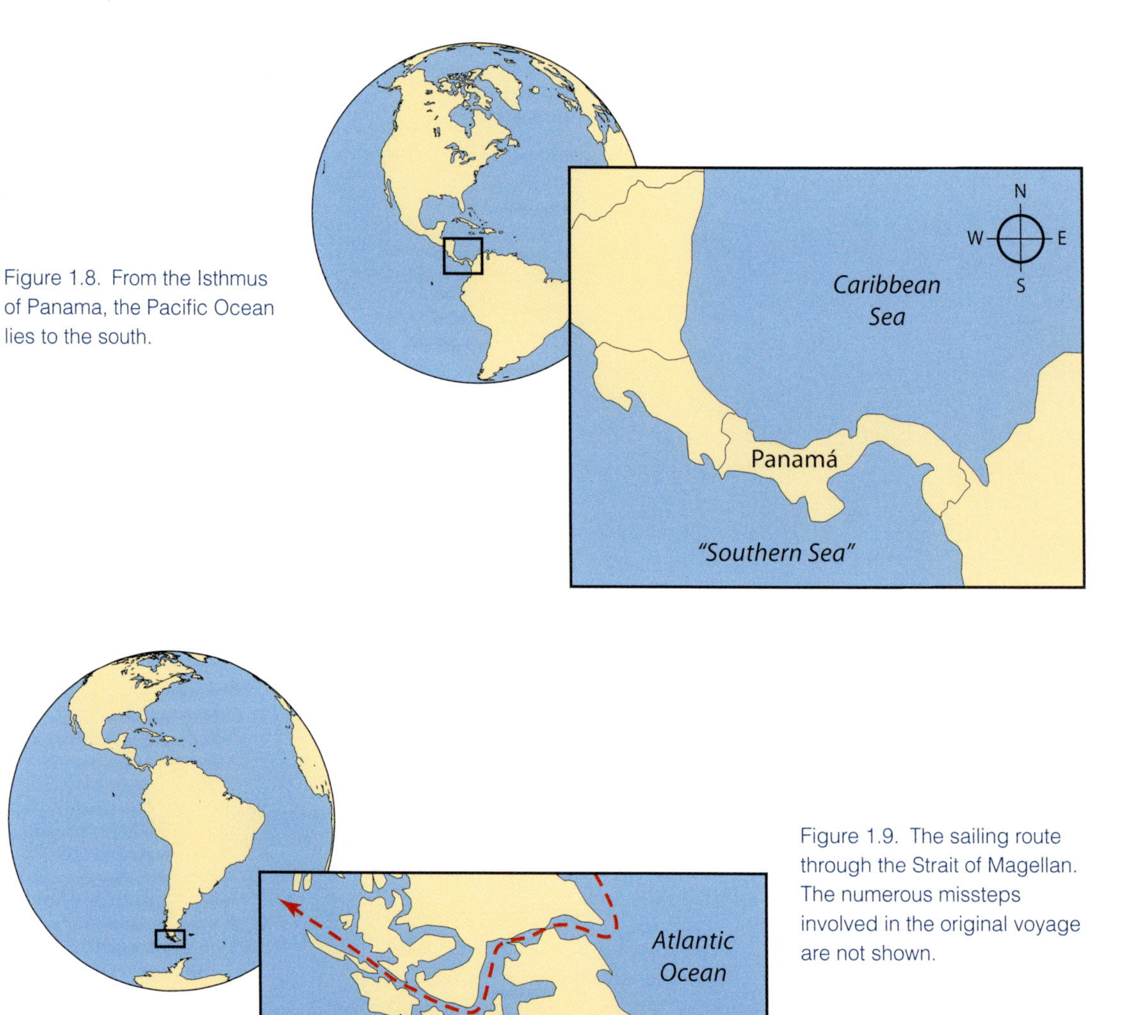

Figure 1.8. From the Isthmus of Panama, the Pacific Ocean lies to the south.

Figure 1.9. The sailing route through the Strait of Magellan. The numerous missteps involved in the original voyage are not shown.

arrived in India later that year, and from there, they retraced Gama's route, arriving back in Spain in 1522.

Magellan's expedition proved that the earth was a sphere: one could continuously sail westward and arrive back at one's starting point. But by 1522 that was almost a foregone conclusion. In fact, when Magellan arrived in the Philippines from the east, his largest problem was not to find a path into the unknown, rather it was to avoid the Portuguese and Dutch who had arrived in the Philippines after using Gama's route from the west. The Portuguese would soon set up trade with China and Japan. Thus, by the mid-1500s, European voyages of discovery had opened the entire globe to ship-borne commerce.

It is interesting to note that success as an explorer had few rewards for those who pioneered new routes. Adept as he was as a sailor, Columbus had limited abilities as a cartographer and administrator. In the face of growing evidence to the contrary, he continued to insist to his dying day that he had discovered the islands of Japan, and that a sea route past them would lead to Asia. The colonies that Columbus attempted to establish on his voyages failed, at least in part because he left his son in charge while he himself searched for the path to China. And Columbus began the atrocious tradition of enslaving local Indians. His record as an administrator was so abominable that at one point Columbus was sent back to Spain in shackles. Balboa fared even worse. Shortly after his discovery of the Southern Sea, Balboa was accused of plotting to become emperor of Peru, and he was beheaded. Magellan, too, did not live to reap the benefits from his exploration. In fact, he did not even complete the circumnavigation for which he is famous. While intervening in a local conflict in the Philippines, he was wounded by poison arrows and died. Only 18 of his original crew of about 250 were on the sole remaining ship that made it back to Spain.

Discovery continued apace after these early explorers. Driven by a lust for gold, silver, and spices, and facilitated by military technology and the death of the natives from smallpox, Spanish, Portuguese, British, and French colonists succeeded where the Polynesians and Vikings had failed, spreading their hegemony across North and South America. The coastline of the New World was quickly mapped, and explorers investigated the remaining pockets of unknown ocean.

Perhaps the greatest of this second generation of explorers was Captain James Cook (1728–79) of the British Navy, who, in three grand voyages, mapped much of the Pacific Ocean. In the process he "discovered" Hawai'i, surveyed the west coast of North America from Oregon to Alaska, and surveyed the east coast of Australia. (It wasn't until 1801 that Matthew Flinders, another British explorer, discovered that Australia was a continent, not just a series of islands.) Cook's success stemmed in part from new technology. The first explorer with access to an accurate clock—chronometer Number 4, invented by John Harrison—Cook could fix his longitude without having to come ashore to make astronomical sightings. In 1775, Cook crossed the Antarctic Circle and circumnavigated Antarctica, probing the boundaries of the ice, but he never got close enough to land to see the last of the continents. Not until 1820 did Fabian Bellingshausen sight Antarctica, but he too could not penetrate the ice surrounding the continent to make landfall. Mankind first set foot on Antarctica in 1838 when Dumont D'Urville landed on a small island off the Antarctic Peninsula. It was thought at that time that Antarctica was two continents, separated by a deep gash and overlaid with ice. The unity of the continent was not definitively established until the International Geophysical Year of 1957–58.

At the other end of the earth, numerous expeditions (mostly by the British) explored the Arctic, searching for the Northwest Passage, a hypothetical sea path around the northern end of North America: a route to counter the Spanish land routes across Central America. Martin Frobisher, Henry Hudson, and William Baffin all captained multiple expeditions, to no avail. The frustration of Arctic exploration is apparent in the fact that Hudson and his son died when they were set adrift by a mutinous crew, who preferred to head home rather than spend another winter in the North. Finally, in 1906, the Norwegian Roald Amundsen and his ship *Gjøa* took advantage of favorable ice conditions and, in an arduous voyage, navigated the Northwest Passage. Rumors of solid land in the midst of the Arctic Ocean ("Crocker Land") persisted until 1926, when Amundsen and his companions, Lincoln Ellsworth, Umberto Nobile, and Hjalmar Riiser-Larsen, flew the lighter-than-air ship *Norge* over the Pole, sighting nothing but ice and water.

From our vantage point in the twenty-first century, it is easy to look back at the heroic era of ocean exploration and think of it as ancient history. But it is important to realize how recent our knowledge is. For example, my parents were children when Amundsen flew over the North Pole, and I was in grade school when we found out that Antarctica was a single continent, not two. We have had only a single lifetime to contemplate the complete map of the earth.

Biology and the Oceans

As informative as it is, that map is far from the full story. Discovery of the oceans has involved much more than just mapping their shape and size. What plants and animals live in the sea? Where are they found and in what abundance? How do they survive? To begin to answer these questions, we need to move beyond maps and trace the history of the study of oceanic biology, a story that intertwines threads of commerce, war, and science. We begin with science and the history of diving.

Diving

Humanity's first contact with life in the seas occurred in shallow water along the coast where people could dive to the bottom and tides periodically uncovered the seafloor. Nearshore organisms played a large role in ancient commerce: fish, shellfish, and seaweeds were used for food, for example, and indigo, a precious purple dye, was extracted from shallow-water marine snails. Scientific description of the often-slimy marine organisms began with Aristotle, who aptly noted that one "must not recoil with childish aversion from the examination of the humbler animals. In all things of nature, there is something of the marvelous."

In contrast to shallow-water plants and animals, which were familiar because of their accessibility, the plants and animals of the deep remained largely unknown to the ancients. Deep-living organisms could be caught occasionally in nets and on hooks, but until the invention of an apparatus for delivering air to a submerged human, scientific observation of these organisms in their native habitat was impossible.

A variety of diving apparatuses have been invented over the years. One of the earliest was devised by Leonardo da Vinci, consisting of a leather helmet that covered the diver's head and a tube that connected the diver to air at the water's surface. As with many of Leonardo's ideas, this diving helmet, although innovative, was not practical. Unless the diver were within a foot or two of the surface, water pressure pushing in on his chest would prevent him from inhaling, and the "dead space" of the tube would mean that even if he could inhale he would simply be re-breathing the same air as it moved back and forth in the tube.

Figure 1.10. Edmund Halley's diving bell. Redrawn from M. W. Denny and J. L. Nelson (2006), *Conversations with Marco Polo* (Xlibris, Philadelphia, PA).

A more practical apparatus was designed in 1691 by Edmund Halley, the Halley of Halley's Comet. A watertight, weighted wooden bell suspended in the water received air periodically from weighted barrels lowered from the surface (figure 1.10). This diving bell was put to use salvaging cannons and other valuable objects from ships sunk in shallow water. The helmet-and-tube apparatus shown in the figure would suffer the same problems faced by da Vinci's helmet, although pressurized air would flow to the diver if his head were higher than the water level in the bell.

A truly practical diving apparatus appeared in the 1800s with the advent of the "hard-hat" diving suit. Originally built in England by Augustus Siebe in 1837, the apparatus consisted of a rigid helmet that enclosed the diver's head and an inflatable canvas suit. Air was pumped from the surface to the helmet, which in essence formed a small, personal version of Halley's diving bell. Updated versions of this apparatus are still used today for shallow-water salvage operations.

Although the hard-hat diving suit allowed divers to spend substantial time underwater, the air hose connected to the surface restricted the divers' mobility and thereby hampered their ability to observe and explore. Jacques Cousteau and Emile Gagnan solved this problem in 1943 with their invention of a regulator that made self-contained underwater breathing apparatus—scuba—practical. Carrying a tank of compressed air on their backs, scuba divers can move freely through the marine environment, and much of what we know about life in the shallow oceans is due to the use of this apparatus. Judicious use of different mixtures of gases allows scuba divers to reach depths of several hundred feet.

Darwin and the *Beagle*

Let us return now to the early 1800s and pick up a thread that combines science and the military. At that time, England possessed the world's preeminent navy, and as a means to facilitate that navy's projection of maritime power, the British busily sent expeditions around the globe to map the world's coastlines. One such expedition was that of the HMS *Beagle*, a 27-meter-long, 10-gun brig charged with mapping portions of the South American coast. The *Beagle*, with a crew of 74, set sail from Plymouth on December 27, 1831, and a young man named Charles Darwin was aboard as an unpaid naturalist.[3]

It was an opportune time for an open-minded naturalist to be sailing around the world. For instance, recent advances in geology were causing scientists to revise their estimates of the age of the earth. Christian doctrine (as calculated by Irish Bishop James Ussher) suggested that the earth came into being at noon on October 23, 4004 BC.[4] But evidence garnered from rock formations led nineteenth-century geologists to believe that the earth was much older, hundreds of thousands, if not millions, of years old. Geologists thus began to think about the origins of mountains and seas in the context of this newly lengthened timeline for the history of the earth. They no longer needed to suppose, for instance, that mountains must have risen in a single catastrophic upthrust as the earth formed. Instead, slow, imperceptible movements over the long millennia could arrive at the same end. Similarly, the lengthened timeline gave biologists new scope for thinking about the origin and evolution of life. Perhaps species need not have been created all at once; the great age of the earth might have allowed them to evolve incrementally.

In this context, Darwin set sail. Over the next five years, he gathered as much information as he could about life on earth. He collected beetles in the Brazilian rain forest and fossil shells in the Andes, and observed finches in the Galapagos Islands, information that, years later, contributed to his conception of the radical idea of evolution by natural selection. Stated in modern terms: chance produces random genetic mutations, but only the better adapted survive to pass the mutation on to their children. In this fashion, organisms change through time. Twenty-three years after his return from the *Beagle*'s voyage, Darwin published *The Origin of Species* in which he laid out his theory of evolution, and biology has never been the same.

During the voyage of the *Beagle*, Darwin also made a major contribution to geology. In the course of visiting islands in the Pacific, he noted a pattern in their structure (figure 1.11). Some islands were mountainous in the middle, with a fringing reef of coral around the periphery. Other islands had smaller central mountains, but were surrounded at some distance by a barrier reef. Still other islands, atolls, consisted of a series of low-lying coral islets spread in a circle around a central lagoon. Why were the atolls always circular or oval, the same shape as fringing and barrier reefs?

[3] February 12, 1809, was an auspicious date for humanity. Both Charles Darwin and Abraham Lincoln were born that day, a few hours and an ocean apart.

[4] Later, John Lightfoot revised Ussher's calculations, and suggested a starting date of October 26, 4004 BC, at 9:00 in the morning. A lot of effort and thought has gone into this subject.

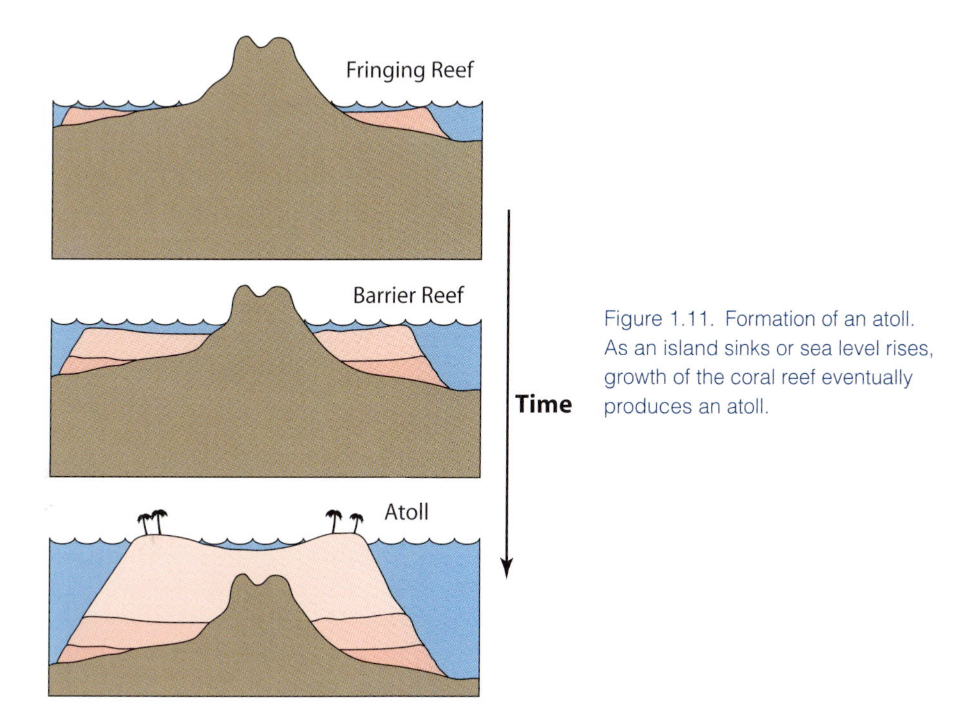

Figure 1.11. Formation of an atoll. As an island sinks or sea level rises, growth of the coral reef eventually produces an atoll.

A simple explanation occurred to Darwin. If a mountainous volcanic island were to slowly sink, or sea level were to slowly rise, the corals of its fringing reef might be able to grow upward fast enough to maintain their position near the sea's surface. In this fashion, the fringing reef would become a barrier reef. Eventually, once the center of the island had completely disappeared into the sea, the remaining reef would persist in the circular shape of an atoll.

Darwin's idea met resistance at the time. After all, given the hundreds of atolls in the Pacific, his theory demanded that numerous Pacific islands had been drowned. And since no one had ever seen an island disappear beneath the waves, Darwin's proposition seemed far-fetched at best.

As we will see, it wasn't until 1952 that Darwin was eventually proven right. Sea level has risen since the last ice age, and islands in the Pacific do indeed commonly sink as the seafloor beneath them cools and shrinks. Today, the explanation of atoll formation, based on Darwin's hypothesis, is an important piece of evidence in our understanding of *plate tectonics*, the scientific field that studies how the face of the earth changes over time. We will return to the atolls later in this chapter when we discuss the testing of the atomic bomb.

The Deep

But first, let's shift gears for a moment and trace the history of our understanding of life in the ocean depths. Due to its inaccessibility, the deep ocean was not studied until about 150 years ago, and the initial impetus came not from biology, but rather from expanding technology. In the 1840s and 1850s, the telegraph revolutionized communications on land, and telegraph companies searched for a practical way to stretch cables across the sea. As a part of this search, they needed information about the seafloor: Was it rock or mud? How deep was it?

To help with this exploration, the U.S. Navy's Depot of Charts and Instruments in 1858 undertook the survey of a potential cable route from the United States to Great Britain, and they found that the Atlantic was far from uniformly deep. Instead, there appeared to be a large hill in the seafloor in the middle of the ocean, a feature they termed the telegraphic plateau, or the "*Dolphin* rise" (after the ship that performed the survey). We now know that the Mid-Atlantic Ridge, as it has become known, plays an important role in plate tectonics.

This survey as well as others indicated that there was structure to the ocean floor, but provided no evidence as to whether there was life in the deep sea. In the wake of Darwin's *Origin of Species,* many thought that the deep oceans hosted only a few simple life forms. If evolution proceeded in response to natural selection, how would anything ever evolve in the cold, unchanging depths of the ocean? Thomas Huxley (an outspoken proponent of Darwin's theory) even proposed that the bottom of the ocean was coated with a primordial ooze (*Urschleim,* to use the German expression) from which all other life had evolved. Indeed, early expeditions to sample the ocean floor showed evidence of gelatinous ooze. Huxley was delighted, and named the supposed creature *Bathybius haeckelii* after his friend, the German biologist, Ernst Haeckel.

Others proposed that life in the deep ocean wasn't just rudimentary, it was impossible. It was common public perception at the time that water (seawater included) was substantially compressible: the more pressure one applied to it, the denser it became. Surely, at the great pressure of the ocean depths water would become so dense that nothing could survive. Some seafarers even thought that in the deep ocean the density of seawater was greater than that of lead. As a result, this reasoning went, lead sounding weights used to measure the depth of the ocean would "float" on this deep, dense water, never actually reaching the seafloor. The depth of the ocean was therefore unknowable.

In fact, seawater is so nearly *incompressible* that even in the deepest reaches of the ocean it is only about 6% denser than it is at the surface (see chapter 3). Nonetheless, the myth of compressibility persisted well into the twentieth century. For example, when the *Titanic* sank in 1912, the relatives of many of those lost were tortured by the thought that the ship had sunk to a depth at which the density of water equaled that of iron, and that as a result, the *Titanic* was doomed to drift forever through the ocean interior.

In that age of telegraphs, *Urschleim*, and compressible seawater, the British government decided in 1872 to combine the search for new deep-sea cable routes with an expedition to answer the question of life in the abyss. The ship chosen was the HMS *Challenger*, an 83-meter-long steam corvette, and over the course of four years she circumnavigated the globe, sounding the depths, measuring ocean temperature, and dredging for biological specimens (figure 1.12).

The *Challenger* expedition revolutionized biological oceanography. Wherever the expedition dredged, the crew found abundant and diverse life. Describing the *Challenger's* samples took more than twenty years and filled 51 large volumes, including 32 dealing with animals and two dealing with plants. The *Challenger's* feats were not limited to biology, however. While sounding the ocean off the island of Guam in the Marianas, by luck the *Challenger* found

Figure 1.12. An engraving from the report of the *Challenger* expedition (vol. 1), showing the gear used for dredging and sounding. The sailor at lower right stands next to a steam winch used to lower instruments into the sea.

the deepest spot in the earth's oceans, a location we now call the Challenger Deep in her honor.[5]

For a while, it appeared that the *Challenger* had also found the *Urschleim.* As with previous expeditions, when the *Challenger's* specimens of ocean-bottom mud were pickled in alcohol and returned to the laboratory, they were coated with a gelatinous ooze. However, a skeptical member of the expedition demonstrated that the ooze was just an artifact caused by the interaction of alcohol with seawater. Huxley was embarrassed, but humbly retracted his proposed *Bathybius.*

[5] There is some controversy on this point. Many current accounts suggest that the Challenger Deep is named for another HMS *Challenger*, a British survey ship that, in 1951, measured a depth of 10,863 m in the same area visited by the original *Challenger*. However, a 1912 publication by the Challenger Society (*Science of the Sea: An Elementary Handbook of Practical Oceanography for Travellers, Sailors, and Yachtsmen*, ed. G. H. Fowler) refers to the area as the "Challenger Deep," so the term was in use well before the second *Challenger* made its soundings.

The *Challenger*'s findings were far-reaching and important, but they were nonetheless indirect. No one on the *Challenger* ever actually saw the deep seafloor, and direct observation of the deep ocean was not attempted until well into the twentieth century.

Below the limits of hard-hat and scuba, a few hundred meters, divers need protection from water pressure and therefore require a closed pressure vessel to observe life in the deep. Over the years, a wide variety of deep-diving vessels have been constructed, ranging from one-man rigid suits to small submarines. Three of these vessels are of particular historic note.

In 1934, William Beebe and Otis Barton devised a plan to transport themselves into the deep ocean. For a pressure vessel, they used a simple, hollow iron sphere, just large enough for the two of them to cram themselves in. This *bathysphere* had a small window that allowed them to look out and an external light to illuminate the deep.

In its maiden voyage, the bathysphere was suspended from a cable and lowered down into the ocean from a barge anchored off Bermuda. Beebe and Barton reached a then record depth of 923 meters and, during their 2.5-hour descent, saw a whole zoo of weird and wonderful ocean creatures. From time to time they would turn off their light, and during those interludes of darkness, they were particularly impressed with the ability of deep-sea animals to luminesce. Barton recalled that "at 2000 feet [depth (610 m), he] made careful count and found there were never less than ten or more lights—pale yellow and pale bluish—in sight at any one time. Fifty feet below [he] saw another pyrotechnic network, this time, at a conservative estimate, covering an extent of two by three feet. [He] could trace mesh after mesh in the darkness, but could not even hazard a guess at the cause."

The bathysphere demonstrated the potential for deep-sea observation, but its lack of mobility was a serious handicap. It went where the barge lowered it, and could not follow the animals Beebe and Barton saw. The next generation of deep submersibles—epitomized by the bathyscaph *Trieste*—solved this problem (figure 1.13). The *Trieste* consisted of a steel pressure vessel similar to that of the bathysphere—just large enough to house two men—suspended below a 22-meter-long, thin-skinned tank filled with gasoline that served as a "balloon" to provide buoyancy. Electric motors attached to the gasoline tank allowed the ship to move under its own power. The *Trieste* was designed and built in 1953 by Auguste Piccard, a Swiss scientist and adventurer.

After a few dives in the Mediterranean Sea, the *Trieste* was purchased by the U.S. Navy, and in 1960 she descended 11,000 meters (36,000 feet) to the near-freezing water at the bottom of the Challenger Deep in the Mariana Trench. Her passengers (Jacques Piccard, Auguste's son, and Don Walsh of the U.S. Navy) were, for a brief few moments, able to observe the living organisms present in the deepest of the deep. In a haunting parallel to humankind's subsequent exploration of the moon, no manned submersible has returned to the Challenger Deep in the 47 years since the *Trieste*'s descent, although an unmanned robot (the Japanese *Kaiko*) has conducted experiments and taken pictures there.

The next generation of deep-sea submersibles after *Trieste* is best represented by *Alvin*. Commissioned by the U.S. Navy in 1964, *Alvin* has been in service

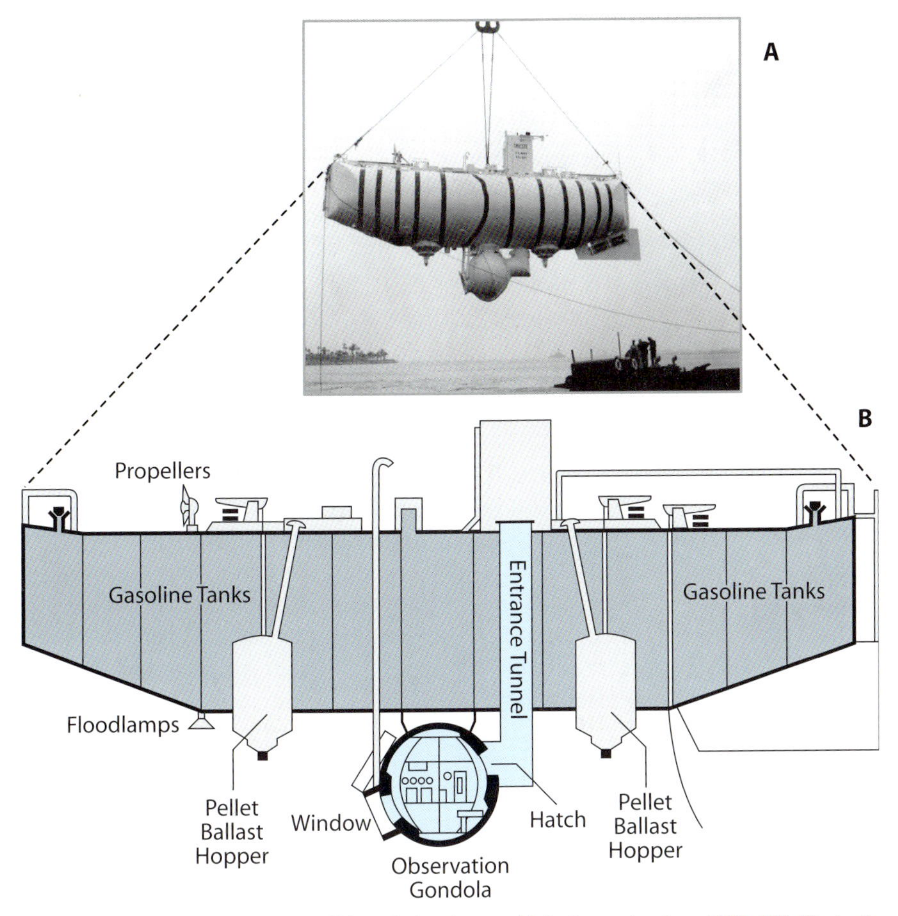

Figure 1.13. (A) The bathyscaph *Trieste* being lowered into the water (ca. 1958–59). Photo # NH 96801, U.S. Navy. (B) A description of the various parts of the vessel. Redrawn from photo # NH 96807, U.S. Navy.

more or less constantly ever since. Capable of carrying three passengers (two scientists and a skipper) to a depth of 4500 meters, *Alvin* has an impressive list of accomplishments, among them: the first collection of organisms from the bizarre communities that grow in conjunction with hydrothermal vents (more about these in a moment), a visit to the sunken *Titanic*, and (reminiscent of a Hollywood script) a dive to find a hydrogen bomb lost on the seafloor off Spain.

Despite its accomplishments, *Alvin* has had its ups and downs. In 1968, cables broke while the vessel was being deployed. Water flowed in through the open hatch, and the submersible sank to the bottom in 1500 meters of water. *Alvin* was salvaged eleven months later, and a bologna sandwich left behind in the vessel by the crew was still fresh. Either the rate of decomposition is slow in the cold, oxygen-poor water 1500 meters down, or heavy metals leaching from the submarine's corroding hardware inhibited bacterial growth. In either case, even *Alvin's* failures contributed to our understanding of oceanic biology. As of 2007, *Alvin* had logged more than 4300 dives.

War and the Oceans

We now turn to the military's role in oceanographic exploration. Throughout history wars have spawned technical innovations, some of which have been useful to science, and the wars of the past century have been particularly beneficial to the study of the oceans.

Take, for example, the use of sound to measure ocean depth. The concept likely originated in 1838 when Professor C. Bonnycastle and R. M. Patterson made a proposal to the U.S. Coast Survey. They suggested that, by firing a gun underwater and listening for the echo, they could measure the distance to the seafloor. Unfortunately, due to the primitive listening technology of the time, their experiment was unsuccessful, and the idea gathered dust until the *Titanic* disaster of 1912 provided impetus to try it again. Perhaps a loud underwater sound emitted by a ship would echo off any iceberg in the ship's path, alerting the crew to the iceberg's presence. Reginald A. Fessenden, a former assistant to Thomas Edison, invented a method for using electricity to make a loud "ping," and in 1914 he used this new technology to detect an echo from an iceberg.

Before the idea could be fully implemented, however, World War I broke out, creating a more pressing need for echolocation. The same sound waves that would echo from an iceberg would echo from the hull of a submarine, allowing a ship on the surface to avoid the submarine or to hunt it. An immense effort ensued to build what is now known as *sonar* (an acronym for "*so*und *na*vigation *a*nd *r*anging"). A practical sonar apparatus was not perfected in time to be play any role in World War I, but the concept was used by the German *Meteor* expedition in 1925 to make 33,000 soundings, thereby measuring the depths of the South Atlantic Ocean. This expedition was the first to reveal how variable the seafloor's topography can be, a topic we will return to in chapter 2. Sonar was put to effective military use in World War II. Wars giveth and wars taketh away. The depth recordings from the *Meteor* expedition were destroyed in the bombing of Berlin during World War II.

After World War II, oceanographers made good use of the many available surplus sonar systems and continuously recording depth sounders, greatly enhancing our knowledge of the shape of the ocean floor. Even the depths of the ocean trenches were accurately measured by recording the time it takes for an echo to return from the bottom. In this case, standard sonar techniques sometimes did not suffice: they couldn't produce a sound loud enough to be heard after the long round trip. Instead, in a repeat of the experiment of Bonnycastle and Patterson, the mapping crew exploded a charge of high explosive in the water near the ship, and timed its echo. In the case of the Challenger Deep, the echo arrived more than 15 seconds after the explosion.

While scientists devised ways for surface ships to find submarines, other scientists struggled to keep submarines unnoticed. For example, U.S. submarines in World War II were equipped with bathythermographs, devices to record simultaneously the depth and temperature of the water outside the sub's hull. The bathythermograph was elegantly simple: a small stylus moved up and down as the temperature shifted and left and right as the depth changed, recording a graph of temperature versus depth onto a small smoked-glass slide (figure 1.14A).

Information from bathythermographs was of immediate practical interest to those aboard a submarine in wartime. An abrupt shift in temperature with

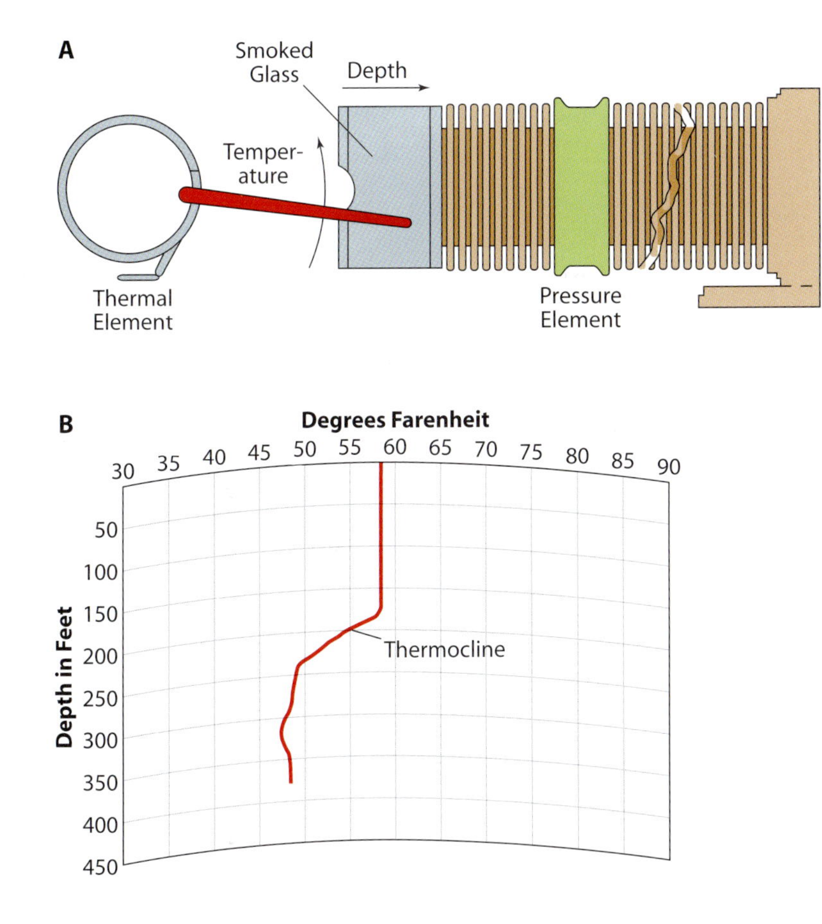

Figure 1.14. (A) A schematic description of a bathythermograph. A smoked-glass slide is attached to a bellows, which moves left or right under the influence of varying pressure. At the same time, a stylus attached to a "thermal element" moves up and down in response to changes in temperature, marking a trace on the smoked-glass slide. (B) A typical bathythermograph trace: water is uniformly warm near the surface, but abruptly decreases in temperature at depth. Redrawn from *Naval Sonar,* 1953, NAVPERS 10884, figures 1-12 and 1-10.

depth indicates a *thermocline*, an oceanic feature below which a submarine can attempt to hide (figure 1.14B). Sound attempting to pass through a thermocline can be deflected, making it difficult for a surface ship to locate the source of an echo. Thermoclines will be discussed in detail in chapter 6.

After the war, Woods Hole Oceanographic Institution (in Massachusetts) obtained the thousands of bathythermograph records from the Atlantic, and Scripps Institution of Oceanography (in California) took charge of those from the Pacific. In each case, the trace of temperature versus depth was accompanied by a record of when and where the trace was taken. Some 60,000 recordings were produced in the Atlantic alone. In this fashion, the wartime effort to hide submarines provided the first large-scale picture of the temperature structure of the ocean's interior. It is interesting to note that the deep submersible *Alvin* was named for Allyn Vine, one of the scientists who perfected the bathythermograph.

Sonar did more than provide a means for measuring the depth of the ocean. During World War II, strange sounds emanating from the sea bedeviled sonar operators. Pops, clicks, gurgles, whistles, and groans came in through their headphones, and no one knew their origin. After the war, considerable effort was expended tracking down these sounds, bringing interesting biology to light. The pops and clicks often came from snapping shrimp, and the gurgles and groans from mating toadfish. Sonar also revealed a concentrated assemblage of small fish, crustaceans, and gelatinous creatures that rise and fall in the ocean depths through the course of the day. This *deep scattering layer* often played havoc with the ability of surface-based sonar to locate objects at depth, but biologists profited from the discovery of these creatures' daily migrations.

Various invasions in World War II required the United States and its allies to land troops on beaches. Waves breaking on the beaches disrupted this maneuver, so the military enlisted the help of Harald Sverdrup and Walter Munk of Scripps Institution of Oceanography to devise methods of predicting wave heights. Much of our current knowledge about the physics of the surf zone can be traced back to their efforts.

And finally, there is the atomic bomb. After World War II, the United States continued to develop its nuclear arsenal, choosing as its test sites several of the Marshall Islands in the Pacific Ocean. As a part of these tests, the nuclear scientists wanted to quantify the effects of their weapons on the structure of the islands, and that required them to know in detail what the islands looked like before the bombs went off. To that end, deep cores were drilled into the atolls of Bikini and Enewetak.

Charles Darwin had predicted what these cores should reveal. If the volcanic islands on which the atolls rested had sunk into the ocean, or sea level had risen, the cores should show the remains of corals to a depth well below where corals can currently live. Once again Darwin was correct. On Enewetak, for instance, the cores were solid coral down to a depth of more than 1400 meters, ten times the depth at which corals can grow. The deepest cores also contained fossilized land snails and pollen, clear evidence that, when alive, that part of the reef had been at the water's surface.

Continental Drift

In another bit of ocean discovery, Darwin, geology, and biology converge. In 1915, Alfred Wegener (1880–1930), a German meteorologist, published a book entitled *The Origin of Continents and Oceans*, in which he made a radical suggestion. Wegener proposed that the continents of earth are not fixed, but rather drift about the globe. As a consequence, Wegener suggested, the oceans change their shape through time.

Wegener cited a variety of evidence to support his brash claim. First, he noted that the outlines of some of the continents seem to fit together. For example, if one could slide North and South America toward Europe and Africa, the land masses would fit together quite nicely (figure 1.15). This observation was not new with Wegener; it had been noted as early as 1596 by the Dutch cartographer Abraham Ortelius. But Wegener supported his proposition with evidence from both geology and biology. For instance, the Appalachian mountains of North America would have been contiguous with the Scottish Highlands if the continents had once been amalgamated, and the geologic structure of the two mountain ranges are indeed identical. Similarly, rocks in South Africa matched those in Brazil. Furthermore, coal deposits in Antarctica contain fossilized tropical ferns, suggesting that, through time, Antarctica has drifted south.

Despite this evidence, Wegener's proposal was met with a skepticism that at times approached scorn: the president of the American Philosophical Society

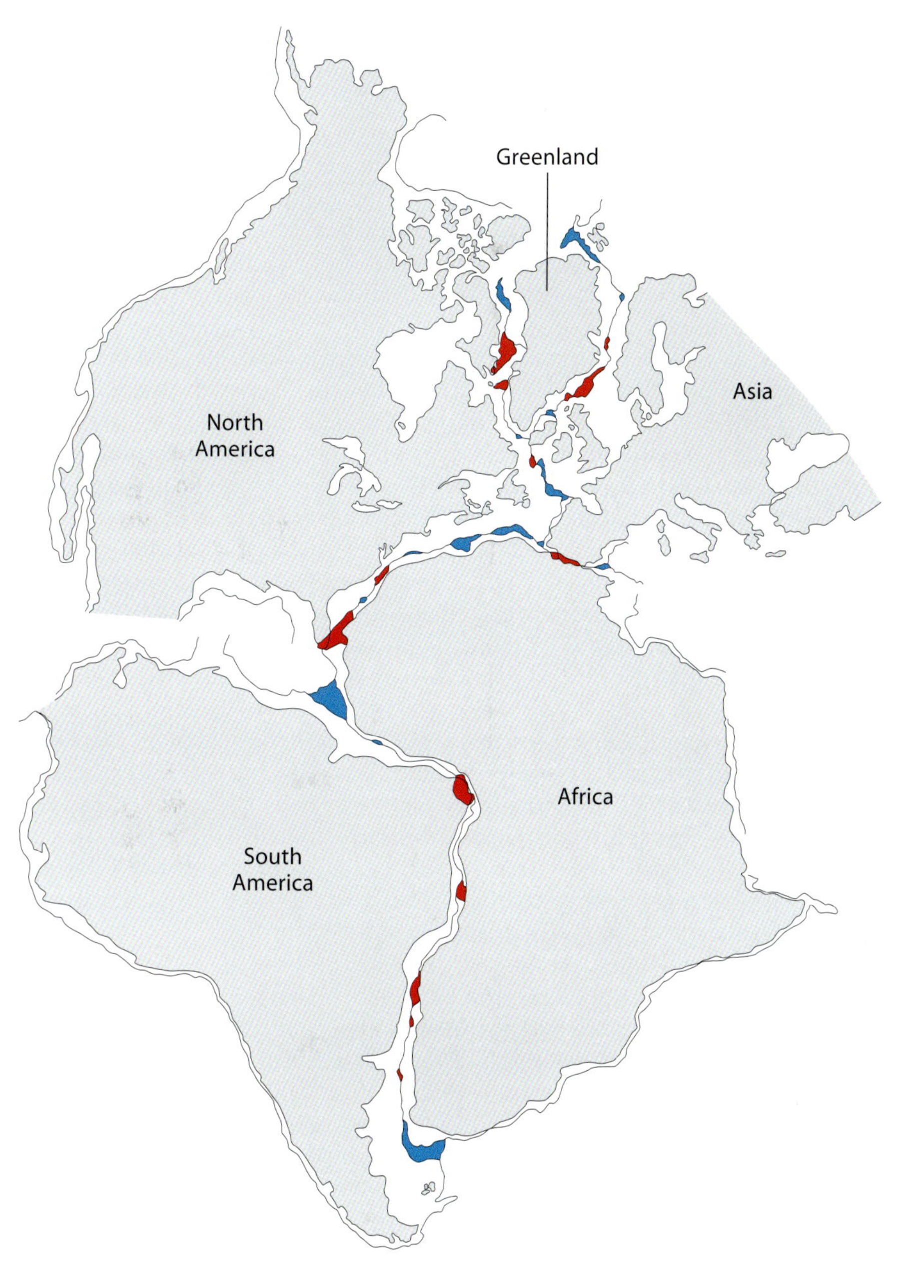

Figure 1.15. The 500-meter depth contours of the continents (their true margins) fit together like pieces of a jigsaw puzzle. Gaps are shown in blue, and areas of overlap (primarily due to the accretion of sediment from rivers) are shown in red. Map redrawn from E. Bullard, J. E. Everett, and A. G. Smith (1965), The fit of the continents around the Atlantic, *Philosophical Transaction of the Royal Society* (A) 258: 41–51.

called continental drift "utter, damned rot." Much of this doubt arose from Wegener's inability to propose a viable mechanism for the continents' motion. Indeed, Wegener's suggested mechanism—that the continents plowed their way through the ocean floor under the impetus of centrifugal and Coriolis forces—was demonstrably wrong. (For an explanation of centrifugal and Coriolis forces, see chapter 7.) Wegener energetically defended his point of view up to his untimely death on a winter expedition to Greenland in 1930. With his passing, the idea of continental drift languished.

Beginning in the 1950s, however, geologists began to reconsider Wegener's ideas. Mounting evidence, which we will consider briefly in chapter 2, suggested that new seafloor was continually being produced along an immense series of undersea mountain ranges, the Mid-Atlantic Ridge being a prime example. If this was true, the Atlantic Ocean was growing, and North and South America were thereby being carried away from Europe and Africa, just as Wegener had proposed.

The idea of a spreading seafloor rapidly gained momentum and led to the current field of plate tectonics, which studies how the face of the earth changes over time. In 1977, as a part of this burgeoning field, the geologist Peter Lonsdale headed an expedition to examine seafloor spreading near the Galápagos Islands. Scientists had proposed that, in areas of active seafloor spreading, seawater percolated through the ocean floor, emerging from the spreading site as a hot vent. Lonsdale attempted to locate sites of such hydrothermal vents by towing an unmanned sled a few feet above the seabed to measure water temperature. Simultaneously, a pair of cameras took pictures of the bottom. As planned, Lonsdale located a potential vent, but then something unexpected happened: the pictures showed a population of large clams clustered around the spreading site.

This discovery, by a geologist, sent a shock through the field of biology. Why would clams be found near seafloor vents? What would they eat? The submersible *Alvin* was quickly brought to the vents, and the story became even more intriguing. An entire community of bizarre animals surrounded the outpouring of superheated water from the spreading seabed, including gigantic tube worms sporting bright red plumes, ghostly white crabs and anemones, and the large clams noted previously.

Over subsequent years, the full story of these deep-sea vent communities emerged. The water spewing forth from the vents (at temperatures up to 400°C) contains a high concentration of hydrogen sulfide, and a few species of bacteria have evolved the ability to use this chemical as a source of food energy. Animals, in turn, use these bacteria as a source of nourishment. Tube worms even house these bacteria within their bodies, using hemoglobin (the chemical that makes our blood red) to deliver the hydrogen sulfide to the bacteria. The hemoglobin gives the worm's plume its bright red color.

It is somehow fitting that the discovery of deep-sea hydrothermal vent communities—one of the most spectacular and unexpected biological finds of the last century—was made by a geologist in quest of information about seafloor spreading. Alfred Wegener would be pleased. It is even more fitting that the discovery was made so close to the Galápagos Islands, the source of much of Darwin's insight into the process of evolution.

The Ocean from Space

On October 4, 1957, the Soviet Union launched humankind's first artificial satellite, Sputnik I. Sputnik did little more than orbit the earth every 98 minutes, transmitting by radio a monotonous beeping, but the small satellite signaled a momentous change in our view of the planet. Three months later, the United States launched its first satellite (Explorer I), and the "space race" began. Within three years of Sputnik's launch, Yuri Gagarin orbited the earth, and then in 1969, Neil Armstrong and Buzz Aldrin set foot on the moon. In a very short period, we had gained the ability to see the oceans as a whole, and scientists have been taking advantage of this lofty perch ever since.

Many ocean attributes can now be measured from space, including the size and direction of surface waves, the direction and speed of surface currents, the temperature of surface water, the color of the water (from which the growth rate of marine plants can be inferred), and the height of the tides. We can track storms that stir the ocean, and measure the concentration of ozone, which protects the ocean from "sunburn." Sensors dropped to the ocean floor or attached to fish or whales can detach themselves, float to the surface, and radio their information via satellite relays. And satellites are the basis for the global positioning system that allows scientists to locate themselves accurately on the planet.

In an intriguing extension of science's reach—and a reversal of the history of exploration on earth—we are currently exploring Mars, but starting, rather than ending, with the big picture. In this case, our first contact with the planet was from space, providing excellent broad-scale maps but few intimate details. Only in the past three decades have we managed to deliver probes to Mars's surface for a close-up view that can begin to answer the question of whether Mars ever had an ocean and if so, whether that ocean ever supported life.

Full Circle

We now return to the perspective from which we started this chapter. Imagine yourself standing on the shore and staring out to sea. You now know the ocean to be many things: a source of food, a path of commerce, and a vast habitat for diverse plants and animals. You can see the ocean as a whole, and you know its shape and its parts.

But as you stare out to sea, questions still arise. Why did Columbus have to sail south from Spain to the Canary Islands to find winds to blow his ships west? The deep waters of the ocean are cold; but why? What do animals eat in the dark ocean depths? Having chronicled the ocean's discovery, it is now time to explore how it works.

A Statement about "Truth"

The information presented in this chapter is, to the best of my knowledge, true. But, as a much abbreviated story, it is far from the whole truth. For example, Basque fishermen may have preceded Columbus to the New World, a fact they concealed to avoid competition from other fishermen. Although we know exactly who first reached the South Pole—a Norwegian expedition led by Roald Amundsen—and when they arrived—December 14, 1911—there is considerable controversy as to who first set foot on the North Pole. And there is much more to the story of continental drift than the trials and tribulations of Alfred Wegener. If you want to progress toward a more complete truth, you will need to read further, and several excellent books are suggested below. This chapter will have done its job if it piques your interest sufficiently to read more.

This same warning applies throughout this book. Nowhere do I approach the judicial standard of "the truth, the whole truth, and nothing but the truth." Instead, my goal is to truthfully describe the "forest" that is oceanography. Perhaps with this view of the forest, you will feel compelled to learn more about the trees.

Further Reading

Bergreen, L. (2003). *Over the Edge of the World: Magellan's Terrifying Circumnavigation of the Globe*. Harper-Collins, New York.

Boorstin, D. J. (1983). *The Discoverers: A History of Man's Search to Know His World and Himself*. Random House, New York.

Diamond, J. (2005). *Guns, Germs, and Steel: The Fates of Human Societies*. W.W. Norton & Co., New York.

Matsen, B. (2005). *Descent: The Heroic Discovery of the Abyss*. Pantheon Books, New York.

Piccard, J., and R. S. Dietz (1961). *Seven Miles Down: The Story of the Bathyscaph* Trieste. Putnam, New York.

Schlee, S. (1973). *The Edge of an Unfamiliar World: A History of Oceanography*. E.P. Dutton & Co., New York.

Sobel, D. (1995). *Longitude: The True Story of a Lone Genius Who Solved the Greatest Scientific Problem of His Time*. Walker Publishing Co., New York.

Wilford, J. N. (2001). *The Mapmakers: The Story of the Great Pioneers in Cartography from Antiquity to the Space Age* (revised edition). Vintage Books, New York.

Ocean Basins

In the last chapter, we briefly traced the history of ocean discovery, but in the process, we gave short shrift to many of the details. We know, for instance, that the ocean is large and deep, but how large? How deep? As a part of learning how the ocean works, we need to understand better its dimensions and its shape. Let's begin by picking up where we left off with the ancient Greeks and delve more deeply into the dimensions of the earth.

Systems of Measurement

When describing an object such as the earth, two basic tools are required. First we need a coordinate system to allow us to locate places. Then we need a unit of length with which to measure distances between those locations. As chronicled in chapter 1, Hipparchus provided us with a workable coordinate system in which locations on earth are described by their latitude and longitude. This grid system is likely to be familiar to you, but it is worth taking a few moments to review the basis for its use.

First, for the system of latitude and longitude to be useful, we need to attach it to the globe. And for that, we need three points fixed on the planet's surface, points that do not all lie on the same straight line. The selection of points is, to a certain extent, arbitrary, but the poles—the two spots where earth's axis intersects the planet's surface—are obvious choices for two of our three required points, and they have been used in this capacity since antiquity.

How does one locate the poles? As a practical matter, you could find them by traveling north or south until you came to a spot where, when you looked directly overhead at night, the star at which you gazed did not move in the course of the evening. In Apsley Cherry-Garrard's terms, the poles are the places where "a man loses his orbit and turns like a joint [of meat] on a spit."

Once you have located the poles, you can then use them to provide information about other locations on the globe. For example, San Francisco, California, is 3/10 of the way from the North Pole to the South Pole. But then, Athens, Greece, is also 3/10 of the way from the North Pole to the South Pole, as is

Figure 2.1. Specification of three points (the North Pole, South Pole, and Greenwich, England) fixes the latitude-longitude coordinate system to the globe, allowing us to locate spots such as San Francisco and Athens.

Seoul, Korea. To be able to discriminate among these cities, two fixed points won't suffice; we need that third point.

Unlike the poles, there is no obvious choice for this third fixed point, and the decision regarding its location is a matter of history rather than logic. In the 1700s, when maps of the world were becoming standardized, the British chose a third point on the globe that fell within Great Britain: the location of a specialized astronomical instrument at the national observatory in Greenwich, England. Other nations used other locations (Paris for France, Cadiz for Spain, for instance), but the prevalence of British charts made Greenwich a de facto standard that was made official by an international conference in 1884.

The utility of the third point at Greenwich is this. We draw a line on earth's surface that extends from one pole to the other and passes through Greenwich (figure 2.1). If we stand on this line and face the North Pole, we define the direction "west" as lying to our left and "east" as lying to our right. We can then say that San Francisco is three tenths of the way from the North Pole to the

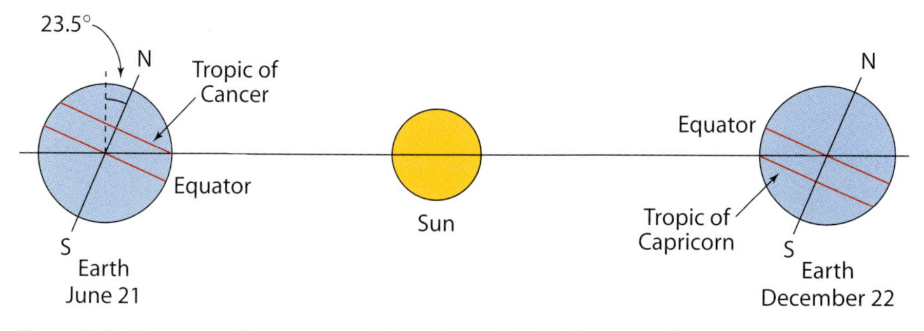

Figure 2.2. At noon on the summer solstice (June 20–22), the sun is directly overhead at the Tropic of Cancer. At noon on the winter solstice (December 21–22), it is directly overhead at the Tropic of Capricorn.

South Pole, and one third of the way around the earth west of the line through Greenwich. In this fashion, we can discriminate between San Francisco and Athens, because Athens is 1/15 of the way around the earth east of Greenwich. These three points, the two poles and Greenwich, thus serve to fix the system of latitude and longitude onto the globe.

A brief review of that system is in order. The line on earth's surface that lies equidistant from the poles is the equator, 0° of latitude. All latitude lines are parallel to the equator, and consequently they are often referred to as *parallels*. Now, the axis about which our planet spins is not perpendicular to the plane of earth's motion around the sun; it is currently tilted by 23.5° (figure 2.2). As a result, the latitude at which the noontime sun is directly overhead varies through the year, from 23.5° South on the winter solstice in the northern hemisphere (December 21 or 22) to 23.5° North on the summer solstice in the northern hemisphere (June 20, 21, or 22).[1] These limiting latitudes mark the boundaries of the tropics: the Tropic of Cancer in the north, the Tropic of Capricorn in the south.

Any line on earth's surface extending from pole to pole is called a *meridian*, and the line that passes through Greenwich is the *prime meridian*, 0° of longitude (figure 2.1). Unlike the lines of latitude, meridians are not parallel; they converge at each of the poles.

In the tradition of Hipparchus and Ptolemy, any circle of latitude or longitude is divided into 360 degrees. Each degree is divided into 60 minutes, and each minute into 60 seconds. Thus, we can precisely locate any point on earth by noting how many degrees, minutes, and seconds it is north or south of the equator (this is its latitude) and how many degrees, minutes, and seconds it is east or west of the prime meridian (this is its longitude). Again, we can use San Francisco as an example. Along the meridian that passes through San Francisco, the city is approximately 1/10 of the way around the earth north of the equator.

[1] It takes the earth approximately 365.24219 days to complete an orbit around the sun (a period known as the "tropical year"). Because this period is not an integral number of days, the seasons (including the solstices) would migrate through the calendar if the "year" were set simply to 365 days. Instead, an extra day is added every four years (leap years), and other adjustments are made every 100 and 400 years to keep the seasons from wandering. The nature of these adjustments causes the date of the solstices to vary slightly.

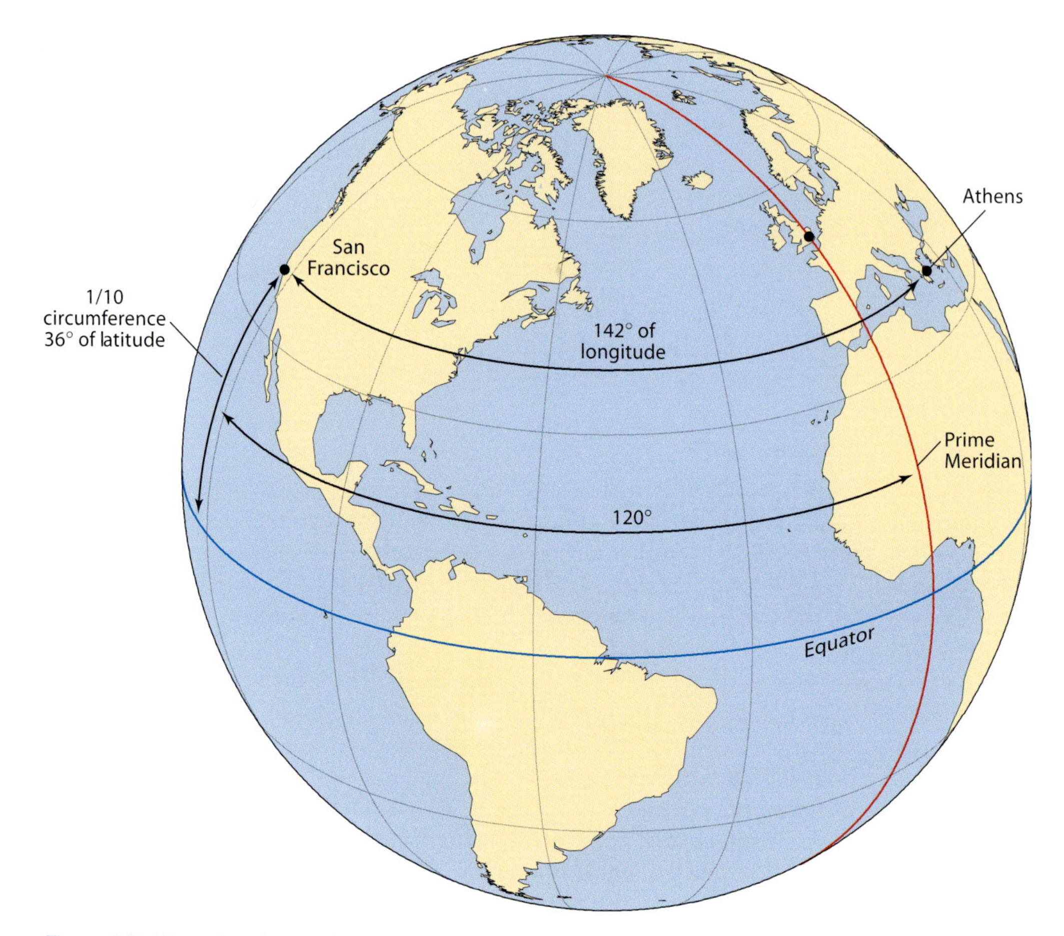

Figure 2.3. Measuring the angular separation between San Francisco and Athens.

Thus, the latitude of San Francisco is about 360°/10 = 36° North. As we have noted, San Francisco is about 1/3 of the way around the earth west of the prime meridian, so its latitude is 360°/3 ≈ 120° West (figure 2.3).

This system is logical and precise, but it provides information in a form that is not a part of everyday experience for those of us who travel primarily by land. For example, if you wanted to travel from San Francisco (36° North, 120° West) to Athens (36° North, 22° East), you could simply progress 142° eastward at latitude 36° North (figure 2.3). But how long a trip is that? How many steps would you have to take to walk from San Francisco to Athens? How much gas would you have to put into your car? For us terrestrial folk, it would be more convenient to measure the journey in terms of distance traveled rather than degrees of longitude.

History provides us with two systems for measuring distances on earth. The first was devised specifically for the use of mariners. Each degree of latitude on a sphere corresponds to a certain distance on the sphere's surface. Thus, we can define distance directly in terms of degrees. With an implicit nod to Ptolemy, we define a *nautical mile* as the distance on earth's surface corresponding to 1/60 of a degree of latitude. Note that this definition does not work for degrees of longitude, except at the equator. The distance around the earth measured parallel to the equator is smaller the farther one is from the equator (figure 2.3). As a consequence, a degree of longitude (being 1/360 of the way around the earth) gets progressively smaller the farther one is from the equator, and this variable distance would not be an appropriate measure for the nautical mile. Because nautical miles are easily translated to degrees of latitude, they are still in common use among sailors. (Actually, the current definition of the nautical mile isn't quite as simple as stated above, but we will postpone for a moment a discussion of the complexity.)

As useful and logical as the nautical mile is, it is not the unit of measure used by scientists. Instead, science employs the *meter*, a measure of distance with its own interesting history. As a spirit of revolution swept through France in the late 1700s, French scientists grasped the opportunity to scrap the nation's archaic system of weights and measures and to substitute a standard system based on sound scientific principles. When it came to the measure of length, there was considerable debate. For example, some argued that the standard of length should be based on a measure of time. The longer the arm of a pendulum, the longer it takes to swing back and forth, and in this proposal, the standard length would be that of a pendulum with a period of precisely one second. But opponents noted that this length varied slightly from place to place on earth due to variations in the strength of gravity. It wouldn't do to have a standard length that was variable. Instead, the opponents argued, the standard of length should be based on the dimensions of the earth, dimensions that presumably would not change. This second group won the debate, and the meter was defined as one ten-millionth the distance along a meridian from the North Pole to the equator,[2] a distance that, coincidently, is very close to the length of a pendulum with a period of one second. With that decided, all the French had to do was to measure the size of the earth, thereby establishing the length of a meridian, and the length of the meter would be fixed.

The method used to quantify the length of the meridian was, in principle, the same used by Eratosthenes 2000 years before. The French chose a section of meridian running from Dunkirk, France, in the north to Barcelona, Spain, in the south. Careful astronomical observations fixed the latitude of these two cities, and it only remained to accurately measure the distance between them to determine the length of a meridian. Two eminent surveyors were charged with the task, and they set out to make the measurements, one working south from Dunkirk, the other north from Barcelona.

It was a hell of a time to conduct a large-scale survey in Europe, but the surveyors persisted through the French Revolution and several skirmishes between France and Spain, doggedly gathering data as governments came and went. Eventually, in 1799, after more than seven years of painstaking work, the meridian was measured and the meter was set. The prescribed distance was etched onto a platinum-iridium bar, held safe in a vault in Sèvres (near Paris) for future use: a standard for the ages.

There are several aspects of the meter that deserve mention. First, the standard meter is the wrong size. A quarter meridian of the earth, which by definition should be 10,000,000 meters long, is actually 10,002,001.23 meters long

[2] Specifically, the meridian running through Paris. After all, this was a Gallic endeavor.

as measured by the standard meter in Sèvres, an error of 0.02%. Furthermore, the meter bar in Sèvres isn't conveniently portable. If you conduct field measurements in Auckland, New Zealand, and want to know if your meter stick is accurate, it would be a bother to have to catch a plane to France to compare your stick to the standard bar. As a practical solution to this problem, the standard length of the meter has been redefined in terms of a portable standard, the wavelength of light. One meter is now defined as exactly 1,650,763.73 times the wavelength in a vacuum of the light produced by the unperturbed transition of electrons between the $2p_{10}$ and $5d_5$ quantum levels of krypton 86. This definition makes somewhat liberal use of the term "practical": one can use the definition as a measure of the meter, but only if one has a physics lab with the equipment for measuring wavelengths of light and a willing physicist to use the equipment. But at least the definition can be applied in theory.

The French survey of the meridian had other unintended consequences, as well. The surveyors were unsure how to cope with the inevitable errors in measurement that arose during the survey, and approached a leading mathematician of the day, Johann Friedrich Gauss, for advice. In the process of exploring this problem, Gauss developed a mathematical model for how errors should be distributed. This bell-shaped distribution comes down to us as the "normal" or "Gaussian" distribution that is central to the field of statistics.

Another intriguing result from the measurement of the meridian was proof that the earth is not a sphere. As the surveyors progressed along the meridian, they found that the distance on earth's surface corresponding to a degree of latitude varied with distance from the equator: the farther north they went, the longer the distance. This effect had been predicted a hundred years before by Sir Isaac Newton. As the earth spins, the resulting centrifugal force causes the planet to bulge along the equator, and this deviation from a spherical shape explains the variation in the length of a degree noted by the French. We will return to the physics of this effect when we discuss centrifugal force in more detail in chapter 7.

In the present context, however, the latitudinal variation in the length of a meridional degree raises a problem in the use of the nautical mile. We noted above that a nautical mile is 1/60 the distance of 1° along a meridian. If this distance varies with latitude, the length of the nautical mile also varies. Indeed, by the definition we have used, the length of a nautical mile varies from about 1843.1 meters at the equator to 1861.4 meters near the poles. As a practical solution to this problem, the International Nautical Mile was defined in 1929 as exactly 1852 meters, the average length across latitudes. The United States is often at odds with the rest of the world, and it did not adopt the international standard until 1954.

Thus, the meter and nautical mile. Of course, if you are reading this in the United States, Liberia, or Myanmar (Burma), neither of these measures is particularly familiar because these countries persist in using yards and statute miles as their common units of length. The yard was originally defined as the length between the nose and outstretched thumb of King Henry I of England, and the statute mile is 1760 yards, distances that are neither logical nor precise. Conversions among meters, yards, nautical miles, and statute miles are given in table 2.1.

As we have seen, meters and nautical miles provide standard units with which to measure distances between points on earth's surface, and we have

Table 2.1 Conversion factors for length

To convert from	to	multiply by
meters	yards	1.09361
yards	meters	0.91440
meters	kilometers	0.001
kilometers	meters	1000
kilometers	statute miles	0.6214
statute miles	kilometers	1.6093
kilometers	nautical miles	0.5400
nautical miles	kilometers	1.8520

briefly explored an example in which meters were used to define the distance between two locations on the same meridian (Barcelona and Dunkirk). Other measurements are not as straightforward. For example, calculating the shortest distance between two points that lie at the same latitude but along different meridians (e.g., San Francisco and Athens) is more difficult. And what if one desires to know the distance between two points at different latitudes *and* longitudes, London and Rio de Janeiro, for instance? In this case, the calculation is far from straightforward. In fact, on the surface of a sphere, a general answer to the basic question, "How far is it from here to there?" isn't as simple as one might expect. It involves principles of spherical trigonometry, and a brief primer is provided in the appendix to this chapter. Nonetheless, with units of measurement in hand and the coordinate system of latitude and longitude to apply them to, we can get on with the business of describing our world in quantitative fashion.

The Dimensions of the Earth

The radius of earth, measured from its center to either of the poles, is 6357 kilometers. As noted above, the distance from earth's center to the equator is slightly more, 6378 kilometers, a difference of 21 kilometers. This difference is a measure of the equatorial bulge predicted by Newton and verified by the French.

As a sphere (well, very nearly a sphere) with an average radius of $R = 6367.5$ kilometers, we can easily calculate that the circumference of earth is $2\pi R =$ 40,008 kilometers, the value so nearly predicted by Eratosthenes. Similarly, we can calculate the surface area of the globe: $4\pi R^2 = 510$ million square kilometers. Of this area, about 29.2% is dry land (149 million square kilometers) and 70.8% is ocean (361 million square kilometers). It is this preponderance of ocean that makes earth appear from space as a "blue planet."

The average height of land above sea level is 875 meters (2871 feet, roughly half a statute mile). This figure may seem surprisingly high to most people. For example, in the United States, Denver (the "Mile-High City") is noted for its elevation, although it is only about 800 meters above the world average. Perhaps Denver seems so high because most people in the United States live along the periphery of the continent, where land is near sea level. The extensive, but thinly populated, high plateaus of places like Tibet and Antarctica raise the average height of the continents above what one might intuitively expect.

The average depth of the ocean is 3794 meters, about 2.5 statute miles. In other words, on average the oceans are more than four times as deep as the continents are high. For those of us who are impressed by the sight of the Rocky Mountains or the Alps, the depth of the ocean sounds impressive. But it is valuable to view the ocean depth in relation to the size of the planet: the average depth of the ocean is only 0.061% of the radius of the earth. In other words, deep as they are, the oceans are merely a thin skin on the globe. It may also be useful to think of ocean depth relative to ocean area. For example, the Pacific Ocean has the same ratio of surface area to thickness as does a thin sheet of typing paper.

In an interesting quirk of history, the average depth of the Pacific Ocean was first deduced in 1856 by Alexander Dallas Bache, a grandson of Benjamin Franklin. From eyewitness reports of an earthquake in Japan and records from tidal stations in California, Bache measured the time it took for a tsunami to cross the Pacific. Knowing from physics that the speed with which tsunamis travel depends on the depth of the ocean, he used this transit time to estimate that the Pacific is about 3600 meters deep. This estimate is not far off the actual value of 4280 meters.

The highest point on the continents is Mount Everest: on the border between Nepal and Tibet; it rises 8848 meters above sea level. In contrast, the deepest point in the ocean—the Challenger Deep—is 11,022 meters below sea level. The deepest of the ocean depths is 24% greater than the highest of the mountain peaks.

The Ocean Habitat

Let's now take a look at the pattern of land and sea on earth (figure 2.4). Several points should leap out at you. First, land and ocean are not evenly distributed across the globe. Nearly 81% of the southern hemisphere is ocean, while only about 61% of the northern hemisphere is ocean. In other words, most of

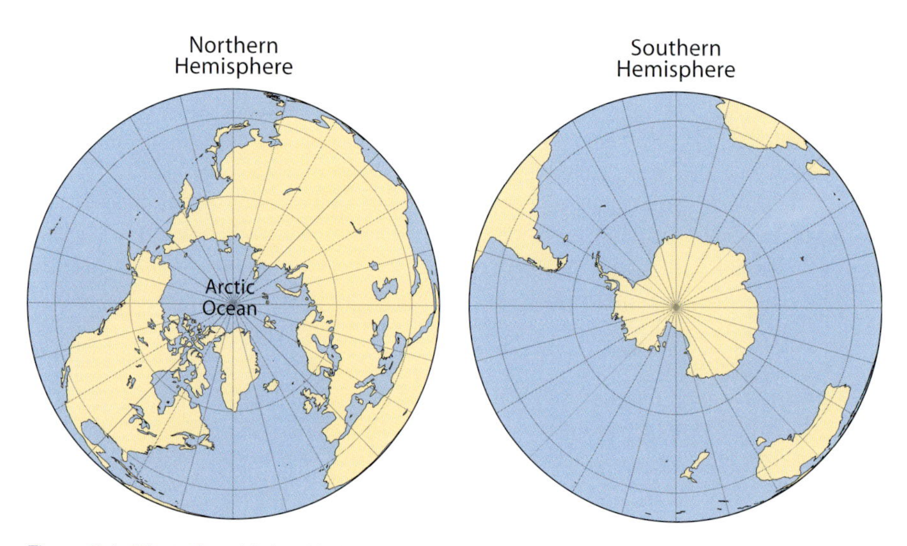

Figure 2.4. Most of earth's land is in the northern hemisphere, most of the ocean in the southern hemisphere.

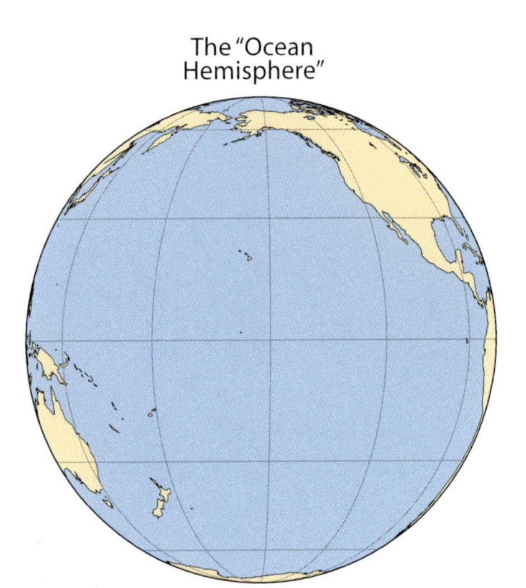

Figure 2.5. The Pacific Ocean occupies nearly an entire hemisphere.

the land is concentrated in the northern hemisphere, while most of the water is concentrated in the southern hemisphere.

The contrast becomes even more stark if we allow ourselves to detach from the traditional north-south division of the earth. If we were to center ourselves not above one pole or the other, but rather above the equatorial Pacific Ocean, we would find ourselves looking down on an "ocean hemisphere" (figure 2.5). The Pacific is so big that it occupies nearly half our world's surface.

When we view the oceans from above in this way, we tend to think in terms of their area. However, for organisms living in the sea, volume is as important as area. How much water is there in the oceans? Multiplying the oceans' surface area (361 million square kilometers) by their average depth (3.8 kilometers), we find that they contain approximately 1.4 billion cubic kilometers of water. Now, a large home aquarium might contain a single cubic meter of water, so the oceans' volume is the equivalent of 1.4 billion billion home aquaria. This is a huge habitat, 220 million aquaria for each of the 6.4 billion people alive on earth today.

It is difficult to compare the size of the marine environment to that of the environment on land because, unlike the ocean, which has a fixed depth, the habitat available to terrestrial organisms has no easily fixed height. If we assume, however, that most plants and animals on land are found within 100 meters of the surface, we can calculate that the volume of the terrestrial habitat is roughly 0.014 billion cubic kilometers. By this calculation, the oceans (which are habitable from surface to seafloor) are 100 times as large as the terrestrial habitat. As terrestrial organisms, this fact is difficult for us humans to absorb. For all the living space we know on land, there is 100 times as much for us to explore below the surface of the ocean.

Ocean Basins

As with any large habitat, to facilitate discussion of the ocean, we need to break it up into smaller parts and give those parts distinctive names. Some of

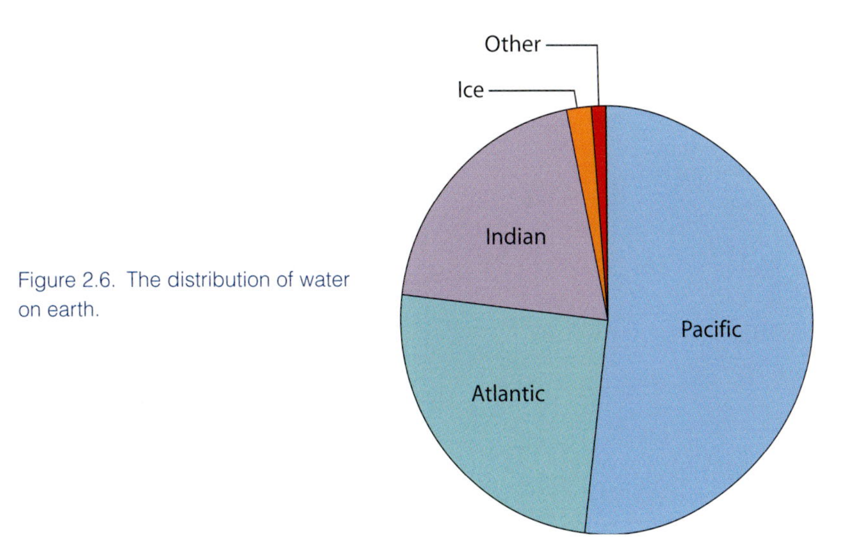

Figure 2.6. The distribution of water on earth.

this categorization we have already accomplished. The world ocean is conventionally divided into three parts: the Pacific, the Atlantic, and the Indian Oceans. The boundaries between these oceans are somewhat arbitrary. For example, the boundary between the Pacific and Atlantic lies along a line drawn south from Cape Horn at the southern tip of South America. Similarly, the Atlantic meets the Indian Ocean along a line south of Cape Agulhas at the southern tip of Africa. The interface between the Pacific and Indian oceans is more complex.

The Pacific Ocean is by far the largest of the three oceans (figure 2.6). More than half (52%) of all water on earth is found in the Pacific. The Atlantic Ocean is next largest, containing about 25% of the world's water, and the Indian Ocean is close behind with 20%.

A quick bit of math reveals that when the water in the Pacific, Atlantic, and Indian Oceans are combined, there isn't much left for the rest of the world. More than 97% of all the water on earth is in its oceans. Furthermore, most of what is left is tied up in ice: about 2% of earth's water resides in the glaciers, ice fields, and snow that coat the polar regions. About 0.5% of all water is ground water, and only the remaining 0.01% is in lakes, rivers, and the atmosphere.

Even as subdivisions of the world ocean, the Pacific and Atlantic are themselves so large that they often need to be subdivided. Thus, the Pacific is commonly split in two at the equator to give the North Pacific and South Pacific. Similarly for the Atlantic and Indian Oceans.

Then there are the polar oceans. The Arctic Ocean is well defined because it is more or less confined by continents and islands (figure 2.4, left). Less well defined is the Southern Ocean, the confluence of the Pacific, Atlantic, and Indian Oceans that surrounds Antarctica (figure 2.4, right). Various boundaries have been used for the Southern Ocean, but a dividing line at latitude 65° South will serve for present purposes.[3]

[3] The Southern Ocean should not be confused with Balboa's "Southern Sea," an archaic name for the Pacific.

A variety of smaller bodies of water, subsections of or adjacent to the oceans, are termed seas: for example, the Norwegian, Baltic, and North Seas adjacent to northern Europe, the Bering and Chukchi Seas between Siberia and Alaska, the South China Sea, and the Caribbean Sea. The boundary lines between seas and oceans are often somewhat vague. Here we treat the seas as subsections of the ocean, so when we speak of "the ocean," the seas are included.

In the early days of oceanic exploration, a well-traveled adventurer was said to have "sailed the seven seas." At the time—prior to the rapid expansion of geographical knowledge in the fifteenth century—the seven known seas were the Adriatic, Black, Caspian, Mediterranean, and Red Seas, along with the Persian Gulf and the Indian Ocean.

The Ocean in Cross Section

So far, we have discussed ocean basins viewed from above, as if we were orbiting in space. Let us now switch perspectives and describe the ocean basins as if we were to drain the water and drive across them. What would we see?

The common perception is that the bottom of the ocean is a featureless flat plain of muddy sediment, the kind of seafloor we see in pictures of the wreck of the *Titanic*. Although there are, indeed, large stretches of flat floor under earth's seas, there is also much more. In fact, a drive across the Pacific or Atlantic basin would not be all that different from a drive across North America. The ocean floor has vast mountain ranges and huge canyons, isolated peaks and deep trenches, all interspersed with Kansas-like areas of plains. In the absence of life and water, it would be difficult to tell from space what was continent and what was ocean basin.

In fact, the similarity between the topography of continents and ocean basins has been a problem for planetary scientists studying Mars. It has been suggested that millions of years ago Mars had an extensive ocean that could have supported life. However, if there ever was an ocean on Mars, it dried up long ago. As a result, as we look at Mars today, we must differentiate its terrain based on topography alone. But, you might think, if Mars ever had an ocean, we should be able to see the basin it left behind. Wouldn't something as big as an ocean basin be obvious?

Apparently not. Certainly there are parts of the Martian surface that are depressed below others, and therefore could have contained an ocean. However, there are many parts of the continents on earth that are similarly depressed, but they are not ocean basins. Detailed topographic maps of Mars made by the Mars Global Surveyor revealed evidence of areas around some of the Martian depressions that resembled the wave-sculpted continental shelves on earth, and on this basis, some planetary scientists proposed that, indeed, Mars once had an ocean. Their evidence was immediately challenged by other scientists, who interpreted the data differently, and the debate continues. In fact, it is proving very difficult to look at the topography of Mars and tell whether it had an ocean, and the issue will be settled only after more scientific probes have successfully landed on the planet and viewed its surface close up.

But I digress. To explore the complexity of earth's ocean basins, let's start on a continent and drive across the basin as proposed above (figure 2.7). As we head out from shore, the first thing we encounter is the *continental shelf*, actually

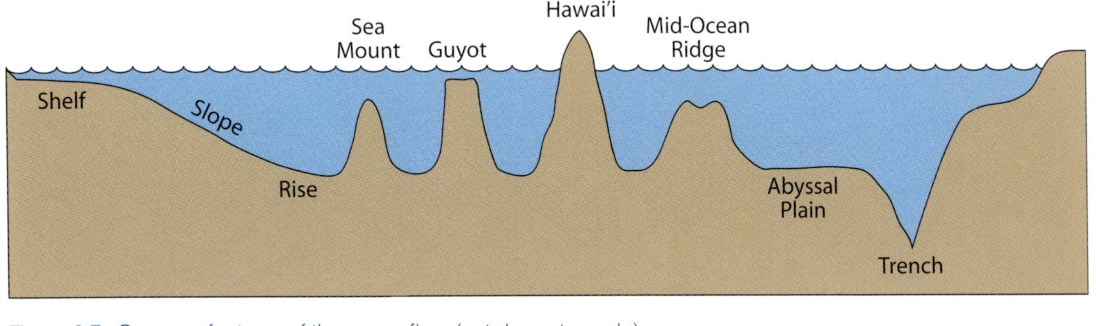

Figure 2.7. Common features of the ocean floor (not drawn to scale).

a submerged part of the continent. The shelf has a gentle slope, about 1 meter down for every 500 meters we travel forward, and on average it extends about 100 kilometers from shore. There is much variation around this theme, however. In places, the shelf is rent by deep canyons. For example, the Monterey Submarine Canyon in Monterey Bay, California, closely resembles in length and depth the Grand Canyon of the Colorado. There are places where the continental shelf extends no farther than a stone's throw from shore and other places where it goes on for more than 200 kilometers.

In the course of earth's history, the continental shelves have been alternately exposed and submerged. During ice ages, when much of earth's water is frozen in glaciers, sea level is several hundred meters lower than it is now, low enough to expose the shelves.

As we reach the edge of the continental shelf, the seafloor begins to slope steeply downward. This is the *continental slope*, the true edge of the continent, with a gradient of about 1 in 25. We can put this slope in perspective by noting that the maximum grade a railroad train can handle is about 1 in 50, half the steepness of the continental slope. You would not want to pedal a bicycle far up this part of the ocean basin. Nonetheless, at a slope of 1 in 25, it takes about 100 kilometers of forward travel for us to descend the 4 kilometers required to reach the average depth of the ocean.

At the foot of the continental slope is the *continental rise*, a more gradual descent that eventually delivers us to the *abyssal plain*, the Kansas of the sea, 4 kilometers below sea level. Traveling across this plain, we would occasionally encounter isolated mountains. Many, called *seamounts*, are not tall enough to reach the surface. Others, called *guyots* (pronounced ghee-OHS), rise close to the surface and have characteristically flat tops.[4] These peaks once extended above sea level, where their tops were worn flat, and they have subsequently subsided or have been submerged by rising sea levels. Still other mountains extend from the seafloor to well above sea level. The Hawai'ian Islands are a prime example of this kind of oceanic mountain. These islands are volcanic in origin, and there are still active volcanoes on Hawai'i, the biggest (and youngest) of the islands. Mauna Loa is the largest, with its peak about 4300 meters above sea level, not a notably tall mountain. But measured from its foot on the ocean floor, Mauna Loa is over 10,500 meters high, nearly 2000 meters higher than Mount Everest.

[4] Guyots are named in honor of Arnold H. Guyot, a Swiss-born American geologist (1807–84).

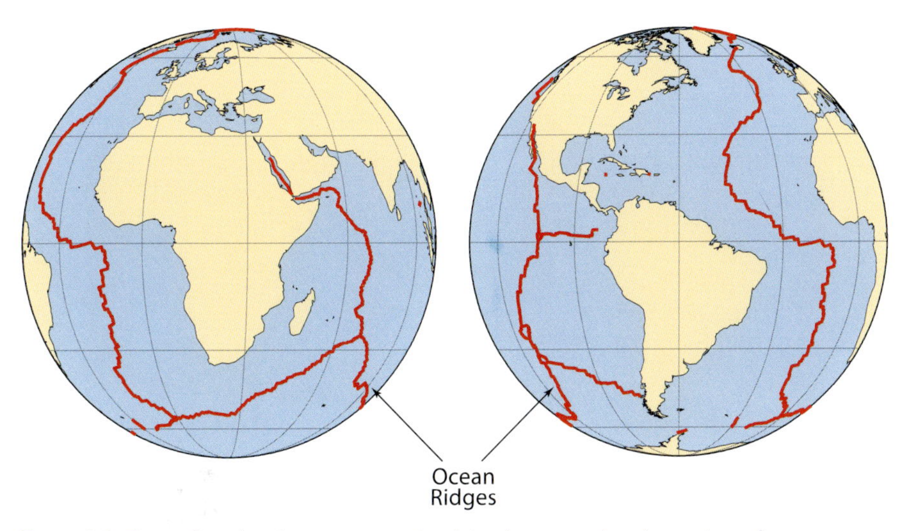

Figure 2.8. Spreading sites between tectonic plates form an extensive system of ocean ridges.

In addition to these isolated peaks, our drive takes us over a huge mountain range in the middle of the ocean basin. These oceanic mountains resemble the Rocky Mountains of North America in the height of their peaks—about 2 kilometers—but they are part of a much longer range. In oceanographic terminology, this mountain range is an example of a *mid-ocean ridge*, and it is part of a continuous range of mountains that extends into all three of the major oceans (figure 2.8). The total length of the combined mid-ocean ridges is 60,000 kilometers, many times the length of any terrestrial mountain range. It is the single largest geological structure on the face of the earth.

After we have picked our way over the mid-ocean ridge, which may, disconcertingly, be located far from the middle of the ocean basin, and have driven across another stretch of abyssal plain, we approach the far side of the basin. In most places, this approach is the reverse of our initial descent: a drive up the continental rise, then a steep grind up the slope, and out onto the continental shelf. In some places, however, before we get to the continental rise and slope, we must first drive down into a deep trench that runs parallel to the continental margin (figure 2.7). For example, if we were to drive east across the Pacific Ocean basin toward Ecuador, Peru, or Chile, we would encounter the Peru-Chile Trench, which descends to a depth of 8000 meters below sea level. There are similar trenches off the coasts of the Philippines, the Aleutian Islands, and Puerto Rico. We have already mentioned the Mariana Trench, which contains the Challenger Deep.

Plate Tectonics, Revisited

In chapter 1, we encountered Alfred Wegener and his proposal that continents drift. We briefly learned that continents do indeed move about on the globe, driven by the spreading of the sea floor. The process of continental drift, now incorporated into the field of plate tectonics, provides an explanation for much

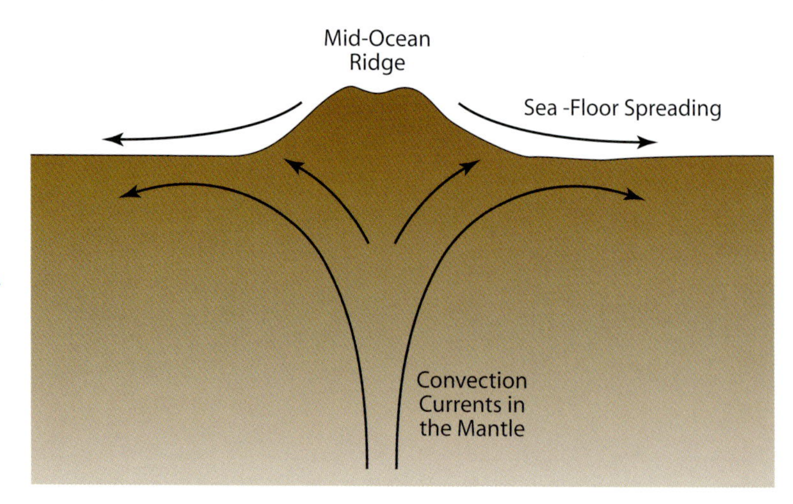

Figure 2.9. Convection currents in the mantle cause seafloor spreading.

of the topography we saw on our drive across the ocean basin. We will not delve into plate tectonics in depth, but an overview is appropriate.

First, we note that the seafloor and the continents are made from different materials. The ocean floor is typically formed of basalt, rock with a density of about 3000 kilograms per cubic meter. In contrast, the continents are made of granite, with a typical density of 2700 kilograms per cubic meter. Because granite is less dense than basalt, the continents "float" on the basalt beneath them, just as ice floats on water. And because continents float, they move at the whim of the basalt below. Thus, the process of continental drift is a secondary result of the manner in which basaltic seafloor spreads.

The mid-ocean ridges are areas of active seafloor spreading (figure 2.9). Driven by convection currents in the earth's mantle, molten basaltic rock is pushed up to the seafloor at the center of the ridge, and then is forced sideways. This divergence of molten rock has several effects. First, it produces the parallel ridges typical of mid-ocean mountain ranges. Second, it causes the ocean floor to spread away from the center of the ridge. It is this spreading that accounts for the gradual increase in the width of the Atlantic Ocean, for instance, and the consequent drift of North America and South America away from Europe and Africa. The seafloor spreads at a rate of 2 to16 centimeters per year, roughly the rate at which your toenails and fingernails grow.

Measurements of the flux of heat through the oceans' floor support this idea of sea-floor spreading. For example, along the Mid-Atlantic Ridge, heat rapidly pours out into the ocean, but farther from the ridge, the flux of heat decreases.

The earth as a whole is not getting any bigger, so if new seafloor is produced at mid-ocean ridges, old seafloor must disappear somewhere else. In general, this "recycling" of seafloor material occurs at continental margins. As the seafloor butts up against a continent, it may be pushed down below the continental margin, a process known as *subduction*. Subduction, in turn, has several effects. As seafloor rock (and any sediment on it) is driven into the mantle, it is heated and becomes molten. Much of this liquid rock then joins the grand *convection cell* that powers seafloor spreading at the mid-ocean ridge. Some of the melted rock, however, finds its way upward to the surface locally, forming volcanoes along the continental margin. The string of volcanoes in the Andes, inland of the west coast of South America, is a prime example.

As seafloor is pushed down below the continents, the margin of the continents can, in reaction, be lifted up. Uplift of this sort is responsible for the Andes.

A wonderful variety of other effects can be explained by plate tectonics. For example, because of the manner in which new seafloor is produced along mid-ocean ridges and old seafloor is recycled by subduction, there is a gradient in the age of the seafloor from ridge to continent. The longer a patch of seafloor has been in existence, the more sediment it can accumulate. Thus, one would expect the depth of sediment to increase from ridge to continent, and this is exactly what we find. The discovery of seafloor recycling also solved an historic paradox that plagued marine geologists for decades. Throughout the early years of the twentieth century, measurements of the rate of sedimentation and the age of the ocean basins suggested that ocean sediments should be very thick. Actual measurements of the thickness of seafloor sediment, however, showed them to be quite thin. Something was clearly amiss, but it wasn't until the theory of plate tectonics came along that the two sets of measurements could be reconciled. Yes, ocean sediments accumulate at a relatively rapid rate, but the process of seafloor spreading does not allow sufficient time for thick sediments to form before the seabed is recycled.

Seafloor basalt is hot when formed at a mid-ocean ridge, and it cools as it moves away from the ridge. As the rock cools, it contracts. As a result, volcanic islands produced near ridges gradually subside as they ride the slowly contracting seafloor conveyor belt. This is one of the mechanisms behind the process Darwin proposed for the formation of coral atolls.

The discussion here of plate tectonics is merely an appetizer for the intellectual banquet of this field. We have left many obvious questions unanswered: How did the continents get there in the beginning? Why does the earth have the number of tectonic plates that it does? Where does the heat come from that powers plate movements? For answers to these and other fascinating questions, you should consult a text dealing specifically with geology.

Horizontal and Vertical Terminology

Just as we have divided the world ocean into smaller areas for ease of description, we can divide the marine habitat into different categories.

One scheme is to divide the territory depending on its distance from shore (figure 2.10). The portions of the shore that are sequentially covered and uncovered by the tides are termed *intertidal* or *littoral*. Just offshore, the waters that cover the continental shelf are termed *neritic*. The vast remainder of the oceans waters are termed *pelagic*. These terms are also commonly applied to organisms. For example, a particular species of fish can be described as intertidal, neritic, or pelagic, depending on where it lives.

We can also divide the marine habitat vertically. Water near the ocean's surface is called *epipelagic*. This zone is also referred to as the *euphotic* ("well-lit") zone because it is the part of the ocean where there is sufficient light to power photosynthesis effectively. We will return to this subject in chapter 4. The depth of the euphotic zone varies depending on the clarity of the water; it is typically 50 to100 meters and seldom exceeds 200 meters.

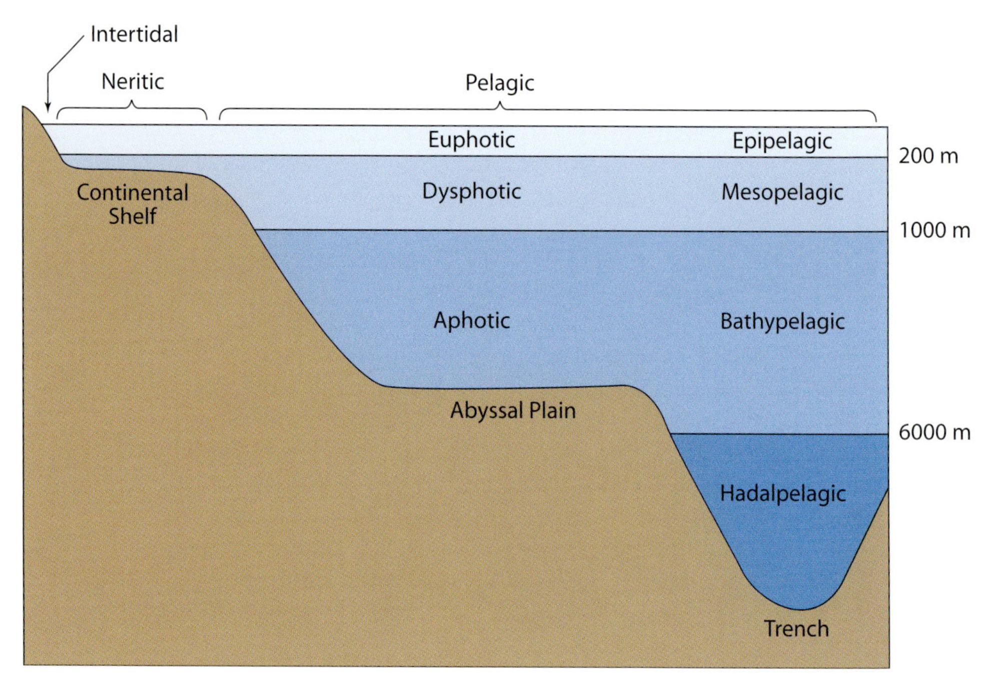

Figure 2.10. The named zones of the ocean.

Below the epipelagic zone is the *mesopelagic* or *dysphotic* ("badly lit") zone. Here there is enough light for animals to see, but not enough for plants to make a living by photosynthesis. Again, the depth of this zone depends on the clarity of the water, but by a depth of 1000 meters, light is generally gone.

Below the mesopelagic zone is the *bathypelagic* or *aphotic* zone, which extends all the way down to the seafloor. This is by far the largest zone in the ocean, comprising fully three quarters of the marine world. It is totally dark—*aphotic* means without light—and as we will see, it is cold.

Occasionally one will see reference to one other zone within the ocean. The term *hadalpelagic* is used to refer to depths below 6000 meters, essentially to the water in the great ocean trenches. The term "hadal" is a reference to Hades, the underworld.

Plants and animals in the sea can be characterized by their habits as well as by their habitats. Organisms that live on or in the seafloor are called *benthic*, whereas organisms that live in the water column above are called *pelagic*. Sea urchins, snails, corals, and clams are all examples of benthic organisms, whereas whales, squid, and most jellyfish are pelagic.

And finally, pelagic organisms can be characterized by their swimming ability. If the plant or animal is a weak swimmer or cannot swim at all, its movement through the ocean is governed by the motion of the water around it. These passive organisms are the *plankton*. We will deal with the phytoplankton (planktonic bacteria and algae) in chapter 4, and the zooplankton (planktonic animals) in chapter 5. In contrast, organisms that can control their movements independently of the water around them are the *nekton*. Fish, squid, dolphins, and whales are prime examples of marine nekton.

Summary

In this chapter, we have quantified the size and shape of the ocean basins. The ocean is the largest habitat on earth, with 71% of earth's surface area and 1.4 billion cubic kilometers of livable volume. The topography of the ocean floor—its mountains and trenches—results in large part from the process of plate tectonics, and the constant spreading of the ocean floor has caused the continents to drift. Although the oceans are but a thin film on the earth's surface, they form a huge and complex habitat nonetheless, and we have given names to its different parts.

And a Warning

This chapter has described the oceans as they are today, but they have not always been in this configuration. Three hundred million years ago, all of earth's continents were bunched together to form one single land mass, which paleogeographers call *Pangaea*. Pangaea stretched from pole to pole and occupied roughly a third of earth's circumference at the equator. Surrounding Pangaea was a single ocean, *Panthallassia*. Then, 200 to 225 million years ago, Pangaea split into two supercontinents, *Laurasia* in the northern hemisphere and *Gondwanaland* in the southern hemisphere, separated by the shallow *Tethys Sea* along the equator. Subsequent splitting, reorientation, and drift led to the geography we see today. As the continents performed their global do-si-do, the shape of the ocean basins shifted drastically, and the dynamics of these ancient oceans must have been substantially different from those of the Pacific, Atlantic, and Indian Oceans of today.

Unfortunately, the intriguing story of the wandering continents and their effects on the ocean is well beyond the scope of this book. It is quite large enough a task for one book to explain the oceans as they are now. With our attention firmly focused on the present, let's turn to seawater, the liquid that fills the ocean basins.

Further Reading

Cox, A., and R. B. Hart (1986). *Plate Tectonics: How It Works.* Blackwell Scientific Publications, Inc., Boston.

Oreskes, N. (2003). *Plate Tectonics: An Insider's History of the Modern Theory of the Earth.* Westview Press, Boulder, CO.

Wilford, J. N. (2001). *The Mapmakers: The Story of the Great Pioneers in Cartography from Antiquity to the Space Age* (revised edition). Vintage Books, New York.

Appendix

Calculating Distances on a Spherical Surface

The task before us is to calculate the shortest distance between any two given points on our planet's surface. For points close together—within a few hundred kilometers—the job is straightforward. At this small scale, the surface of the earth can be closely approximated by a flat surface, and we can plot the shortest distance from point A to point B by drawing a straight line on a conventional map. At larger scales, however, we encounter a fundamental problem in mapmaking: any attempt to represent the spherical earth on a flat map results in some distortion. For instance, a Mercator projection of the earth's surface (a standard method often used for wall maps) grossly exaggerates the size of areas near the poles (see figure 9.10). Different mapmaking methods have different advantages and disadvantages, but nothing can completely substitute for examining the earth in its true, spherical form.

Before tackling the measurement of distances on a sphere, let's review the measurement of distances on a circle. Consider a circle of radius R as shown in figure 2.A1. If we take a string R long and stretch it along the circumference of the circle (thereby measuring an arc length), the ends of the string subtend a certain angle measured relative to the center of the circle, as shown. This angle is, by definition, one *radian*. In other words, the ratio of distance along the circumference of a circle to the circle's radius is the central angle, measured in radians:

$$\text{radians} \equiv \frac{\text{distance along circumference}}{\text{radius}} \qquad \text{(Eq. 2.1)}$$

For example, because the circumference of a circle is 2π times its radius, there are 2π radians in a complete circle. This fact allows us to convert between degrees and radians. There are 360° in a circle, so:

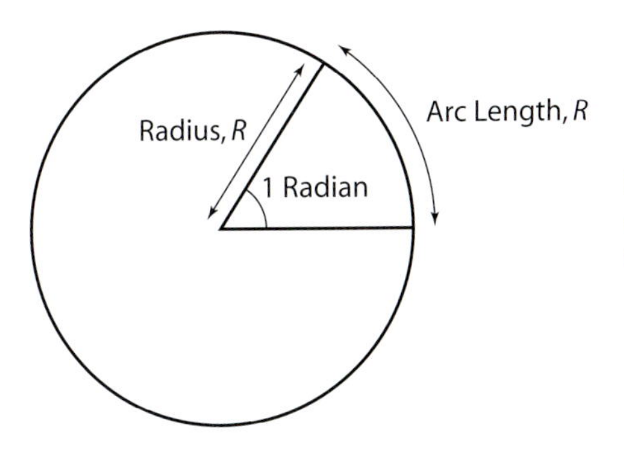

Figure 2.A1. Calculating distances on a circle. Radians are the ratio of arc length to radius.

$$\text{radians} = \text{degrees} \times \frac{2\pi}{360} \qquad \text{(Eq. 2.2)}$$

We can make practical use of the definition of radians by rearranging equation 2.1:

$$\text{distance along circumference} = \text{radians} \times \text{radius} \qquad \text{(Eq. 2.3)}$$

So, if we know the radius of a circle, we can calculate the distance that corresponds to any given central angle.

We need one more basic concept before we can get on with our calculation. If you take a sphere and slice straight through it, the edge of the slice is a circle. If your slice goes through the center of the sphere, its edge is a *great circle*, the largest possible circle for that sphere.

For any given sphere, there are an infinite number of great circles. For instance, on earth, the equator is a great circle, as is each meridian. Actually, meridians are a special case. There are many different meridians that connect earth's poles, but this is possible only because the poles are at exact opposite ends of the earth. For any two points on earth's surface *not* exactly at opposite ends, there is a unique great circle that connects them, and a path along this great circle is the shortest route from one point to the other. For instance, the great circle that connects San Francisco and Frankfurt, Germany, passes over the high Artic. This is the route that airliners fly between these cities, although it is certainly not the route one would intuitively pick from looking at a flat map.

Now consider the situation shown in figure 2.A2, in which I have picked two points at random on earth's surface (London and Rio de Janeiro). We desire to know the length of the great circle path that links them.

We begin by drawing a "triangle" that connects London, Rio, and the North Pole. Each side of this triangle is a segment of a great circle; sides *1* and *2* because they lie on meridians, and side *3* because we define it to be so: it is the great circle route we desire to measure.

For each segment of a great circle, let's use as a proxy for its length the angle the segment subtends with the center of the earth. Side *1* has angular length a, equal to the difference in latitude between Rio and the North Pole. Rio lies 22.27° south of the equator, so its angular distance from the North Pole is $22.27° + 90.00° = 112.27°$ (1.959 radians).

Figure 2.A2. Calculating distances on the globe, using an example discussed in the text.

London lies 51.30° north of the equator, so side *2* has angular length b equal to 90.00° − 51.30° = 38.70° (0.675 radians). Side *3* has angular length c, which we do not yet know.

At this point, we have a triangle. We know the length of two of its sides and are attempting to calculate the length of the remaining side. To do so, what else do we need to know? It turns out that if we can measure the angle between sides *1* and *2*, we can solve the problem. And because sides *1* and *2* are meridians, measuring the angle between them is easy. Rio's longitude is 42.43° (0.741 radians) West, and London's longitude is 0.07° (0.001 radians) West, both angles measured from the prime meridian, which passes through Greenwich. The difference in longitude between London and Rio is thus 0.740 radians, shown as angle C in the figure.

With these values in hand, we can now make use of a fundamental equation from spherical trigonometry (the analogue of the side-angle-side relationship for plane geometry):

$$c = \arccos[\cos(a)\cos(b) + \sin(a)\sin(b)\cos(C)] \qquad \text{(Eq. 2.4)}$$

Table 2.A1. Values used to calculate the great circle distance from London to Rio

$\cos(a)=\cos(1.959 \text{ radians})=-0.3785$	$\cos(b)=\cos(0.675 \text{ radians})=0.7807$
$\sin(a)=\sin(1.959 \text{ radians})=0.9256$	$\sin(b)=\sin(0.675 \text{ radians})=0.6249$
$\cos(C)=\cos(0.740 \text{ radians})=0.7385$	

In other words, we know two angular distances, a and b, and angle C. Taking the sines and cosines of these angles as appropriate, we plug into equation 2.4 to calculate the quantity inside the square brackets. The arc cosine of this quantity (the angle whose cosine equals this quantity) is the angular distance c, the value we desire. Values for this particular example are shown in table 2.A1. Thus, the quantity in the square brackets of equation 2.4 is 0.1317, and the arc-$\cos(0.1317)=1.4388$ radians.

To finish the calculation, we make use of equation 2.3. Multiplying angle c (1.4388 radians) by the average radius of the earth (6367.5 kilometers) gives us the great circle distance from London to Rio: 9163 kilometers.

The same set of calculations can be carried out for any two points on earth. If both points are in the southern hemisphere, it would be simpler to form the triangle using the South (rather than the North) Pole.

Seawater

Seawater is one of the most familiar substances on earth and, at the same time, one of the most mysterious. In the last chapter, we calculated that there are 1.4 billion cubic kilometers of the stuff, but as we will see, it is extremely difficult to specify precisely what it is. Seawater is both the crystal blue liquid of advertisements for tropical islands and the gun-metal grey menace of a storm at sea. It is even the cause for the normally staid editors of the *Handbook of Chemistry and Physics* to compose (perhaps inadvertently) a short poem for their index:

> Seawater, *see* Water, sea.

If we are to understand how the ocean works, we need to know some basic facts about the substance that fills the ocean basins. Let's begin with pure water and work our way up to the complexities of seawater.

Pure Water

The Chemistry of Pure Water

Pure water is H_2O: two hydrogen atoms bonded to an oxygen atom (figure 3.1). Hydrogen atoms have an atomic mass of 1 dalton each, and oxygen is 16 times as massive, so the molecular weight of water is 18 daltons.[1]

The structure of the electron orbitals in an oxygen atom is such that atoms bonding to oxygen attach themselves at points corresponding approximately to the vertices of a tetrahedron (figure 3.2). Thus, when two hydrogen atoms bond to oxygen to form water, the hydrogen atoms lie at an angle of about 104.5° to each other, not at opposite ends of the oxygen (an angle of 180°) as one might expect.

[1] One dalton is 1/12 the mass of a typical carbon atom, ^{12}C.

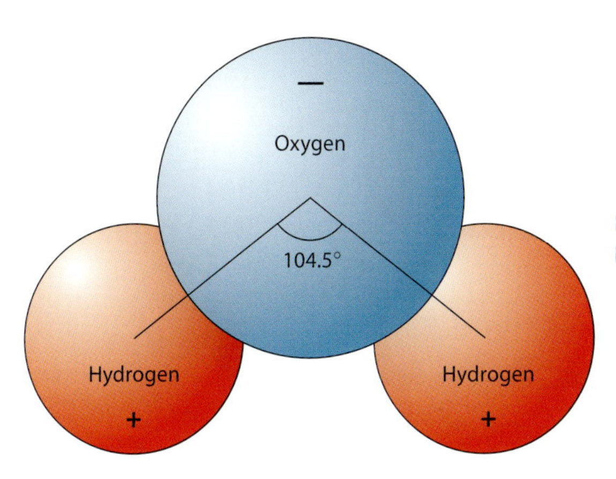

Figure 3.1. A water molecule is formed from two hydrogen atoms covalently bonded to an oxygen atom.

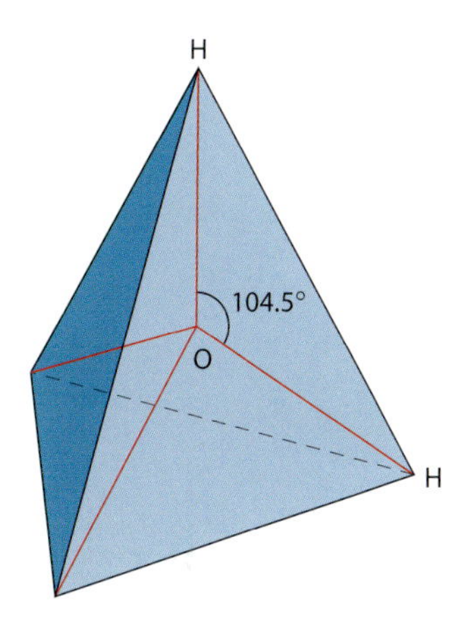

Figure 3.2. Electron orbitals of an oxygen atom are arranged approximately as a tetrahedron, a shape that allows the maximum separation among the four orbitals. This tetrahedral shape establishes the angle at which hydrogen atoms attach (104.5°), resulting in the polarity of water molecules.

This angle has profound implications. The bonds between hydrogen and oxygen are covalent; that is, the hydrogen and oxygen share two electrons to form a relatively stable configuration. However, oxygen has a higher affinity for electrons than does hydrogen—oxygen is more electronegative—so the sharing of electrons in the covalent bond is unequal: the shared electrons are drawn more toward the oxygen atom. As a result, the oxygen end of the water molecule is slightly negatively charged due to the extra charge from the unequally shared electrons, and the hydrogen end of the molecule is slightly positively charged by default (figure 3.1). This separation of charges is referred to in chemistry as a *dipole*, and because it forms a dipole, water is a *polar molecule*.[2]

Liquidity

The most important effect of water's polar nature is that it sticks to itself. The positively charged hydrogen of one water molecule is attracted to the negatively

[2] The bond angle for an exact tetrahedron is 109.5°, but the distribution of charges that leads to the polarity of the water molecule also causes the molecule to "bend" slightly away from a tetrahedral configuration such that the actual bond angle is the 104.5° cited earlier. If you enjoy math, you may take some pleasure in using equation 2.4 (in the appendix to chapter 2) to calculate the tetrahedral bond angle.

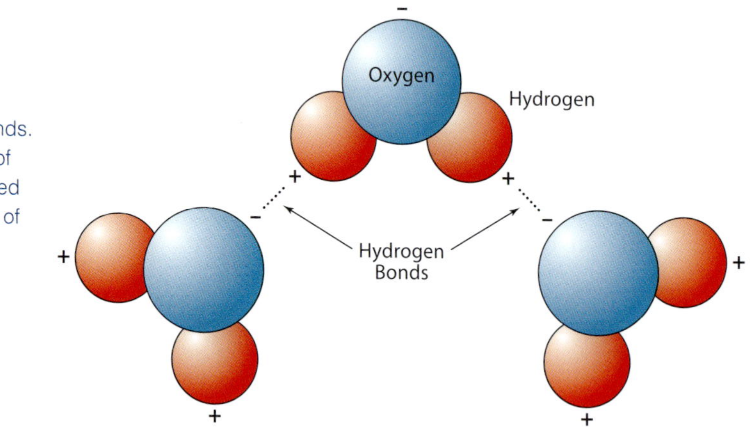

Figure 3.3. Hydrogen bonds. The net negative charge of an oxygen atom is attracted to the net positive charge of the hydrogen atom of an adjacent molecule.

charged oxygen of a neighboring molecule, and the two form a weak chemical interaction known as a *hydrogen bond* (figure 3.3). In liquid water, each molecule can be hydrogen bonded to as many as four other water molecules.

To grasp the importance of water's tendency to stick to itself, let's digress for a moment to review the differences between gases, liquids, and solids. As with all molecules, gas molecules move at a speed proportional to temperature (a phenomenon known as thermal agitation), and as they rattle around they occasionally encounter each other. However, molecules in a gas are minimally attracted to each other, so when they collide, they simply rebound. This freedom of motion is the basis for the familiar properties of gases: their variable volume and their tendency to expand when heated and contract when cooled. These properties are discussed in more detail in chapter 8.

If we lower the temperature of a gas, however, there comes a point at which the slight attraction of one molecule for another outweighs the tendency to rebound. At this temperature, the gas condenses to form a liquid, a tenuous state. Molecules in a liquid still have some freedom to move about, but this freedom is much restricted: the liquid can change its shape, but its volume is more or less fixed. Often we think about the process of condensation from a different perspective: if we start with a liquid and raise its temperature, when we reach the temperature of condensation, the liquid boils. Thus, the condensation temperature is the same as the boiling temperature.

If we lower the temperature of a liquid, eventually another critical point is reached when the tendency of a molecule to shift its position due to thermal agitation is less than the attraction between molecules. At this temperature, molecules become fixed in place, and the liquid freezes into a solid.

The temperature at which particular substances boil and freeze depends on the properties of their constituent molecules, particularly on their strength of attraction. And it is in this context that the polarity of water molecules is so important. Because of its tendency to form hydrogen bonds, water is a liquid at room temperature. At sea level, water doesn't boil until the temperature is above 100°C, and it doesn't freeze until the temperature is below 0°C.

These temperatures are notable in two respects. First, both the boiling and freezing temperatures of water are unusually high. For example, methane (CH_4) has a molecular weight of 16 daltons, very similar to that of water, but because methane is not polar and does not form hydrogen bonds, it boils at a frigid −162°C. Furthermore, methane freezes at −183°C, only 21°C below its boiling point. In contrast, water is liquid across an exceptionally wide range of 100°C.

It would be difficult to overstate the importance of these properties. Life as we know it requires liquid water. If water were liquid over a smaller range of temperatures, the probability would be much reduced that a planet, by chance, would have the requisite conditions for the evolution of life. In many respects, the fact that I am here writing this book and you are here reading it is due to the hydrogen bonds between water molecules.

Heat Capacity

Water is exceptional in other ways as well. For example, we can measure how much heat energy is required to raise the temperature of 1 kilogram of any substance 1°C. This property is known at the *specific heat capacity* of the substance (*specific heat* for short), and it varies widely among materials. Gold has the lowest specific heat of any element; it takes only 130 joules of energy[3] to raise the temperature of 1 kilogram of gold 1°C. This is one reason gold is so highly prized for making jewelry. Because little heat is required to raise gold to body temperature, a gold ring or necklace feels warm. At the other end of the spectrum is hydrogen gas, which has a specific heat of 14,000 J/(kg°C). Liquids tend to fall in the middle of this range. For example, liquid methane has a specific heat of approximately 2410 J/(kg°C). The specific heat of water is almost twice as large, about 4200 J/(kg°C). Among room-temperature liquids, only ammonia has a higher specific heat (4520 J/[kg°C]).

It is also useful to compare the specific heat of water to that of two other substances, substances common on earth. The specific heat of water is about four times that of air (1006 J/[kg°C]). This difference is striking enough, but remember that specific heat is measured on the basis of mass. Per volume, water is much more massive than air (see the discussion of density below). Consequently, at sea level it takes about 3500 times as much heat to raise the temperature of a given volume of water as it does to raise the temperature of the same volume of air. The specific heat of water is also much higher than that of the rocks from which the continents are constructed. The specific heat of granite is approximately 800 J/(kg°C), only about a fifth that of water.

Perhaps it will help to give these numbers some tangibility. One watt of power is an energy expenditure rate of one joule per second. Thus, a 100-watt light bulb gives off 100 joules of energy every second, some of it as light. This energy tends to heat up objects in the bulb's vicinity, an effect you probably have noticed when you replace a bulb and find that you forgot to turn off the switch: your hand quickly gets hot as the new bulb lights up. Water has a specific heat of 4200 J/(kg°C), so if you place a 100-watt bulb in contact with 1 kilogram of water (1 liter, about a quart), it takes 42 seconds for the bulb to raise the temperature of that kilogram just 1°C. At this rate, it would require

[3] One joule (symbolized by J) is the energy required to move an object 1 meter against a force of 1 newton. One newton, in turn, is the force required to accelerate 1 kilogram at 1 meter per second squared.

about an hour to boil a single liter of initially-room-temperature water, testimony to the high heat capacity of water.

For our purposes, it isn't so much the heat one has to put into water to raise its temperature that is important, it is the heat one can get out of water as it cools. For example, the large amount of heat that sunlight injects into ocean surface water in the tropical Atlantic helps to keep Great Britain and Scandinavia relatively warm when surface water is delivered to the North Atlantic by ocean currents. We will return to this effect in chapter 6.

Latent Heat

Consider what happens when water evaporates. As noted above, molecules in room temperature water move about as they are thermally agitated, but they are not able to break completely away from their neighbors. This is why water is a liquid at room temperature. At any given moment, however, the speed with which a water molecule moves varies from one molecule to the next. Most molecules move within a range of speeds too low to break hydrogen bonds, but a few molecules move faster. If one of these energetic molecules arrives at the water's surface, it can escape into the air, a process known as evaporation or vaporization.

Note, however, what must happen as this molecule escapes. When it is in water, the molecule is bound to its neighbors, an energetically favorable condition, and these hydrogen bonds must be broken as the molecule separates from the liquid and enters the gaseous state. Energy must be supplied to break these bonds, and this energy is supplied by the heat of the liquid water. This bond-breaking energy, in turn, imparts potential energy to the evaporating molecule as it enters the air. This potential energy remains latent ("hidden") in the molecule as long as it is in its gaseous form, hence the energy required to evaporate molecules is known as the *latent heat of vaporization*. Because this latent heat is carried away as water evaporates, the liquid water left behind is cooled. This is the basis for one form of temperature regulation in your body. When you overheat, you sweat, and evaporation of this sweat cools you down. The cooling is quite effective. For each kilogram of water that evaporates, approximately 2,454,000 joules of heat energy are drained from your body. The latent heat of vaporization of water is the greatest of any substance, a fact that has had interesting historical consequences. For example, one of the difficulties faced by the rebellious British colonies in New England after 1776 was the lack of imported salt. The colonists could boil seawater to manufacture salt, but even with a whole continent of trees to use as fuel, they were frustrated by the heat of evaporation. One had to chop several cords of wood to produce a single bushel of salt.

There is a flip side to this process, though. Once it has vaporized, a molecule of water has the potential to form new hydrogen bonds. If a water-vapor molecule strikes the surface of liquid water, it can form bonds to its new neighbors and condense. Just as heat energy was taken from liquid water to break these bonds when the molecule evaporated, heat is returned to the liquid when the molecule condenses. For each kilogram of water vapor that condenses, 2,454,000 joules of energy are delivered to the liquid water, heating it up.

Density

Density is the ratio of mass to volume; it varies widely among materials, and within a material it varies with temperature. The density of pure water is about 1000 kilograms per cubic meter, roughly 830 times the density of room temperature air (1.2 kilograms per cubic meter).

Most materials expand as they are heated. That is, as the temperature rises their mass takes up a larger volume, and as a consequence, an increase in temperature results in a decrease in density. Over a small range of temperatures, pure water is an exception to this rule. As the temperature of liquid water rises from 0°C to 3.98°C, small ice-like groups of water molecules melt, allowing for more ordered packing. As a result, the volume of the material decreases and its density goes up slightly (figure 3.4). Maximum density is reached at 3.98°C, and at higher temperatures water behaves like a more typical material: its density decreases as temperature rises. This unusual behavior provides the basis for defining the kilogram, the standard unit of mass: 1 kilogram is the mass of one liter of pure water at the temperature of its maximum density.

Because the coldest water is not the densest, plants and animals are afforded a wintertime haven in freshwater lakes. In fall and winter, heat is lost at the surface of a lake, and its temperature decreases. As a result, surface water becomes denser than the warmer water below, and the surface water sinks to the bottom. However, this process can only continue until the surface reaches 3.98°C. Below that temperature, as the surface continues to cool, water becomes less dense and it stays at the surface. The net result is that liquid water at the bottom of the lake never gets colder than 3.98°C. This is good news for the fish, frogs, turtles, and other animals that use this relatively warm bottom water as a refuge against winter's cold. Only if the entire lake freezes can the bottom get colder than 3.98°C.

Pure water freezes when it reaches 0°C, and the reorganization of water molecules during freezing causes ice to expand. As a consequence, ice at 0°C has a density of 917 kilograms per cubic meter, much lower than liquid water at the same temperature (999.8 kilograms per cubic meter). This reduction in density is why ice floats.

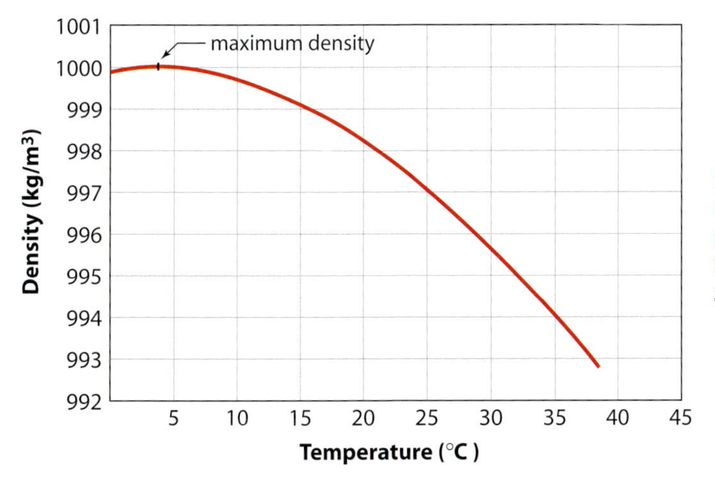

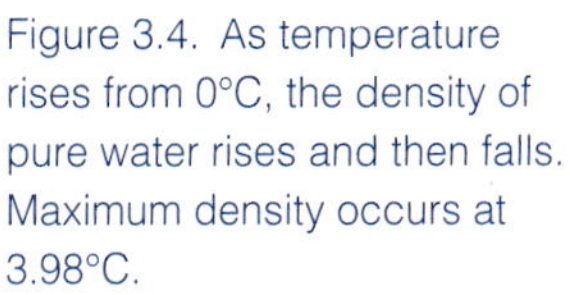
Figure 3.4. As temperature rises from 0°C, the density of pure water rises and then falls. Maximum density occurs at 3.98°C.

If one applies hydrostatic pressure to a sample of water, the sample's mass is squeezed into a smaller volume. Thus, an increase in pressure results in an increase in density. This is the basic effect that led people to suppose that water in the deep sea could be denser than steel or lead (chapter 1). For a given pressure, however, the change in volume of water is actually very small. The pressure at the average depth of the ocean (4 kilometers) is sufficient to increase the density of water by only about 2%.

Optical Transparency

There are many other properties of pure water that are interesting and important: its electrical conductivity, thermal conductivity, and surface tension, to name just a few. But there is only one other property that we need mention in the context of learning how the ocean works: water is quite transparent to visible light. In itself, this is not a notable property—many, if not most, pure liquids are transparent—but it is an exceedingly important property nonetheless. Because water is so transparent, visible light can penetrate several hundred meters into the ocean, allowing planktonic algae to photosynthesize and animals to see.

The transparency of water depends on the color of light. For instance, red light is absorbed more readily than blue light. We will return to the optical properties of liquid water in chapter 4, and we will briefly touch on the optical properties of water vapor in chapter 10.

Speed of Sound

The sonic properties of seawater are not a primary factor in how the ocean works, so by rights we should not discuss them here. But the speed of sound is just too interesting not to mention.

The speed of sound quantifies the velocity with which sound waves travel. In water, it increases with either an increase in density or an increase in temperature, which in the ocean leads to a strange phenomenon. As we will see in chapter 6, water at the ocean's surface is typically warm, and therefore has a high speed of sound. With increasing depth, the water gets colder, which causes the speed of sound to decrease. Below a depth of 600 to 1200 meters, however, the increase in density as water is squeezed by hydrostatic pressure results in an increase in the speed of sound that more than offsets the decrease due to colder temperatures. In other words, in the ocean, the speed of sound is minimal at a depth of 600 to 1200 meters.

Imagine, now, what would happen if a whale were to dive to this depth and let out a cry. If the sound were directed sideways but slightly downward, its speed would increase as the sound wave moved into water with higher density, and this increase in speed would cause the wave to bend upward (figure 3.5). Similarly, if the sound traveled sideways but slightly upward, it too would travel faster, and consequently the wave would bend downward. This tendency for waves to bend toward areas with a lower speed of transmission is known as *refraction*, and it works for light waves as well as for sound. For example, telephone messages are often carried by pulses of light in glass optical fibers, where

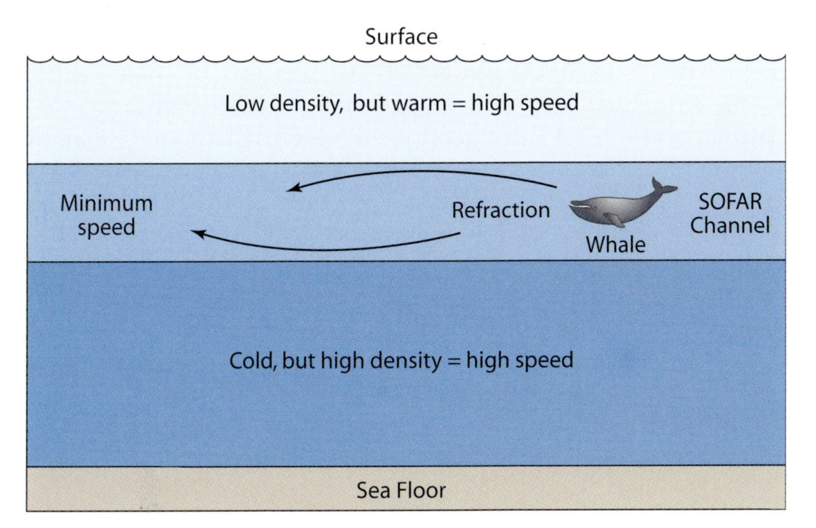

Figure 3.5. The SOFAR channel, a region of minimal sound speed. Refraction of sound confines it to the channel.

the speed of light is slower than in air. As a result of the difference in speed between glass and air, light entering the end of such a fiber stays in the fiber. If it tries to move out into air, the speed of the light wave increases, and the wave is refracted back into the glass. In this fashion, a glass fiber acts as a "light pipe," effectively channeling light where the phone company wants it, with minimal loss during transmission.

The same channeling effect operates on the whale's cry in the ocean. Because sound released near the depth of minimum speed tends to stay at that depth, it is confined to a "sound pipe," and as a result, it can travel thousands of kilometers and still be heard. This oceanic sound pipe is known as the SOFAR channel, an acronym for *so*und *f*ixing *a*nd *r*anging, reference to a use of the phenomenon by the U.S. Navy. If a pilot had to ditch his plane at sea, he lowered a small explosive into the SOFAR channel and detonated it. The sound could then be picked up by remote hydrophones, and the pilot's location determined by triangulation. The Navy also used the SOFAR channel to snoop on foreign activity. During the cold war, the Navy submerged hydrophones in the SOFAR channel off the California coast and was able to listen to the sounds made by the propellers of ships leaving Vladivostok Harbor in Russia.

Seawater

So far, we have discussed the physical properties of pure water, H_2O. It is now time to take the next step, and explore the properties of seawater, the substance that really concerns us. We begin by asking a simple question.

What Is Seawater?

As elementary as it sounds, this question is exceedingly difficult to answer. Simply put, seawater is pure water with some other stuff added. In every 1000 grams of seawater, there are approximately 965 grams of pure H_2O and 35 grams of dissolved material, collectively known as *salt*. The problem arises when we attempt to define the composition of this salt.

First, the term "salt" can itself be misleading. Traditionally, salt was what was left when seawater (or any other brine) evaporated. Its composition was not specified. About 86% of the salt in seawater is sodium chloride—table salt—so in this respect, the term "salt" matches one's expectations. There is, however, that other 14%, and the composition of this remainder includes virtually every known element. It is safe to say they we still don't completely know the composition and concentration of salts in seawater, and new—and sometimes important—discoveries are made every year. For example, a tiny amount of iron is dissolved in seawater, and we will see in chapter 10 that the availability of iron has important effects on the ability of algae to grow. But it has only been in the last twenty years that scientists have been able to accurately measure the concentration of iron in seawater. Other trace elements are also important in living things. Cobalt, for instance, is required for the synthesis of vitamin B12 and selenium for the synthesis of thyroid hormones.

Table 3.1 provides a list of some of the major constituents of sea salt. Note that there is considerably more mass of chloride than there is of sodium. This might appear odd, given that sodium chloride, the primary salt in seawater, contains an equal number of sodium and chlorine atoms. However, each chloride ion weighs 35 daltons, more massive than its partner sodium ion, which weighs 23 daltons. Thus, with equal number of sodium and chlorine atoms, there is more mass of chloride than sodium. For future reference, note that bicarbonate, at the bottom of this list of abundant salts, is nonetheless present in substantial amounts. Bicarbonate is essentially a hydrated form of carbon dioxide, a molecule important in the process of photosynthesis. We will return to the role of bicarbonate in chapter 4.

Please remember that table 3.1 is a list only of abundant ions, it is not a complete list of the components of seawater. To give you a feel for how incomplete this short list is, consider the miniscule concentration of gold in seawater, about 8 millionths of a gram per cubic meter. At this concentration, gold ranks forty-ninth in the overall list of elements present, six places behind yttrium. But, there are a billion cubic meters in each cubic kilometer of seawater, and 1.4 billion cubic kilometers in the ocean. Thus, the ocean as a whole contains 11 million *tons* of gold. This for an element that doesn't even come close to making the list shown here.

Table 3.1 Common components of sea salt

Ion	Symbol	Percent by weight	Grams per kg seawater
chloride	Cl^-	55.07	18.98
sodium	Na^+	30.62	10.56
sulfate	SO_4^{2-}	7.72	2.65
magnesium	Mg^{2+}	3.68	1.27
calcium	Ca^{2+}	1.17	0.40
potassium	K^+	1.10	0.38
bicarbonate	HCO_3^-	0.41	0.14

Before you head off to make your fortune mining gold from the sea, you might want to estimate the production expenses. It would take far more money to extract gold from the other elements in sea salt than the gold itself is worth.

Salinity

The concentration of salt in seawater defines the water's *salinity*. Loosely defined, salinity is the number of grams of dissolved solids in one liter of seawater. Salinity is traditionally expressed as parts per thousand, symbolized by ‰. As noted above, if you were to collect one kilogram of seawater and wait until all the water had evaporated, there would be 35 grams of solid matter left—about 6 teaspoons—indicating that the salinity of the sample was 35‰. This method of measuring salinity is both messy and time consuming, and more convenient (and more accurate) methods have taken its place. Currently, salinity is quantified by measuring the ratio of the electrical conductivity of a sample of seawater relative to that of a standard concentration of potassium chloride. As a ratio of conductivities, this measure of salinity technically has no units, but it is often reported in *practical salinity units* (psu). A salinity of 35 psu is very close to a salinity of 35‰.

The measurement of salinity via conductivity can easily be incorporated into instruments used at sea. As the measuring apparatus is lowered into the ocean, seawater is pumped through a small chamber, and its conductivity is measured and recorded, often to an accuracy of 0.005 psu or better.

Salinity varies from place to place in the ocean, especially in semi-confined seas. For example, in the warm Mediterranean Sea, more water evaporates than is replaced by freshwater input from rain and rivers, and the salinity is consequently high, about 40‰. In contrast, the cold Baltic Sea has a relatively low rate of evaporation and it is fed by many rivers. As a result, its salinity can be as low as 10‰.

Just as salinity varies from place to place, one might suppose that the chemical composition of sea salts would vary. For instance, as rocks weather, some of their chemicals might be leached into a river, and the input of these chemicals could skew the composition of seawater near the river's mouth. This can indeed occur, but the effect is quite local. In fact, the chemical composition of seawater is surprisingly constant throughout the vast majority of the world's oceans. Seawater from the North Atlantic is very similar to seawater from the South Pacific, despite how different they might look in the travel brochures. This suggests that the oceans are relatively well mixed. In other words, the time it takes for ocean water to be stirred about among the ocean basins is short compared to the time it takes for a substantial amount of new salt to be added. The mixing time is about 300 years for the deep Atlantic, and about 600 years for the deep Pacific.

The chemical constancy of seawater is one of the important and enduring results of the British *Challenger* expedition. On the basis of his analysis of 77 widely scattered samples collected by the expedition, William Dittmar (a professor at Anderson's College in Glasgow) concluded in 1884 that seawater is seawater the world over; only its concentration changes to any appreciable extent.

The Physical Consequences of Salinity

The salinity of seawater affects the water's colligative properties: its boiling point, freezing point, and the osmotic pressure it can exert. Of these properties, the most relevant for our purposes is the freezing point. At a typical salinity of 35‰, seawater does not freeze until its temperature is –1.89°C, nearly two degrees below the freezing point of freshwater.

In addition to affecting the temperature at which seawater freezes, salt also affects the *way* in which it freezes. Given opportunity during the process of freezing, the pure water in seawater separates itself from the salts. In essence, as water molecules bind tightly to form ice, they push any salt atoms out of the solid matrix. Thus, an ice floe formed from seawater contains mostly freshwater ice, and the salts are extruded into the sea.

You may want to test this effect at home. Gather some seawater or mix a bit of table salt into some freshwater (6 teaspoons per liter), and put a cup of the liquid in the freezer of your refrigerator. Leave it overnight. In the morning, chip off a chunk of the resulting ice and put it on your tongue. What do you taste?

Actually, what you will taste is salt. When frozen quickly in the confined volume of a cup, seawater freezes with much of the salts included. But this is not how ice forms in the ocean. When water at sea cools to its freezing point, initially small flat crystals of ice begin to form, giving the sea surface a "greasy" appearance. These crystals then amalgamate, first to form frazil ice, and then to form slush ice. The aggregation of slush produces thin sheets, known as pancake ice, which eventually thicken to form floes. All this takes substantial time—days to weeks—allowing the ice matrix to extrude a concentrated brine and causing the ice itself to be mostly fresh. In polar ice sheets, the process of brine formation can continue slowly for years. Another experiment at home will convince you that ice can indeed evolve given sufficient time. Ice cream, which arrives from the store smooth and creamy, eventually becomes granular and crunchy in the freezer as large crystals of pure ice grow and extrude the sugar and flavorings. Salt extrusion by sea ice will be an important component of our discussion of ocean circulation in chapter 6.

When salts are dissolved in pure water, the density of the resulting solution is increased. For example, seawater with a salinity of 35‰ has a density of about 1028 kilograms per cubic meter. In other words, a cubic meter of seawater weighs 28 kilograms more than a cubic meter of freshwater. To a rough approximation, an increase of 1‰ in salinity results in an increase of 0.8 kilograms per cubic meter (figure 3.6).

Density is important for our purposes primarily because it determines whether a bit of seawater will float or sink. Just as lower density oil floats on higher density water, low salinity (hence, low density) seawater floats on higher salinity, higher density seawater.

The presence of dissolved salts also affects the way in which seawater's density varies with temperature. Recall that pure water reaches a maximum density at 3.98°C, and becomes less dense as the temperature is lowered below this point. Not so with seawater. For salinities greater than 27‰, seawater becomes steadily denser as the temperature is lowered, all the way to the freezing point (figure 3.7).

So far, we have listed two important properties of seawater that change with changes in salinity: density and the freezing point. Other pertinent properties

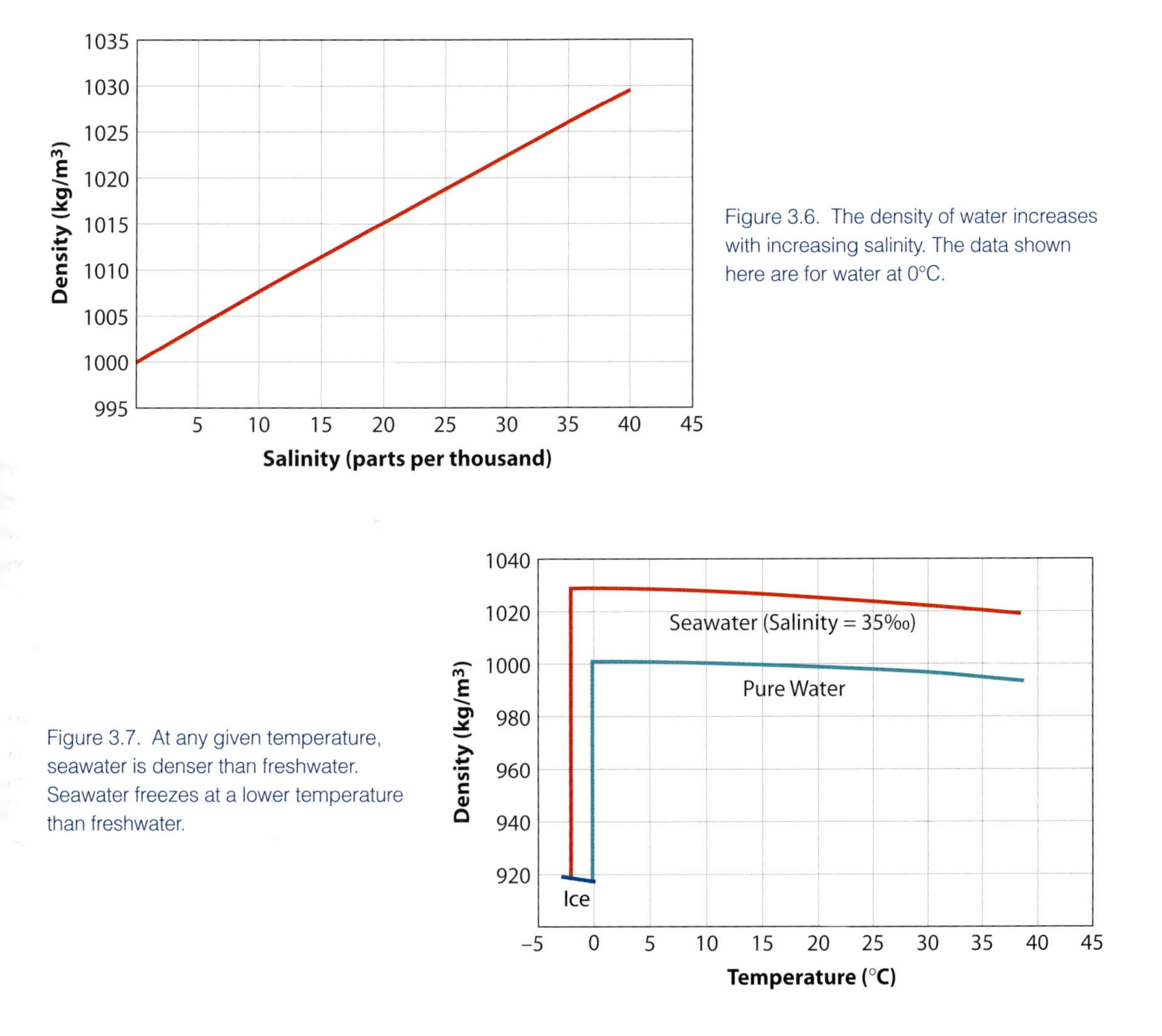

Figure 3.6. The density of water increases with increasing salinity. The data shown here are for water at 0°C.

Figure 3.7. At any given temperature, seawater is denser than freshwater. Seawater freezes at a lower temperature than freshwater.

do not change appreciably, including latent heat, light transmission, and specific heat.

The Organic Continuum

Scooped out of the ocean and poured into a bottle, seawater appears homogeneous. Indeed, for the small molecules that make up the bulk of the salts, one cubic millimeter of seawater in a sample is the same as the next. But the concentration of organic molecules produced by plants and animals often varies from one small sample to another. For example, the "jelly" in the bell of a jellyfish is more than 95% seawater. When a jellyfish dies, this gel gradually breaks down, disintegrating into small bits and becoming even more dilute. But until completely broken down into its constituent molecules, the bits of gel (and the seawater they encompass) maintain some structural integrity and do not mix readily with water around them. As a result, water that appears homogeneous to large organisms like us might actually differ from place to place depending on whether the measurement was taken inside or outside a bit of jellyfish jelly or any of a wide variety of other organic gels. This small-scale variability can have biological importance. Bacteria, for instance, can eat jellyfish jelly, and life might be quite different inside than outside even a dilute bit of gel that you and I cannot see. For this reason, oceanographers have come to think of seawater as an "organic continuum." Salts and small biological molecules grade into chains of organic molecules, which in turn grade into aggregations of chains, and finally, into gels. Only when viewed in bulk—a few cubic centimeters or more—is seawater homogeneous.

Where Does Seawater Come From?

The earth formed some 4.6 billion years ago, coalescing from a cloud of debris left over after the explosion of a star that preceded our sun. This debris included substantial amounts of both hydrogen and oxygen. Thus, the raw materials for water have been here from the very start, and an immense amount of water is trapped within the rock of the planet's mantle and crust. This trapped water has been, and continues to be, slowly released into the atmosphere as water vapor works its way to the planet's surface, a process known as *outgassing*.

Initially, earth was quite hot, and water released into the atmosphere remained in the vapor phase. However, about 4 billion years ago earth's atmosphere cooled below the boiling point of water, and vapor in the atmosphere began to condense. Subsequent millennia of outgassing and condensation filled the oceans.

Other gases are also released from rocks, sulfur, nitrogen, carbon dioxide, and chlorine among them. As the oceans filled with water, outgassed chlorine combined with sodium (leached by rain from the continents) to form salt, and the combination of salt and condensed water is, to a good approximation, seawater. The remaining constituents of seawater have similar origins. Because salts were added continuously as the oceans filled, the salinity of seawater has remained nearly constant throughout earth's history.

As soon as earth cooled sufficiently for water to condense, the oceans filled at a rapid pace, and by 2 billion years ago, the volume of the ocean was very nearly the same as it is now. Thus, the ocean we know today is very similar in volume and composition to the ocean that existed when multicellular life evolved half a billion years ago.

It is worth placing this timeline in context. Current estimates suggest that the observable universe is approximately 13.7 billion years old. Thus, the oceans on earth have been present for almost a third of the universe's history, and their composition and volume have been much like they are now for fully a sixth of that long, long time.

Summary

Pure water is an unusual substance: it is liquid at room temperature, nearly transparent to light, and capable of storing a large amount of heat. The addition of salts modifies some of the properties of water. For example, seawater has a higher density and a lower freezing point than freshwater. The presence of salt also affects the way in which water freezes: brine is extruded and fresh ice left behind. The importance of these physical properties will begin to become apparent in chapter 6, but first we need to learn about plants in the sea and how they start the food chain.

Further Reading

Chaplin, M. (2006). Water Structure and Behaviour. http://www.lsbu.ac.uk/water/anmlies.html.

De Podesta, M. (2002). *Understanding the Properties of Matter* (2nd edition). Taylor and Francis, New York.

Denny, M. W. (1993). *Air and Water: The Physics of Life's Media*. Princeton University Press, Princeton, NJ.

Kurlansky, M. (2002). *Salt: A World History*. Penguin Books, Hammondsworth, England.

Open University (1989). *Seawater: Its Composition, Properties and Behaviour*. Pergamon Press, London.

Pilson, M.E.Q. (1998). *An Introduction to the Chemistry of the Sea*. Prentice-Hall Inc., Upper Saddle River, NJ.

Thomas, D. N. (2004). *Frozen Oceans: The Floating World of Pack Ice*. Firefly Books, Buffalo, NY.

Photosynthesis and Primary Production

Life on earth is powered by the sun. In the ocean, for example, plants harvest energy from sunlight and use that energy to grow and reproduce. Animals eat the plants and are themselves eaten, passing the sun's energy up the food chain. Although this basic concept is easily grasped, to comprehend the vast quantity of energy involved takes some work.

The sun is an average star, located about 150 million kilometers from earth. Every second, the sun fuses a billion tons of hydrogen to form helium, releasing sufficient energy to maintain its surface at a white-hot temperature of 5500°C. In the process, the sun radiates four hundred million billion billion joules of light energy into space each second, and about a billionth of this power—1.8×10^{17} watts—falls on earth. The oceans occupy 71% of earth's surface, and as a result, they intercept 71% of the sunlight, a total of about 1.2×10^{17} watts.

To make sense of this number, we need something with which to compare it. Humankind's current rate of energy consumption—the combined total of all the energy produced from oil, gas, coal, nuclear, and hydroelectric sources across the entire globe—is only about 1.5×10^{11} watts. Thus, energy from the sun hits the oceans at a rate 816,000 times as large as the entire energy consumption of humankind.

Some of the sun's energy is reflected back into space. Some is absorbed by water and thereby heats the ocean, a phenomenon we will cover in depth in chapter 6. And as noted above, some of the sun's light is harvested by living organisms. Marine organisms capture only about 3% of the overall energy hitting the sea, but the captured energy is still 24,000 times as much as all humankind consumes. The way in which marine plants use this immense energy bonanza is the topic of this chapter.

Introducing the Phytoplankton

Before we return to the theme of energy, let's first introduce the players. There are about 100,000 species of marine plants,[1] so it would be impractical to start with species *A* and work our way down to species *Z*. Instead, let's explore various categories of marine photosynthesizers. We start with the largest categories and progress to smaller ones.

All living things can be divided into two groups: *autotrophs*, organisms that grow by using energy from nonliving sources, and *heterotrophs*, organisms that grow by eating autotrophs and other heterotrophs. You and I are heterotrophs: the food we eat is derived from plants and animals that were once alive. In contrast, trees are autotrophs: all they require is air, water, sunlight, and a few minerals. In this chapter, we concern ourselves only with the autotrophs of the sea. We will deal with the heterotrophs in chapter 5.

Autotrophs themselves can be divided into two categories: *chemoautotrophs* and *photoautotrophs*. Some bacteria grow using the energy stored in chemicals such as hydrogen sulfide, and they comprise the chemoautotrophs. These bacteria form the base of the food chain in areas such as the deep-sea hydrothermal vents mentioned in chapter 1. In this chapter, however, we are concerned with the photoautotrophs, the organisms that "eat light."

In turn, marine photoautotrophs come in two varieties. As defined in chapter 2, those that are attached to the seafloor are termed *benthic*; those that are suspended in the water column are termed *planktonic*. The seaweeds one encounters on a rocky shore at low tide are a prime example of benthic photoautotrophs. But, as we will see, benthic plants face a severe problem in the ocean. If the water above them is deep, more than a few tens of meters, it absorbs much of the available sunlight before it can reach the seaweeds, rendering survival difficult or impossible. Most photoautotrophs in the ocean have solved this problem by floating free above the seafloor. These free-floating organisms, called *phytoplankton*, are the central focus of this chapter.

Again we progress by division. All organisms can be divided into two general categories. The cells of some have relatively complex internal structures. In particular, they each have a well-defined cell nucleus (a membrane-bound structure that contains their genetic material) as well as other membrane-bound organelles (such as *mitochondria* and *chloroplasts*, described later in this chapter). Organisms with this relatively complex cell structure are the *eukaryotes*, from the Greek for "true nucleus." In contrast, the cells of other organisms lack membrane-bound nuclei and organelles—their genetic material is not separate from other cellular contents—and these simpler organisms are the *prokaryotes*.[2] In accordance with this categorization, we divide the phytoplankton into eukaryotic and prokaryotic camps (table 4.1).

[1] I use the term "plant" in its generic sense to refer to any organism that uses photosynthesis. Technically, the term should only be used to refer to members of the Plantae. This excludes algae, which are protists, and cyanobacteria, which are bacteria. However, "protist" sounds far too pedantic and "bacteria" could be misleading, so at the risk of offending botanists, I will stick with the generic term.

[2] Recent work suggests that earth's organisms should be divided into three categories or "domains:" eukaryotes, bacteria, and archaea (tiny single-celled organisms that superficially resemble

Table 4.1 Major players among the phytoplankton

		Typical Size (micrometers)	Minimum Doubling Time (days, at 20 °C)
Cyanobacteria	Prokaryotic		
Trichodesmium		5–15	3–5
Prochlorococcus		0.5–1	1.5
Haptophytes	Eukaryotic	3–10	0.3–1
Dinoflagellates	Eukaryotic	10–100	0.3–1
Diatoms (centric)	Eukaryotic	20–300	0.3–1
Diatoms (pennate)	Eukaryotic	2–25	0.3–1

Phytoplankton can also be categorized by size. Traditionally, plankton have been collected by dragging a net through the water, and the smallest practical mesh size for such an apparatus is about 20 micrometers, a fraction of the thickness of a human hair. Thus, the traditional method of sampling phytoplankton captures organisms with a maximum dimension greater than 20 micrometers, and these organisms are classified as *net plankton*. For many years, all phytoplankton small enough to escape from plankton nets were referred to as *nanoplankton*, and they were thought to have sizes between 2 and 20 micrometers. More recently, however, oceanographers have discovered even smaller phytoplankton, and the term *ultraplankton* has been coined to describe organisms less than 2 micrometers long.

The Cyanobacteria

Let's begin our survey of the phytoplankton with the small, prokaryotic organisms. Here, finally, there is some simplicity. All of the prokaryotic phytoplankton fall within one class of bacteria: the *cyanobacteria*. This ancient lineage emerged about 2.8 billion years ago, and these organisms were the first on earth to evolve the type of light-harvesting chemistry currently used by all the phytoplankton discussed here. As we will see, in the process of capturing light energy this chemical system releases oxygen; that is, it is *oxygenic*. In fact, cyanobacteria are in large part responsible for shaping the composition of earth's atmosphere. Until cyanobacteria evolved, there was no free oxygen in the atmosphere. The current atmosphere is 20% O_2, all of it produced by oxygenic photosynthesis.

Cyanobacteria are found worldwide and there are many varieties, three of which bear mention here. First, there is *Trichodesmium*. In addition to captur-

bacteria). Traditionally, bacteria and archaea have been lumped together as prokaryotes, but new discoveries regarding the genetic makeup and biochemistry of archaea imply that they are, in fact, sufficiently different from both eukaryotes and bacteria to deserve their own domain. Unfortunately, our understanding of marine archaea is currently evolving too fast to allow informed discussion of their role in how the ocean works. Consequently, the treatment here is confined to the eukaryotes and bacteria, and application of the traditional term "prokaryotes" to bacteria should not cause a problem.

ing light energy, *Trichodesmium* converts atmospheric nitrogen into ammonia, a nutrient ("fertilizer") on which many phytoplankton rely. *Trichodesmium* is relatively large for a cyanobacterium—5 to 15 micrometers in length—and is found most commonly in tropical and subtropical seas where the water temperature is greater than 25°C. *Trichodesmium* is a single-celled organism, but often attaches end to end, forming filaments that can be a few millimeters long. These filaments can in turn attach to each other, forming larger aggregations that are visible to the naked eye. When winds are calm, these aggregations may congregate at the water's surface, where they are referred to by mariners as "sea sawdust." Both Charles Darwin and Captain Cook noted blooms of *Trichodesmium* in tropical seas. Like other cyanobacteria, individual cells may have flagella, tiny corkscrews used for propulsion. Some species of *Trichodesmium* produce a toxin that repels many potential grazers.

The second and third notable genera of cyanobacteria, *Prochlorococcus* and *Synechococcus*, are extremely small—individuals are only 0.6 to 0.8 micrometers in diameter—and as a result they escaped detection for decades. With the advent of techniques that allow oceanographers to collect and study such small organisms, *Prochlorococcus* and, to a lesser degree, *Synechococcus* have emerged as important players in oceanic photosynthesis: they dominate the phytoplankton in clear, nutrient-poor waters. Recent findings suggest that *Prochlorococcus* and at least one species of *Synechococcus* are incapable of using nitrate, another nutrient, as a source of nitrogen for growth, a limitation that may have important implications for the global pattern of marine productivity (chapter 10).

Cyanobacteria use a variety of pigments in their light-harvesting apparatuses, and therefore assume a variety of colors. Many are bluish, hence the "cyan" in their name. In others, the blue is mixed with green, and in older texts, cyanobacteria are often called *blue-green algae*. Still others, such as *Trichodesmium*, are red, and the Red Sea may owe its name to dense blooms of *Trichodesmium*.

Cyanobacteria are commonly eaten and digested by single-celled herbivores (see chapter 5). However, on a few momentous occasions deep in evolutionary history, a heterotrophic cell engulfed a cyanobacterium, but instead of digesting its prey, the heterotroph kept it alive. This arrangement was advantageous for both the engulfer and the engulfee. By maintaining the cyanobacterium intact, the heterotrophic cell in effect converted itself into an autotroph. The cyanobacterium still captured light energy and used it for growth, but now some of the products of photosynthesis could diffuse out of the cyanobacterium and feed the heterotroph as well. Conversely, the heterotroph protected the cyanobacterium from being eaten by other cells, and the cyanobacterium also used some of the heterotroph's waste products as nutrients. A symbiotic relationship between the heterotroph and the cyanobacterium resulted, and the cyanobacterium eventually evolved into a chloroplast, the photosynthetic organelle in extant eukaryotic organisms. Because the chloroplast lives inside the heterotroph, this symbiosis is known as *endosymbiosis*.

Endosymbiotic events have occurred several times through the course of evolution and have given rise to the major lines of eukaryotic phytoplankton. The details of who engulfed whom are interesting but complex, and we will not delve into them here. Instead, we make do with brief vignettes of a few of the important marine lineages that stemmed from these endosymbiotic events. We introduce the eukaryotic phytoplankton in order of increasing size.

The Haptophytes

Haptophytes are single-celled eukaryotic phytoplankton sporting a pair of flagella, whip-like structures (different from bacterial flagella) that can propel the cell through the water. Both flagella emerge from one end of the cell, and they flank a small, threadlike structure called a haptonema. The haptonema is roughly the size and shape of a flagellum, but it does not contribute to propulsion.

The most familiar haptophytes are the coccolithophorids, which get their name from their unique armor (figure 4.1). Typically, a coccolithophorid cell is covered with small disks, which in some species resemble hubcaps. These *coccoliths* are constructed from calcium carbonate, the same chemical from which chalk, limestone, and marble are made. In fact, the famous white cliffs of Dover, in England, are made of chalk formed, in part, from coccoliths. Coccoliths are formed internally and then extruded onto the cell surface—as often as every 15 minutes—to defend the cell against ingestion by predators. They may also play a role in photosynthesis, a curious connection discussed later in this chapter.

Coccolithophorids are found worldwide. They are larger than cyanobacteria: typical coccolithophorids are 3–10 micrometers in diameter, although a few species can reach lengths of 100 micrometers.

Other haptophytes, notably *Phaeocystis*, do not produce coccoliths. *Phaeocystis* often forms massive blooms in polar seas.

Dinoflagellates

Dinoflagellates are single-celled, eukaryotic organisms, characterized by the presence of an unusual pair of flagella (figure 4.2). Typically, one flagellum emerges from the end of the cell and pushes or pulls the cell through the water; the other flagellum is wrapped around the cell's midsection. Most dinoflagellates have armor (a *theca*) made from cellulose. Roughly half the species of dinoflagellates are photosynthetic, and therefore qualify as phytoplankton; the other half lack chloroplasts and are heterotrophic. We discuss the heterotrophic dinoflagellates in chapter 5.

Dinoflagellates occur worldwide, and they are at times responsible for the phenomena known as "red tides." Under the proper conditions, a coastal dinoflagellate population can grow so rapidly that its concentration turns the water a muddy reddish brown. Suspension-feeding animals such as oysters,

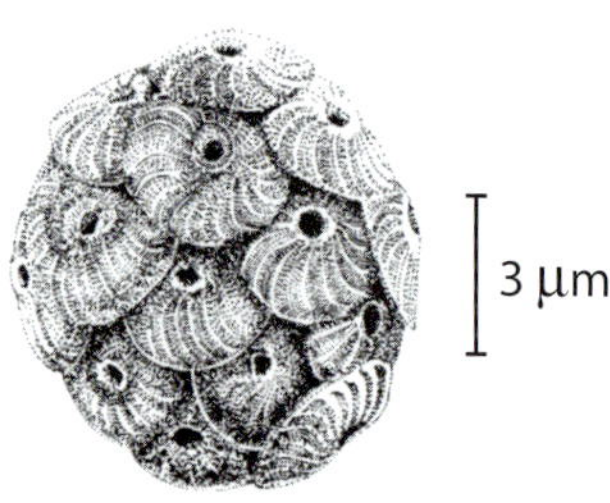

Figure 4.1. A typical coccolithophore. Redrawn from J. S. Levinton (2001), *Marine Biology: Function, Biodiversity, Ecology* (Oxford University Press, Oxford), figure 7.2.

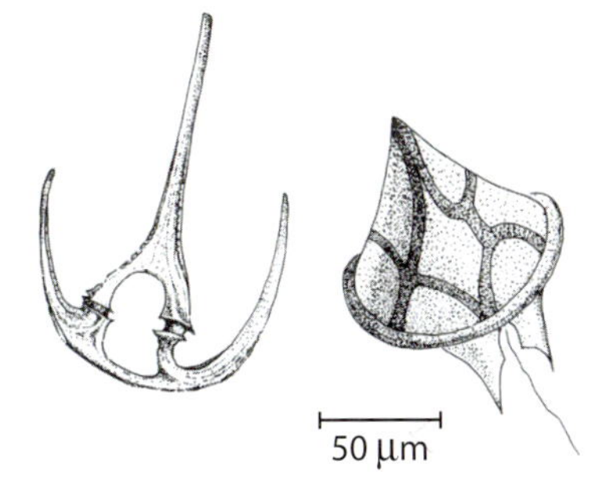

Figure 4.2. Typical dinoflagellates. Redrawn from J. S. Levinton (2001), *Marine Biology: Function, Biodiversity, Ecology* (Oxford University Press, Oxford), figure 7.2.

mussels, and clams feast on the abundant dinoflagellates, in the process accumulating toxins produced by the phytoplankton. These toxins have no effect on shellfish, but they are deadly to most fish, birds, and mammals. If you were to eat a tainted mussel, you would soon feel a tingling in your lips and mouth, a sensation that would spread to your limbs. Over the course of several hours, the tingling would progress to numbness, paralysis, and potentially death. The affliction is known as *paralytic shellfish poisoning*.

Red tides have been known since antiquity. There is possible mention of a red tide in the Bible (Exodus 7:20–21), and early explorers noted red tides and their effects. For example, in 1793 George Vancouver and his crew explored the west coast of North America, sampling animals and plants as they went. On June 15, four men became sick and one died after eating mussels that had fed on a red tide. A few years later, 100 members of a Russian expedition died in the same area. In 1832, during his voyage on the *Beagle*, Darwin reported a dramatic red tide off the coast of Chile, but fortunately for science he did not eat the local shellfish.

When disturbed, some species of dinoflagellate produce a brief flash of blue-green light. The same chemistry used by fireflies produces this dinoflagellate *bioluminescence*, which can be spectacular. When dinoflagellates are in low concentration, their bioluminescence appears as "sparks" in the water, and swimming in their midst at night gives the impression of cruising in space among myriad blinking stars. In high concentrations, the light produced by dinoflagellates can be surprisingly intense. I remember once traveling at night in a friend's small boat. As we sailed through a dense red tide, I thought he had turned on a search light under the hull, the brilliant blue light was so bright. Rumor has it that the U.S. Navy explored the use of bioluminescence as a means for tracking Soviet submarines, the thought being that a sufficiently sensitive satellite sensor could see the submarines' wakes.

No one is really sure why dinoflagellates produce light, although there are several theories. In one scenario, the flash of light released as a dinoflagellate is attacked by a predator acts as a burglar alarm to attract the attention of a still larger predator, who, in theory, then gobbles up the dinoflagellate's attacker. In another scenario, the flash of light startles the attacker, somehow allowing the dinoflagellate to escape.

Dinoflagellates are considerably larger than cyanobacteria and coccolithophorids. A typical dinoflagellate cell is about 80 micrometers in maximum length, and some large species can reach 500 micrometers. Dinoflagellates are often the dominant type of large phytoplankton in nutrient-poor tropical and subtropical waters.

Although most dinoflagellates are planktonic, a few species have taken up symbiotic residence in other organisms. For example, species of *Symbiodinium* are found in many corals and sea anemones, and other species are found in foraminifera and radiolaria, marine amoebas described in chapter 5.

The Diatoms

And finally there are the diatoms. These ubiquitous single-celled phytoplankton were the last of the major phytoplankton lineages to evolve, emerging at about the same time as the mammals, roughly 100 million years ago.

Unlike all other phytoplankton discussed so far, the adult forms of diatoms do not have flagella and therefore cannot actively move themselves through the water. (Diatoms at times undergo sexual reproduction, and their sperm do have flagella.) The cells are covered in armor made from amorphous silica—opal—and they come in two general forms. One type, known as the *pennate* form, is bilaterally symmetrical (figure 4.3A). That is, its left side mirrors its right side. The other type, the *centric* form, is typified by cells that look like small cookie tins or Petri plates (figure 4.3B). Pennate diatoms are commonly benthic, although small pennate diatoms are found in the plankton; centric diatoms are commonly planktonic.

Diatoms' opal armor (the *frustule*) poses a problem when the cells reproduce. As each cell divides, it takes with it one half of the frustule, either a top or a bottom. A new half frustule then forms inside the old one. That's fine for the cell that got the frustule's top, its newly formed bottom is the same size as its predecessor. However, the other daughter cell forms its new half frustule inside the smaller half of the old frustule (the old bottom), and thus this new frustule is smaller than either half frustule of the parent cell. In this fashion, a fraction of diatom cells gets smaller and smaller with each generation. The problem is

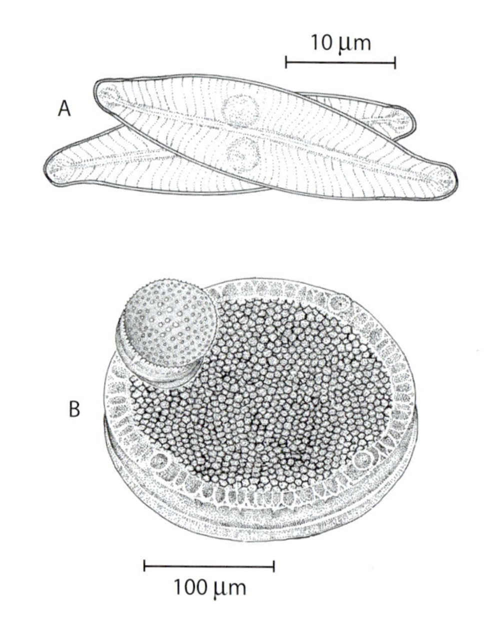

Figure 4.3. Pennate (A) and centric (B) diatoms.

solved when, at a certain small size, a daughter cell sheds *both* its parental frustules, undergoes sexual reproduction, grows back to the size of a full-fledged diatom, and then covers itself with two new frustule halves. Life is complex in opal houses.

Diatoms are among the largest marine phytoplankton. Large centric diatoms can be 300 micrometers in diameter. Small centric diatoms often increase their effective length by attaching themselves end to end to form long chains.

Whereas dinoflagellates are the most common large phytoplankton in clear nutrient-poor tropical waters, diatoms are the most common large phytoplankton in those tropical and temperate areas where nutrient concentrations are high, such as the upwelling areas we will discuss in chapter 9. Furthermore, diatoms are typically the dominant large phytoplankton at high latitudes. It is clear that diatoms are a major player in marine primary production: by some estimates they account for a third to a half of all the solar energy stored by phytoplankton.

One final quirk of diatoms deserves note. They are capable of producing domoic acid, a neurotoxin that has been blamed for the deaths of a variety of sea mammals, dolphins and seals for instance.

Synthesis

There are other types of marine phytoplankton, some with names only a botanist could love: cryptomonads, glaucocystophytes, and chlorarachniophytes, for instance. But we have introduced the major players, and we now pause to discuss four general characteristics that tie the phytoplankton together.

First, phytoplankton can reproduce quickly. When conditions are favorable—warm water with plenty of light and nutrients—the phytoplankton discussed here can divide every 8 to 24 hours. The sole exception to this generality is *Trichodesmium*, which can take 3–5 days to reproduce.

The ability to reproduce quickly leads to the second characteristic: phytoplankton are present in incredible abundance. For example, by rough estimate, there are 10^{24} cyanobacteria in the sea. This number is so large it is difficult to grasp, but a simple calculation can put it in perspective. A typical cyanobacterium (*Prochlorococcus,* for instance) is about 1 micrometer long, about a sixth the width of a strand of spider silk. If you stacked all the cyanobacteria in the ocean end to end, starting on the floor of your house, they would extend up to the ceiling and past it to the roof. They would then extend up past the top of Mt. Everest to the edge of the atmosphere and on to the sun. And back. *Three million times.*

Third, although there are species that strain the definition of "single-celled" (e.g., chains of diatoms and filaments of *Trichodesmium*), the phytoplankton described here are single-celled organisms.

And finally, by human standards, phytoplankton are small. Even the largest chains of cyanobacteria and diatoms are only a few millimeters long. But we should not allow the diminutive stature of phytoplankton to obscure the diversity of their sizes. A large diatom cell 300 micrometers in diameter is 600 times the length of a small cyanobacterium, and 216 million times it volume. In other words, a diatom is to a cyanobacterium as a whale is to a mouse. This diversity of sizes becomes important when we next consider the rates at which phytoplankton sink.

Sinking Speed

A recurring theme in this description of phytoplankton has been the presence of armor: diatoms coat themselves with opal, and coccolithophores cover themselves with limestone. As you might expect, this armor makes these cells denser than the water around them—up to 1100 kilograms per cubic meter for phytoplankton versus 1028 kilograms per cubic meter for seawater—and as a consequence, phytoplankton tend to sink. Due to the concentration of material in their cytoplasm, even those phytoplankton that do not have armor are denser than seawater, and they, too, sink. And to a phytoplankton cell, sinking is a life-threatening problem. If the cell sinks too far below the surface, light becomes too dim for photosynthesis, and the cell can no longer function.

Although few manage to avoid this problem altogether, phytoplankton use several mechanisms to minimize the rate at which they sink. First, it helps that phytoplankton are small. Physics dictates that the rate at which an organism sinks increases with an increase in size. A spherical coccolithophore 5 micrometers in diameter sinks at a speed of about 14 centimeters per day. A dinoflagellate ten times as large sinks fourteen times faster, 2 meters per day. And a large centric diatom 300 micrometers in diameter sinks roughly 18 meters per day. Clearly, the problem of sinking is much more dire for large phytoplankton than for small ones (figure 4.4).

It may not be intuitive that sinking rate should increase with size. If you are feeling inquisitive, the appendix to this chapter presents a brief explanation of the physics involved.

One might suppose that phytoplankton could use their flagella to swim upward, thereby offsetting the rate at which they sink. Three considerations render this method largely ineffective. First, swimming is advantageous only if reliably directed upward toward the light at the water's surface. But surface waters of the ocean are often mixed by the wind, and this stirring causes phytoplankton to

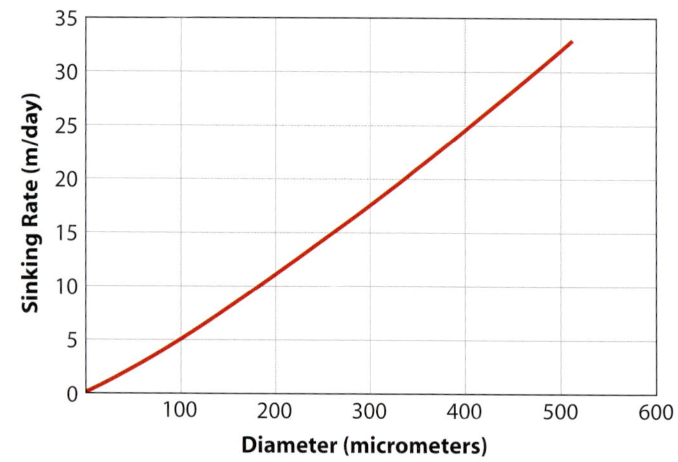

Figure 4.4. Measured sinking rates for phytoplankton of different sizes. The larger the cell, the faster it sinks.

tumble and roll. These gyrations make it difficult for the cell to propel itself reliably in any particular direction, rendering swimming an ineffective means of staying afloat. Imagine walking from one side of a room to the other with the floor rapidly rotating under your feet in an unpredictable fashion, and you will gain some idea of the problems faced by a phytoplankton cell. Second, swimming has a cost. If a phytoplankton cell has to swim upward continuously to stay in the light, the cost of this swimming cuts into any energy profit the cell might turn from absorbing sunlight. Finally, the adult forms of some phytoplankton, such as the diatoms, lack flagella and therefore cannot swim at all. So, in general, swimming can't keep phytoplankton afloat.

Alternatively, phytoplankton can resist sinking by adjusting their buoyancy. By reducing their density to that of the water around them, they can avoid sinking. Several mechanisms have evolved to this end, each with an energetic cost. Some phytoplankton reduce their density by exchanging heavy molecules for lighter ones. For example, an ammonium ion (NH_4^+) has the same charge as a sodium ion (Na^+) but weighs less: 18 versus 23 daltons. By exchanging some of its sodium for ammonium, the cell thus reduces its weight while maintaining the proper internal charge. Other phytoplankton reduce their density by producing droplets of oil. Unlike most other biological molecules, oil is less dense than seawater, and its accumulation in the cell acts much as a life preserver would for you and me. The cyanobacterium *Trichodesmium* produces small gas vesicles that similarly act to provide buoyancy.

Still other phytoplankton (some diatoms, for instance) reduce their rate of sinking by producing long spines that extend out from the cell's armor into the water. By increasing the area of contact between cell and water, these spines increase the viscous interaction (a type of friction) between the cell and its surroundings, thereby slowing the cell's rate of sinking. Of course, the spines also add to the weight of the cell, so this strategy is not as effective as it might first appear. Indeed, spines may have evolved primarily as a means for phytoplankton to avoid being eaten. Lastly, phytoplankton cope with sinking by reproducing. This nonintuitive strategy requires a bit of explanation.

Let's release a phytoplankton cell near the surface of the ocean, and follow its path. In still water, the cell sinks steadily toward the seafloor, as shown in figure 4.5A. In contrast, in water stirred by the wind, the cell performs a random walk, as shown in figure 4.5B: on average, the cell sinks, but at the whim of the turbulent flow, it wanders about in the process.

Now, let's repeat this experiment, but instead of following a single cell we release a large number of cells at a point near the surface and follow the group through time. In still water, the group behaves the same as a single cell: all the cells sink steadily. However, water in the surface ocean is seldom still. Instead, it is stirred through interaction with wind and waves. In water that is thus turbulently mixed, the center of the group—the average position of all the cells—sinks steadily downward (figure 4.5C), but at the same time, the water's mixing causes the group to spread out as individual cells wander at random. Some cells in the group are mixed lower in the water than average; likewise, a few are mixed higher. If, by chance, these latter lucky few happen to reproduce while they are high in the water column, they start the whole process over again. Serendipitously reproducing near the surface, they form a group similar to the one originally released. In this fashion, mixing and reproduction combine to counter the effects of sinking.

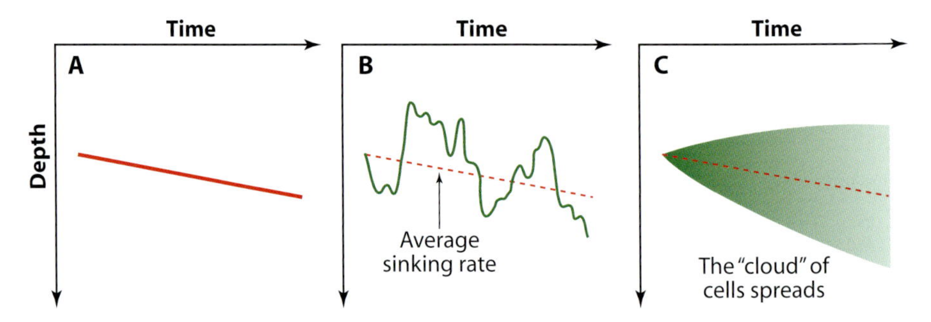

Figure 4.5. Sinking in still and turbulent water. (A) In still water, phytoplankton sink at a constant speed, so depth increases linearly with time. (B) In turbulently mixed water, a sinking cell wanders randomly relative to the average sinking rate. (C) Mixing causes a group of cells to spread over time.

The whole scheme is chancy. The faster cells sink, the more often they must reproduce to ensure that, by chance, enough cells reproduce high in the water column to keep the population afloat. Given the maximal rate at which typical phytoplankton reproduce—once or twice per day—cells must be smaller than roughly 150 micrometers in diameter to sink slowly enough to survive by this strategy.

Though a bit strange, this reproductive tactic has a decided advantage: it does not require any expenditure of energy beyond what the phytoplankton cells would be expending anyway. Thus, in comparison to swimming, forming spines, pumping ions, or accumulating lipids or gas, this reproductive strategy is cost effective, especially for small phytoplankton.

The mathematics of this reproductive strategy were laid out in a paper that appeared in 1949, and reading it always makes me smile. There is something inherently pleasing about an article written by Riley, Stommel, and Bumpus.[3]

Note that all the mechanisms used to cope with sinking work best when phytoplankton cells are small, which is one reason there are few large phytoplankton in the ocean. The major exceptions to this rule are a few genera of large seaweeds (*Sargassum*, for instance) that have evolved small gas bladders that allow them to float at the water's surface. Individual *Sargassum* can grow to nearly a meter in length and weigh several kilograms. At times, these seaweeds are abundant—the Sargasso Sea, east of Bermuda in the Atlantic Ocean, is named for the *Sargassum* that aggregates there in great windrows—but such abundance is rare. For the oceans as a whole, small, single-celled phytoplankton are most important.

Energy

Let us now return to the theme that opened this chapter: the delivery of energy to the oceans in the form of sunlight.

Energy is commonly defined as the capacity to do work. For instance, water penned up behind a dam has gravitational potential energy, and it can do work

[3] G. A. Riley, H. Stommel, and D. F. Bumpus (1949), Quantitative ecology of the plankton of the western North Atlantic, *Bulletin of the Bingham Oceanographic Collection* 12(3): 1–169.

by flowing downhill past a waterwheel or through the turbine of an electrical generator. A battery stores electrical potential energy and can do work as electrons flow down an electrical potential through a motor. A wrecking ball has kinetic energy—the energy of mass in motion—and can do work when it slams into a building.

In this chapter, we are concerned with energy that does biological work. Examples of biological work include growth and reproduction (processes that involve building new tissue), locomotion, feeding, and defense. Although they may include other types of energy as well, each of these examples involves *chemical energy*, the energy stored in chemical compounds.

Specifically, chemical energy is the energy stored in the bonds between atoms. For example, substantial energy is required to break the covalent bonds that hold two hydrogen atoms to an oxygen atom in a water molecule. You can gain a feel for this energy by performing the following experiment. Some bicycles come equipped with a generator that powers a headlight. As you pedal the bike, the generator spins, and current flows through the light. Now, take the bicycle, jack it up so that you can pedal in place, and run the wires from the generator into a glass of water. Then, as you pedal the bike, the energy you expend creates an electrical current that splits hydrogen from oxygen in the water, and the gases bubble up. You will find that you have to pedal laboriously for hours to split all the water in just one small glass.

In the process, however, you would have stored a lot of energy in the hydrogen and oxygen produced, and you could release this energy by igniting the gases and allowing them to burn. Consider the U.S. space shuttle. The large, bulbous tank to which the shuttle is strapped during takeoff contains liquefied hydrogen and oxygen, and the energy released when they burn and exit through a rocket nozzle provides much of the thrust that lifts the shuttle into space.

In biology, we are concerned primarily not with the energy stored in hydrogen and oxygen, but rather with the energy stored in *organic* compounds. This term requires explanation. Organic chemistry is the chemistry of carbon-containing molecules, and just about any molecule with some combination of carbon, hydrogen, and oxygen is considered organic.[4] Sugar, starch, and cellulose are organic molecules. Proteins and fats are organic molecules. In contrast, carbon dioxide is not an organic molecule (no hydrogen); neither is water (no carbon). If a molecule is not organic, it is *inorganic*.

The distinction between organic and inorganic was simpler before farmers and grocers co-opted the term "organic" for foods grown without pesticides or antibiotics. For present purposes, we use "organic" only in its chemical sense.

Living things build organic molecules from inorganic precursors, typically by combining carbon dioxide (CO_2) with the hydrogen from water (H_2O) to form a *carbohydrate*. There are many types of carbohydrates, each a variation on the theme: CH_2O. This basic carbohydrate structure is then added onto and twisted around to form all the other molecules of life. Combining carbon dioxide and hydrogen takes energy, however, and therein lies an important story.

[4] There are organic molecules that contain just carbon and hydrogen (e.g., methane, some terpenes), but they are relatively rare in living organisms.

Primary Production

The conversion of inorganic material to organic material is called *primary production*: "production" because an energy-bearing organic molecule is produced and "primary" because this molecule is constructed from inorganic materials rather than from some preexisting organic source. The rate at which organic material is produced is termed *productivity*. Demonstrating that organic molecules contain a substantial amount of energy is easy. Wood, for instance, is formed from organic molecules (primarily cellulose, a carbohydrate), and heat energy is released when these molecules combine with oxygen: that is, when wood burns. Energy released by burning wood can heat your house or power a steam engine.

The process of converting inorganic carbon-bearing molecules to organic molecules is also referred to as *carbon fixation*. Thus, when carbon dioxide combines with the hydrogen from water to form a carbohydrate, the carbon has been "fixed." This is not meant to imply that prior to its incorporation into a carbohydrate inorganic carbon was broken; rather it conveys the fact that, once fixed, carbon is held in place within an organic molecule.

What energy powers primary production? We touched on this subject when we classified the autotrophs earlier in this chapter. In most cases, the energy comes from the sun, but in some cases, the energy comes from inorganic molecules.

Chemosynthesis

Bacteria excel at scavenging energy from any available source, and some have even evolved the capability to capture energy from molecules that are toxic to other organisms. For example, at hydrothermal vents, heat from the earth's mantle causes extremely hot seawater to percolate past rock in the seafloor. In its journey through the crust, this seawater becomes saturated with hydrogen sulfide, H_2S, the same compound that makes rotten eggs smell rotten. Bacteria that live near these vents use this hydrogen sulfide as a source of energy.

To see how this works, we briefly digress to define two terms: *oxidation* and *reduction*. When chemical compounds are constructed or taken apart, electrons are often exchanged. If, in the course of a chemical reaction, a molecule gives up an electron, that molecule is *oxidized*. On the other hand, if in the course of a reaction a molecule accepts an electron, it is *reduced*. Usually, the donation and acceptance of electrons are coupled in a reduction-oxidation reaction, commonly known as a redox reaction. For example, when hydrogen burns with oxygen to form water (H_2O), each hydrogen atom gives up an electron (and is thereby oxidized), while the oxygen atom accepts the two electrons (and is thereby reduced).[5] The oxidation states of various relevant elements are given

[5] As noted previously, the exchange of electrons between hydrogen and oxygen in a water molecule is incomplete. The bonds holding H and O atoms together are covalent and electrons are shared. But recall that the sharing is not equal; the electronegative oxygen atom dominates the exchange. Thus, the hydrogen atoms are effectively oxidized and the oxygen atom is effectively reduced.

Table 4.2 The oxidation-reduction states of some elements common in the ocean

Element	Most Reduced	Intermediate	Most Oxidized
Carbon, C	CH_4 (methane)	C (elemental carbon)	CO_2 (carbon dioxide) CO_3^{2-} (carbonate ion)
Nitrogen, N	NH_3 (ammonia)	N_2 (elemental nitrogen)	NO_2^- (nitrite ion) NO_3^- (nitrate ion)
Sulfur, S	H_2S (hydrogen sulfide)	S (elemental sulfur)	SO_4^{2-} (sulfate ion)
Iron, Fe	Fe (elemental iron)	FeO (ferrous oxide)	Fe_2O_3 (ferric oxide, "rust")

in table 4.2, to which you may wish to refer from time to time as the subject of oxidation and reduction is raised.

In hydrogen sulfide, sulfur is present in a highly reduced form, having accepted electrons from two hydrogen atoms. Bacteria obtain energy from H_2S by oxidizing the sulfur to form elemental sulfur (S) or sulfate (SO_4^{2-}). This oxidation is chemically similar to burning wood in air, another common oxidation process, but it is carried out at the temperature of the surrounding seawater under the control of enzymes in the bacteria. In this fashion, energy released from hydrogen sulfide is trapped in the chemical bonds of organic molecules rather than being lost as heat. The process by which the energy of hydrogen sulfide is used to fix carbon is one form of chemosynthesis.

Now, hydrogen sulfide is highly toxic to most organisms, and the ability to use it in chemosynthesis is unique to a few bacteria. However, a variety of animals have evolved symbiotic relationships with these chemosynthetic bacteria, allowing complex communities of organisms to flourish around hydrothermal vents.

In addition, hydrogen sulfide emerges from a variety of ocean cold seeps, underwater springs fed by water that has seeped through layers of decomposing vegetation. Complex communities also form around these seeps, based on the chemosynthetic abilities of bacteria.

Photosynthesis

Chemosynthesis, although intriguing, is a minor factor in the overall energy budget of life in the oceans. The vast majority of energy used by marine organisms comes from sunlight, which is transferred to organic molecules through the process of *photosynthesis*. To understand photosynthesis, we first need to understand the properties of light.

A beam of light is composed of individual packets of energy called *photons*. The amount of energy contained in a photon depends on its wavelength, and the range of photon energies corresponds to the colors arrayed in the spectrum of light: a long-wavelength photon (red light) has less energy than a medium-wavelength photon (green light), which in turn has less energy than a short-wavelength photon (blue light) (figure 4.6). By a fortunate cosmic coincidence, the energy contained in a photon of visible light approximates the energy contained in the chemical bonds between atoms. For example, it takes about 6×10^{-19} joules of energy to break the bond that holds one carbon atom to another in an organic molecule. A single photon of red light has an energy of approximately

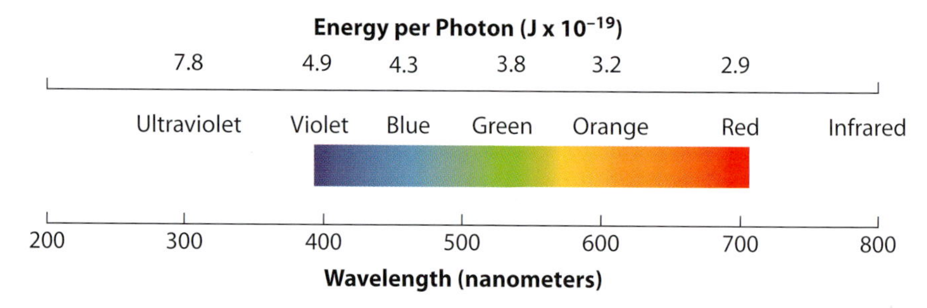

Figure 4.6. The longer the wavelength of light, the lower the energy of photons.

3×10^{-19} joules. Thus, if an organism absorbs two photons of red light, it gains enough energy to rearrange the connection of one carbon atom to its neighbors. A photon of blue light has about a third again as much energy as a photon of red light, slightly more than 4×10^{-19} joules.

Phytoplankton (like other photosynthetic organisms) have evolved complex chemical machinery to absorb photons of visible light and use their energy to fix carbon. In cyanobacteria, the photosynthetic machinery is located on specialized *thylakoid membranes* in the cytoplasm. In eukaryotic cells, thylakoid membranes are confined within chloroplasts, small, membrane-bound organelles within the cell, remnants of cyanobacteria obtained through endosymbiosis.

Although the specific steps in photosynthesis are not our concern here, a general overview will prove useful. In the photosynthetic machinery, photons are absorbed by a variety of colored molecules known as *photopigments*. With very few exceptions, chlorophyll is one of these pigments, and it selectively absorbs light with wavelengths in the blue and red (figure 4.7). Once chlorophyll has absorbed the blue and red constituents from a beam of white light, the remaining light is greenish, and for this reason, many plants appear green. Other photopigments (known as *accessory pigments*) are often present in phytoplankton, allowing cells to absorb additional photons of different wavelengths, especially those readily transmitted by water. The more nearly black

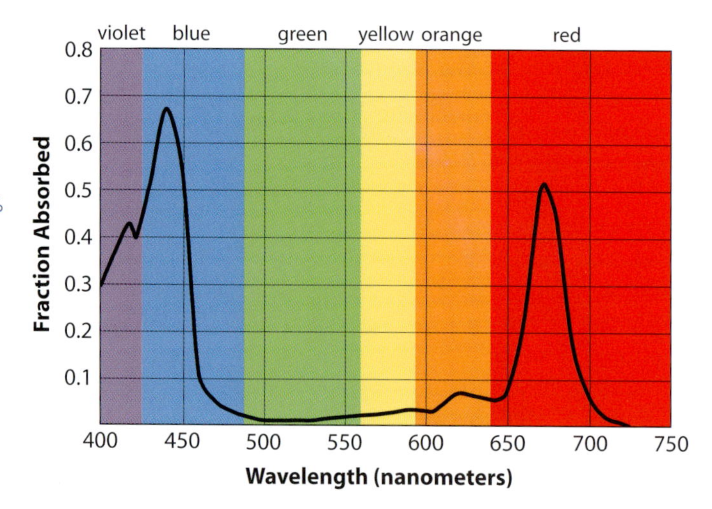

Figure 4.7. The absorption spectrum for chlorophyll *a*. Chlorophyll absorbs nearly 70% of the blue light that hits it, and more than 50% of the red light, but it absorbs very little blue-green, green, yellow, or orange light.

the phytoplankton, the larger the fraction of available light it absorbs for photosynthesis.

Energy absorbed from photons is transferred to electrons within the thylakoid membranes, and these high-energy electrons then pass from one specialized molecule to another. Some of these molecules contain iron, a fact you should store away for use in chapter 10. As the electrons travel through the membrane, they gradually give up their energy, and this energy is put to three uses. First, it is used to split water into hydrogen and oxygen. The oxygen rapidly diffuses out of the chloroplast and exits the cell. (This is the process that maintains the oxygen in the atmosphere.) The hydrogen is available to be combined with carbon dioxide. The remaining energy from the electrons is then used in the formation of two intermediate molecules. One of these molecules, adenosine triphosphate (ATP), serves as an energy source for future reactions, and the other, reduced nicotinamide adenine dinucleotide phosphate (NADPH), is a carrier of hydrogen for these reactions, as well as a source of energy for future reactions.

These two molecules—ATP and NADPH—then power the next step in photosynthesis: a biochemical cycle in which carbon dioxide combines with hydrogen from water to form the sugar *glucose*, a carbohydrate.

It may help to summarize this process in a chemical equation:

$$\text{carbon dioxide} + \text{water} \xrightarrow{\text{light energy}} \text{glucose} + \text{oxygen} \qquad \text{(Eq. 4.1)}$$

Or, to be more precise,

$$6CO_2 + 6H_2O \xrightarrow{\text{light energy}} C_6H_{12}O_6 + 6O_2 \qquad \text{(Eq. 4.2)}$$

In words, six carbon dioxide molecules combine with six water molecules to produce one molecule of glucose ($C_6H_{12}O_6$) and six molecules of oxygen. A total of 54 photons must be absorbed to produce one glucose molecule. Glucose then serves as a source of both energy and fixed carbon for other reactions in the cell.

About 33% of the light energy absorbed by a chloroplast is stored in glucose, which actually represents a notably high efficiency. For instance, when our muscles contract, only about 25% of the energy stored in glucose is released as mechanical energy; the remaining 75% is wasted as heat. This is why you tend to overheat during intense exertion. Efficiency is even less when energy passes from plants to animals: new tissue typically contains only about 10% of the energy ingested as food (see chapter 5). And all these biological efficiencies far exceed those in some manmade machinery, such as the classic steam engine. The conversion of energy stored in coal to the energy of moving railcars is only about 1% efficient. In this context, then, the 33% efficiency of chloroplasts is indeed notable.

In summary, photosynthesis uses the energy of light to fix carbon dioxide into glucose, giving off oxygen in the process. The transformation involves complex biochemical machinery, and some parts of the machinery require iron to function.

Respiration

To this point, we have focused primarily on the process that transforms solar energy into glucose stored in photosynthesizing cells. What happens to this energy after it has been stored? In a nutshell, it powers all the processes that allow the cell to survive and reproduce. It provides energy to build armor around the cell and to power flagella. It provides energy needed to construct new cellular material as phytoplankton grow and ultimately divide.

The energy stored in glucose is recovered through the process of *respiration*. In its barest essentials, respiration is photosynthesis run in reverse:

$$\text{glucose} + \text{oxygen} \xrightarrow{\text{energy}} \text{carbon dioxide} + \text{water} \qquad \text{(Eq. 4.3)}$$

$$C_6H_{12}O_6 + 6O_2 \xrightarrow{\text{energy}} 6CO_2 + 6H_2O \qquad \text{(Eq. 4.4)}$$

This is the formula for oxidizing glucose, and if you were to grind glucose to a fine powder, spray the powder into a volume of pure oxygen, and light the mixture, you would indeed produce carbon dioxide, water, and a lot of energy. In fact, the mixture would explode. In this sort of uncontrolled combustion, the energy of burning glucose is released as heat. Cells, however, have evolved the capacity to "burn" glucose at room temperature, gradually releasing its energy as the molecule is carefully deconstructed. Much of the energy released is transferred to ATP, the same molecule used in photosynthesis. ATP then powers a myriad of other reactions in the cell, including the construction of new cellular machinery.

In eukaryotes, energy is extracted from glucose, in large part in specialized membrane-bound organelles in the cell called *mitochondria*. Just as eukaryotic cells gained their chloroplasts by endosymbiotically engulfing cyanobacteria, eukaryotes gained their mitochondria by endosymbiotically engulfing other bacteria. As with chloroplasts, some of the biochemical machinery in mitochondria requires iron.

Note that all cells perform respiration. When you eat a candy bar, your cells respire to release energy from the sugar you ingest. To support the reactions of respiration, you breathe in oxygen, and after respiration has occurred, you exhale the resulting carbon dioxide and water vapor. Accustomed to thinking of this process in the context of animals, we sometimes forget that it also happens in plants. Phytoplankton respire more or less continuously. During the day, when light is available and chloroplasts are active, some of the glucose produced by photosynthesis is respired immediately, but some is also stored. At night, glucose stored during the day is respired to keep the cell alive until dawn.

In summary, respiration recovers energy from glucose. In the process, oxygen is consumed and carbon dioxide is released. All cells, including phytoplankton cells, respire.

Building Other Molecules

Living cells contain a mind-boggling array of molecules: sugars, fats, proteins, and genetic materials (RNA and DNA). Energy stored in glucose powers the production of these molecules, and the carbon fixed by photosynthesis provides some of the basic building blocks. But production requires other raw materials as well. Glucose contains only carbon, hydrogen, and oxygen, whereas the simplest proteins contain carbon, hydrogen, oxygen, and nitrogen. Thus, at the very least, in addition to carbon and water, a phytoplankton cell needs nitrogen to maintain its cellular machinery and grow. Often, phytoplankton acquire this nitrogen from the seawater around the cell in the form of ammonia (NH_3) or nitrate (NO_3^-). Other cellular constituents require additional elements. For example, genetic materials, as well as molecules involved in energy transfer within the cell, require phosphate ions, PO_4^{3-}. Recall, for instance, that ATP, an important intermediate molecule in photosynthesis, is adenosine tri*phosphate*.

In our future discussions of primary production, it will be useful to keep these additional cellular requirements in mind. To that end, we can write a general equation for the net chemical reaction involved in a typical growing phytoplankton cell:

$$106CO_2 + 122H_2O + 16HNO_3 + H_3PO_4 \rightarrow$$
$$138O_2 + (CH_2O)_{106}\,(NH_3)_{16}\,H_3PO_4 \qquad \text{(Eq. 4.5)}$$

That is, 106 carbon dioxide molecules combine with 122 water molecules, 16 nitric acid molecules, and 1 molecule of phosphoric acid to produce 138 oxygen molecules and a quantity of average cellular "stuff," $(CH_2O)_{106}\,(NH_3)_{16}\,H_3PO_4$. Now, in a cell you would never find a molecule with the chemical formula $(CH_2O)_{106}\,(NH_3)_{16}\,H_3PO_4$. But if you took all the diverse molecules in the cell and averaged across them, you would find that, for every 106 carbon atoms, there are approximately 16 nitrogen atoms and 1 phosphorus atom. In summary, in addition to carbon and water, phytoplankton also need a ready source of nitrogen and occasionally a bit of phosphorus, usually in the form of phosphate.

In 1934, A. C. Redfield (a professor at Harvard and Woods Hole Oceanographic Institution) worked out the average composition of phytoplankton cells, as shown above, and the ratio of the number of nitrogen atoms to the number of phosphorus atoms (16:1) is commonly known as the *Redfield ratio*. This ratio is often taken as an indication that phytoplankton consume more nitrogen than phosphorus, and as we will see, this assumption is a critical factor in predicting when and where primary production can occur. It is important to note, however, that the Redfield ratio is an average value. Nothing intrinsic in the biochemistry of a phytoplankton cell requires nitrogen and phosphorus in a ratio of exactly 16:1, and individual cells often deviate substantially from this ratio.

Measuring Production

We have seen that phytoplankton both photosynthesize and respire, often at the same time. In other words, living cells use energy at the same time they acquire it, and we will profit from a method of bookkeeping to keep track of what is going on.

Gross and Net Production

The simplest way of quantifying the rate of photosynthetic carbon fixation (and thereby, the rate of energy influx) is to track the rate at which carbon molecules enter a phytoplankton cell. This is a measure of the cell's *gross productivity*. An analogy can be made to personal finances: your gross income is the amount of money you have earned, without subtraction of expenses.

As any wage earner can attest, however, one's gross income can be misleading. Before you can buy that new car that caught your eye, you must first pay taxes and rent and buy groceries. At the end of the month, it is the difference between your gross income and your expenses that remains as *net* income in your bank account:

$$\text{net income} = \text{gross income} - \text{expenses} \qquad \text{(Eq. 4.6)}$$

The same idea applies to phytoplankton. The carbon accumulated by a cell (its *net production*) equals its gross production (the carbon that enters the cell) minus its respiration (the carbon, in the form of carbon dioxide, released from the cell):

$$\text{net production} = \text{gross production} - \text{respiration} \qquad \text{(Eq. 4.7)}$$

Or, rearranging this equation, we conclude that

$$\text{gross production} = \text{net production} + \text{respiration} \qquad \text{(Eq. 4.8)}$$

This equation provides the basis for one of the classic methods of measuring productivity in the ocean.

The Light-Bottle, Dark-Bottle Experiment

To measure the gross rate of photosynthesis, we need to track the flow of carbon into phytoplankton cells. We could do this directly (using radioactive tracers, see below), but tracking the flow of carbon indirectly is often simpler. From our understanding of photosynthesis, we know that, for every atom of carbon fixed, one molecule of oxygen is released. Thus, if we can measure the rate at which oxygen is released by phytoplankton, we have a measure of carbon fixation.

Or we would, except that respiration uses some of this oxygen. For each carbon atom respired, an oxygen molecule is taken up. Therefore, as a cell both respires and photosynthesizes, the rate at which oxygen is released into seawater is actually a measure of the *net* rate of photosynthesis. To measure the gross rate of photosynthesis, we also need to know the rate of respiration.

This is the impetus for the "light-bottle, dark-bottle" method, introduced in 1918 by Haakon Gran, a Norwegian oceanographer. Two bottles are filled completely with seawater containing phytoplankton, and both are tightly stoppered. One bottle has transparent walls that allow light to enter and phytoplankton to photosynthesize. The oxygen concentration in this bottle goes up through time at a rate governed by the net rate of photosynthesis, and this change in oxygen concentration is monitored. The other bottle has opaque walls that prevent light from reaching the phytoplankton. As a result, only respiration occurs in this bottle, and its oxygen concentration decreases over time, providing a measure of the rate of respiration. One assumes that respiration occurs at equal rates in the two bottles. (There is evidence that this is reasonable.) Knowing both the rate of net productivity and the rate of respiration, we can then add the two together to estimate the gross rate of production.

This method is conceptually elegant and easily implemented. Originally, a chemical technique known as a Winkler titration measured changes in oxygen concentration, but in recent years, electronic dissolved oxygen meters have streamlined the process. Care must be taken to avoid experimental artifacts. For example, light and dark bottles must be kept at the same temperature, preferably the original temperature of the seawater used, and phytoplankton in the light bottle must be subjected to the same light intensity they would experience in the sea. These criteria are satisfied most easily by actually submerging the two bottles back in the ocean during the experiment. By suspending bottles at different depths, one can measure the variation in photosynthetic rate with distance below the surface.

A host of other possible artifacts arise, and oceanographers will happily debate for hours the merits of different strategies for coping with these problems. But these fine points should not concern us here. The light-bottle, dark-bottle method provides a conceptual model for determining the rate of photosynthesis.

There is one final point we need to consider in this measurement. A light-bottle, dark-bottle experiment tells us the amount of carbon fixed in the volume of seawater contained in our bottles. Eventually, we would like to use this information to calculate carbon fixation over the entire ocean, but to do so we must complete a few intermediate steps. First, we generalize the results of a single bottle at a given depth in the ocean to a larger volume of water surrounding the bottle. For example, let's assume that 1 milligram of carbon (gross) is fixed per day in a 1 liter bottle at a depth of 10 meters. There are 1000 liters in a cubic meter, so we estimate that 1000 milligrams (1 gram) of carbon are fixed per day in a cube of water one meter on a side centered at this depth. We could, in theory, repeat this calculation for bottles located at every meter of depth below the surface (figure 4.8). If we added up all the carbon fixed in all the cubic-meter volumes from the surface to the seafloor, we would have a measure of the total rate of carbon fixation associated with one square meter of ocean surface. In practice, productivity is measured at only a few selected depths, and these data are interpolated to infer the production at other depths, but the overall idea is the same. The final result is a measure of productivity—grams of carbon fixed per unit time—per square meter of ocean surface. Averaged over the entire ocean, the net productivity of phytoplankton is approximately 140 grams of carbon per square meter per year.

What should we make of this figure? For the animals that rely on phytoplankton as a source of food, is this productivity comfortingly large or distressingly small? Fortunately, the measurement of oceanic productivity on a per-area basis allows for ready comparison to other systems. Perhaps the best comparison is to a system with which we are intimately familiar: the primary producers on land. Averaged across all the jungles, forests, tundra, and grasslands

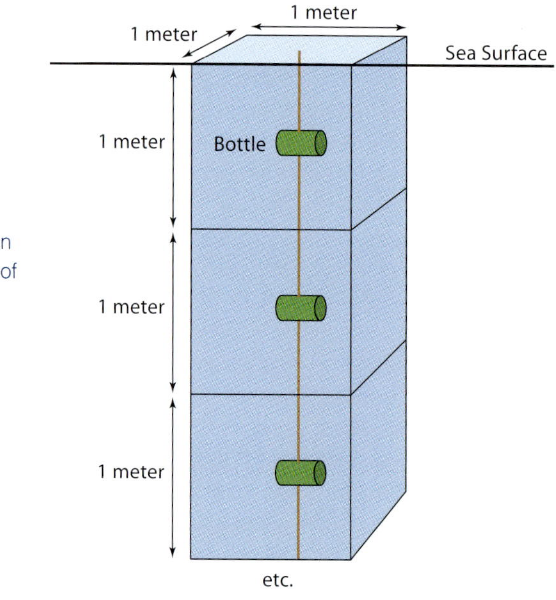

Figure 4.8. Schematic depiction of an experiment to measure the total rate of carbon fixation per square meter of ocean surface.

that grow on the ice-free area of the continents, the net production of terrestrial plants is approximately 426 grams of carbon per square meter per year, three times that of the ocean. Thus, when compared area to area, oceans are much less productive than land, and the reason why is the subject of much of the rest of this text. There is, however, much more ocean than there is ice-free land, and the total yearly net production of the oceans (48.5 billion metric tons of carbon[6]) is nearly equal to that of the continents (56.4 billion metric tons).

It is worth noting that measurement of productivity, whether oceanic or terrestrial, on a per-area basis acknowledges the role sunlight plays in providing the energy for production: the amount of sunlight available is directly proportional to the area on which the light impinges.

Radiometric Assays

The light-bottle, dark-bottle method measures the rate of photosynthesis indirectly by measuring the rate of oxygen release. An alternative method, devised in the 1950s, uses radioactivity to measure directly the rate of carbon uptake by phytoplankton. In the carbon dioxide of a sample of seawater, ^{14}C, a radioactive form of carbon, replaces ^{12}C, the standard carbon. Phytoplankton are added to this seawater, illuminated, and allowed to photosynthesize. The phytoplankton are then filtered out of the seawater, and the amount of radioactive carbon they have absorbed is measured. Because radioactive carbon can be detected in minute amounts, this method provides a sensitive measure of the rate of carbon uptake, and thereby, of gross primary production.

[6] A metric ton is 1000 kilograms, approximately 2205 pounds. Thus, the total yearly net production of the oceans is 48.5×10^{12} kilograms.

The method is not without its drawbacks. For example, if phytoplankton are left in the experimental seawater too long, some of the radioactive carbon they absorb and incorporate into glucose can be respired and released back into the water as carbon dioxide. In this case, the radiometric assay more closely measures net production than gross production. However, when attention is paid to this and other details, radioactive uptake provides accurate estimates of productivity, and the method is widely used.

Measurements from Space

As we discuss below, under some conditions, the rate of primary production is proportional to the concentration of phytoplankton and therefore to the concentration of chlorophyll in the water. Because chlorophyll reflects green light, its concentration can be estimated from the color of a sample of water: the greener the water, the more concentrated the phytoplankton. This simple notion sets the stage for measurements of primary production from space. Some satellites have sensitive cameras tuned to the particular colors reflected by phytoplankton chlorophyll. A picture of an area of the ocean's surface can then be used to estimate the concentration of phytoplankton, thereby providing a measure of primary production. For the record, the estimates of global net production cited earlier were made using satellite-based measurements.

Terminology

In our discussion of primary production throughout the rest of this book, it is usually net primary production—production that can be passed from one link in the food chain to another—that matters. Repetition of this term becomes tiresome, however. To streamline the prose, unless otherwise noted, when I refer to "production," I mean "net primary production."

Control of Primary Production

We now shift from the measurement of production to focus on some of the factors that control the rate at which photosynthesis stores energy. We will continue this discussion in the next two chapters.

At a given time of day, a certain amount of light energy impinges on each square meter of ocean each second. This rate of delivery establishes the maximum rate at which photosynthesis can proceed: the more intense the light, the more photons available to the phytoplankton, and the more carbon they can fix. In reality, however, only a small fraction of the light impinging on the ocean can be used by phytoplankton; the water itself reflects or absorbs much of the light that strikes it. In essence, seawater competes with phytoplankton for the sun's energy. As a result, the rate of photosynthesis depends on the optical properties of seawater. The more absorptive the water, the less light available for carbon fixation.

By the same token, the overall rate of photosynthesis depends on the number of phytoplankton present. The more phytoplankton there are, the better they can compete with water, the more light they can trap, and the faster they can photosynthesize. As noted at the beginning of this chapter, at an average oceanic concentration, phytoplankton themselves absorb about 3% of the light hitting the ocean, a much lower percentage than is absorbed by land plants (about 15%).

There are also factors intrinsic to the phytoplankton themselves. Some species, packed with chloroplasts, absorb relatively large numbers of photons and grow fast. Other species have fewer chloroplasts, absorb less of the light available to them, and grow slowly.

And finally, the growth of phytoplankton can be limited by the availability of raw materials. If there is insufficient carbon, nitrogen, or phosphorus available in the seawater, cells cannot grow even if supplied with abundant light.

Thus, several factors might control the maximum potential rate of production of a given volume of seawater, and we examine each of these factors in turn. Whether production reaches this potential is a matter of ecology: the interaction of phytoplankton with other organisms. These ecological constraints are discussed in the next chapter.

The Availability of Light

If there were no light shining on the ocean, there could be no photosynthesis. The more light there is, the more energy phytoplankton can potentially capture. But the full story is not quite that simple. Let's perform the following experiment. Using the light-bottle, dark-bottle method in the lab, we measure the gross rate of photosynthesis for a given mass of phytoplankton at a given light intensity, and by systematically varying the light intensity between measurements, we ascertain the variation in gross productivity of the sample as a function of light availability. From this experiment, we obtain a graph known as a photosynthesis-irradiance curve, or PI curve for short (figure 4.9A).

At low light levels, the curve behaves as you might suppose: the more light shining on the phytoplankton, the faster they fix carbon. But above a certain light intensity, the process begins to level off. In this region of the curve, the rate of carbon fixation is independent of light intensity. At very high intensities, the rate of fixation may even decrease. How can we account for this behavior?

At low light intensities, the availability of photons limits photosynthesis. Plenty of photopigments are poised to capture light, and the photosynthetic machinery easily processes the energy as photons are captured. As light becomes more intense, however, the light-harvesting machinery becomes busier and busier, until it is processing photons as fast as it can. Beyond this intensity, the photosynthetic machinery—not the availability of photons—limits photosynthetic rate, and production is insensitive to variation in light intensity. This accounts for the plateau portion of the PI curve. At very high light intensities, the rate of energy absorption may actually harm the photosynthetic machinery, causing the rate of fixation to decrease.

Two values of particular interest can be garnered from this curve. The first is the *saturation light intensity*, the intensity at which the phytoplankton reach maximal photosynthetic rate. Below this intensity, phytoplankton cells are limited by light; above this intensity, they are limited by the processing ability of their chloroplasts. In the hypothetical case shown here, the phytoplankton

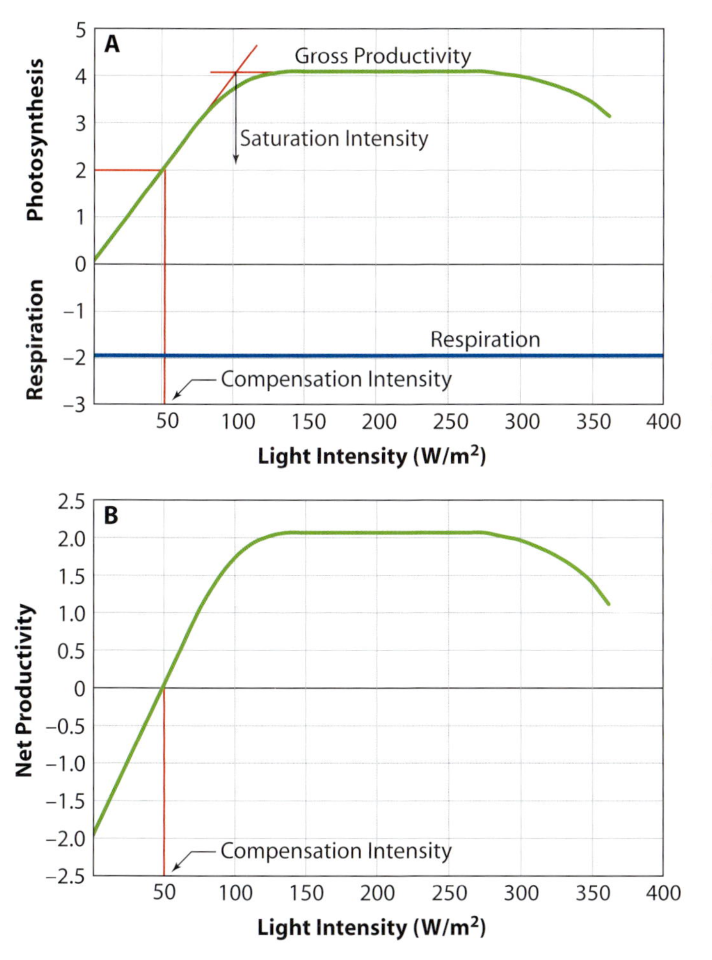

Figure 4.9. Photosynthesis-irradiance curves. (A) In this graph, respiration is shown as "negative productivity." Rates of photosynthesis and respiration can be expressed in a variety of units, such as milliliters of oxygen per gram of phytoplankton per time. In the schematic example shown here, the units are arbitrary. (B) The information of (A) recalculated as net productivity.

reach saturation at a light intensity of 100 watts per square meter, about a tenth of full sunlight.

To calculate the second value, we first recall that phytoplankton both photosynthesize and respire. In other words, as a cell cranks out oxygen as a by-product of photosynthesis (equation 4.2), it uses some of that oxygen to oxidize glucose (equation 4.4). The rate of respiration appears on the PI curve as a line of net oxygen consumption, in effect, a line of negative fixation. When the rate of gross photosynthesis equals the rate of respiration, the cell's energy account breaks even. The specific light intensity at which breaking even occurs is the *compensation intensity*. In the hypothetical case shown here, the rate of fixation equals that of respiration when the light intensity is 50 watts per meter squared. Alternatively, we can plot net (rather than gross) photosynthesis versus irradiance (figure 4.9B). Net photosynthesis becomes positive at the compensation intensity.

The photosynthesis-irradiance curves shown in figure 4.9 were measured in the laboratory, but their information can be applied to phytoplankton in the ocean. We need only to know the variation in light intensity with depth, information easily obtained. As a beam of light passes through water, a certain fraction of its photons are attenuated (absorbed or scattered) by the water for each

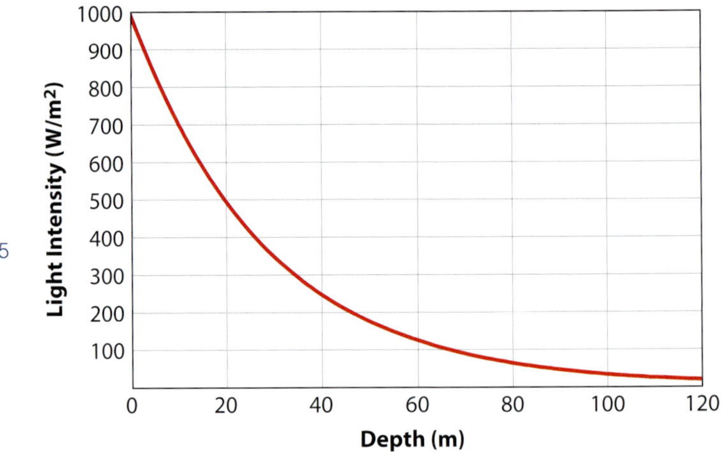

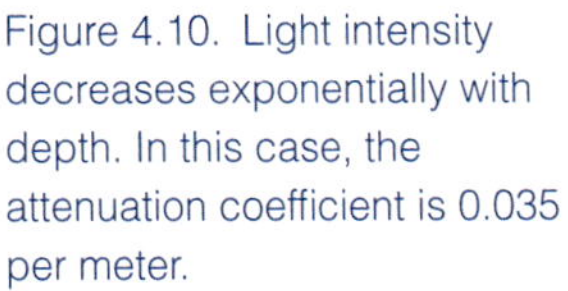
Figure 4.10. Light intensity decreases exponentially with depth. In this case, the attenuation coefficient is 0.035 per meter.

distance traveled. For example, in crystal clear water, about 3.5% of photons are attenuated per meter. In other words, if 1000 photons enter the water, 35 of them are absorbed or scattered and 965 of them are left after the beam has traveled 1 meter. In the next meter, 3.5% of the 965 (34 photons) are absorbed or scattered, leaving 931 photons after an excursion of 2 meters. And so forth.

This process, in which a set fraction of light is attenuated per distance traveled, results in an exponential decrease in the intensity of light, a behavior depicted in figure 4.10. This curve is described by a simple equation:

$$I(d) = I(0)e^{-kd} \qquad \text{(Eq. 4.9)}$$

The light intensity at depth d, $I(d)$, is equal to the light intensity at the water's surface, $I(0)$, multiplied by the value e (the base of natural logarithms, ≈ 2.718) raised to the power $-kd$, where d is again depth below the surface (measured in meters). The coefficient k in this equation is known as the *extinction coefficient*. In the example we have just presented, $k = 0.035$ per meter because 3.5% of the light is extinguished for each meter of depth.[7]

Extinction coefficients in the ocean vary considerably from one water mass to another. In the clear, blue waters of the Caribbean, k is 0.06 per meter. In the more turbid coastal waters off California, k might be 0.12 per meter or higher. The higher the extinction coefficient, the more rapidly light is absorbed by water, and the darker the environment is at any given depth. The euphotic zone is typically 50 to 100 meters thick in clear oceanic water, less where the water is turbid.

So far, we have treated the absorption of photons by water as if that process were independent of wavelength. In reality, the extinction coefficient is sensitive to the wavelength of light: red light is absorbed most readily by seawater, with violet and blue light close behind. Blue-green light is absorbed least. The wavelength-specific absorption of light in typically clear oceanic water is shown in figure 4.11. At the surface, red, blue-green, and blue light are present in the

[7] Note that when k (0.035 per meter) is multiplied by d (in meters), the resulting quantity has no units: it is a pure number, as all exponents must be.

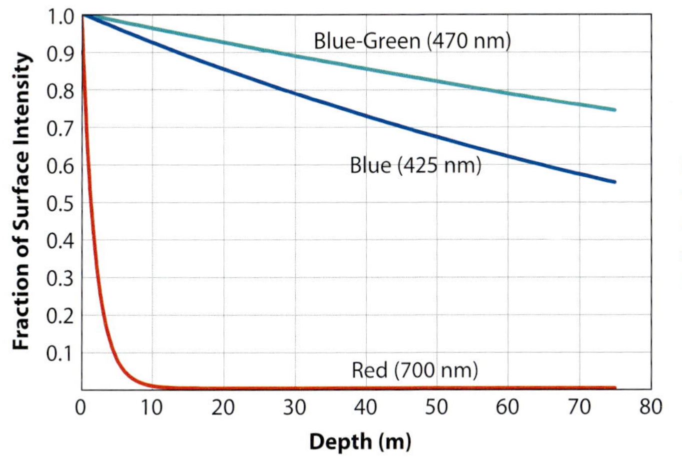

Figure 4.11. Different wavelengths of light are attenuated differently; red light is attenuated most, blue-green light least.

ratio found in sunlight, and the light is white. At a depth of 20 meters, nearly all the red light has been absorbed, along with about 15% of the blue light, and the remaining light is decidedly blue-green. By a depth of 50 meters, all the red and roughly 35% of the blue light are gone, with primarily blue-green light remaining. In this fashion, both the intensity and the color of light change with depth. The presence of any fine particles in the water (silt, for instance) will increase the rate at which blue light is attenuated.

The wavelength-specific attenuation by seawater can be important for photosynthesis. Recall that chlorophyll absorbs only red and blue light, wavelengths that are attenuated rapidly by seawater. As a result, many marine plants have evolved accessory pigments to assist chlorophyll by trapping the blue-green light most readily available.

The change in quality of light with depth is familiar to scuba divers. Dive shops often sell bright red and orange dive gear, which looks flashy in the shop. However, once this gear is submerged a few meters, little red or orange light remains, and the gear appears as a bland gray.

With our knowledge of light extinction, we can now translate the photosynthesis-irradiance curve of figure 4.9 into terms relevant to the ocean. Given the extinction coefficient for a body of water, each intensity on the PI curve can be matched with a given depth. For example, if we combine the PI curve of figure 4.9 with the irradiance-versus-depth curve of figure 4.10, we see that the compensation intensity for this set of phytoplankton (100 micromoles of photons per square meter per second) is reached at a depth of about 65 meters (figure 4.12). In other words, at this depth below the surface, these phytoplankton just break even. This depth is known as the *compensation depth*. If the extinction coefficient were higher, the compensation intensity would be reached at a shallower depth, and phytoplankton would have to live higher in the water column to turn an energy profit.

The exponential nature of light extinction introduces an interesting wrinkle into the calculation of compensation depth. Note that at any point on the curve of figure 4.10, if you move up a meter toward the surface, the light gets substantially brighter, whereas if you move down a meter, the light gets only a little bit darker. Translating this phenomenon into terms of carbon fixation, we find that, if a cell moves up a meter in the water column, its rate of carbon

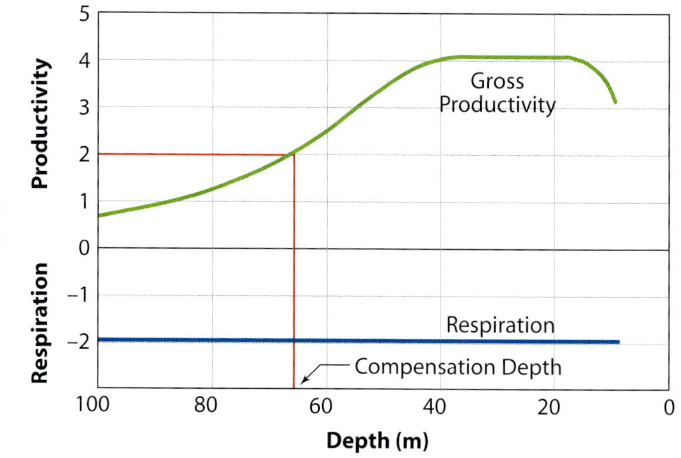

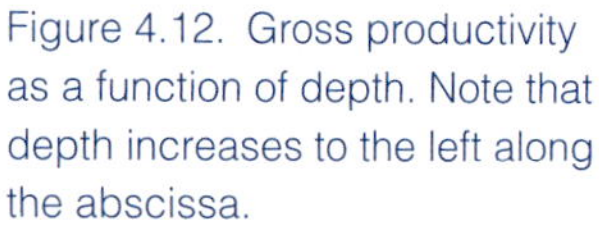
Figure 4.12. Gross productivity as a function of depth. Note that depth increases to the left along the abscissa.

fixation increases more than its rate would decrease if it moved a meter down. In other words, if a cell were to be at the compensation depth *on average*, but moved up and down around this average position, it would more than break even.[8] This scenario might indeed occur if the water column were mixed by the wind.

Now, if cells have a net positive carbon fixation when they are, on average, at the compensation depth, that depth is no longer of much importance. Instead, we should look for the average depth of a phytoplankton cell that just breaks even. This *critical depth* is deeper than the compensation depth, often by a substantial amount. In the North Atlantic, for example, the critical depth can be four to seven times the compensation depth.

In our consideration of compensation and critical depths, we have been concerned with phytoplankton receiving too little light. However, phytoplankton may be inhibited or harmed by too much light as well (figure 4.9), and the high light levels near the water's surface may be deleterious. In general, the highest concentrations of phytoplankton are found a few meters below the surface.

Standing Crop and Growth Rate

We have seen that at certain times and places, light limits photosynthesis. In other circumstances, the capacity of phytoplankton to process the sun's energy—not the availability of energy itself—limits primary production. Phytoplankton's ability to process light has two components. First, the higher the concentration of phytoplankton cells (the higher the *standing crop*), the more light they can intercept, and the more carbon they can fix. Thus, if all other factors are held constant, primary productivity is directly proportional to the con-

[8] This is an excellent example of *Jensen's inequality*, a mathematical theorem that describes how the average behaves in systems that vary nonlinearly. For a readable review of the biological consequences of Jensen's inequality, you may wish to consult J. J. Ruel and M. P. Ayres (1999), Jensen's inequality predicts effects of environmental variation, *Trends in Ecology and Evolution* 14: 361–366.

centration of phytoplankton. However, productivity also depends on the rate of individual cell growth. A low concentration of rapidly growing cells can fix the same amount of carbon per time as a high concentration of sluggishly growing cells.

This combination of factors complicates satellite measurement of productivity. Satellites sense the color of surface water, a measure of chlorophyll concentration and thereby of phytoplankton concentration. However, if the abundant phytoplankton grow slowly, a high concentration may not lead to a high rate of production. Because a single chlorophyll concentration can correspond to different rates of production, it is necessary to "ground truth" satellite measurements. Careful sampling at various places and times reveals the types of phytoplankton that contribute to ocean color, allowing oceanographers to adjust their estimates of productivity.

The Availability of Carbon

Assume for the moment that light is available at saturating intensity and phytoplankton are numerous. With plants and photons abundant, the availability of carbon could potentially limit photosynthesis. Indeed, this is sometimes the case for plants in air, where carbon dioxide forms only 0.026% of the gas molecules in the atmosphere. In units of concentration, there are only 0.36 milligrams of carbon in each cubic meter of air at sea level and even less at higher altitudes. This raises a problem for terrestrial plants, which in response have to keep the pores in their surfaces (the stomata) open to the air to absorb the scarce carbon. In the process, they lose water.

This tradeoff—water for carbon—has heavily influenced the evolution of plants on land. For example, plants such as cacti open their stomata and absorb carbon dioxide in the cool night air to minimize evaporation, and then close the stomata and use stored carbon dioxide for photosynthesis during the heat of the day. In this respect, humankind's penchant for dumping carbon dioxide into the atmosphere might be a boon for terrestrial plants. As the concentration of carbon dioxide increases in the atmosphere, plants, able to take up carbon with less loss of water, presumably will grow faster. However, at least for the immediate future, terrestrial plants will continue to struggle to absorb sufficient carbon dioxide from the atmosphere.

Plants in water do not have this problem of carbon dioxide supply. To understand why, we need to briefly examine the chemistry of carbon dioxide in water. Carbon dioxide readily dissolves in water, and once dissolved, it immediately undergoes a chemical reaction with the water around it. First, it combines with water to form carbonic acid (H_2CO_3), which then splits (dissociates) to yield a hydrogen ion (H^+) and a bicarbonate ion (HCO_3^-). This process can be written as a chemical equation:

$$CO_2 + H_2O \rightleftharpoons H_2CO_3 \rightleftharpoons H^+ + HCO_3^- \qquad \text{(Eq. 4.10)}$$

The double-headed arrows indicate that each reaction in this process can proceed in either direction. Thus, the reaction can run backward from what we have described: a bicarbonate ion can combine with a hydrogen ion to form carbonic acid, which can then split to form carbon dioxide and water. However, in the ocean this equation currently leans toward bicarbonate production. If you take a volume of seawater (with no carbon in it) and shake it up in air, some of the carbon dioxide in the air dissolves, and most of that dissolved CO_2 immediately converts to bicarbonate. As the dissolved CO_2 converts to bicarbonate, however, the concentration of carbon dioxide in the water decreases, allowing more CO_2 to dissolve. This process continues until the water reaches a saturating level of bicarbonate. For seawater in its current state (pH=8.05),[9] saturation is reached when there are approximately 0.028 grams of carbon per cubic meter, a concentration 78 times that in air. Thus, if phytoplankton can get their carbon from bicarbonate instead of from carbon dioxide, seawater provides more raw material for photosynthesis than air does.

This is a substantial "if," however. The chemistry of photosynthesis specifically requires carbon dioxide; it cannot directly use bicarbonate as its carbon source. However, what could have been a major problem has, in fact, been solved in the course of evolution. Most phytoplankton deploy high concentrations of the enzyme *carbonic anhydrase* in their cell walls, which, as its name implies, strips water off bicarbonate to produce CO_2. The presence of carbonic anhydrase thus allows phytoplankton to effectively tap into the huge reservoir of oceanic bicarbonate. As a result, carbon seldom, if ever, limits marine photosynthesis.

Coccolithophorids demonstrate another method of reclaiming CO_2 from bicarbonate. As noted previously, these haptophytes construct calcareous coccoliths inside the cell, which are then extruded onto the cell surface. The chemical reaction by which the coccoliths are formed is:

$$2HCO_3^- + Ca^{2+} \rightleftharpoons CaCO_3 + H_2O + CO_2 \qquad \text{(Eq. 4.11)}$$

In other words, when bicarbonate combines with a calcium ion to produce calcium carbonate, carbon dioxide is released, which can then be used in photosynthesis.

To sum up this section, it is best to repeat its major conclusion. Because of the manner in which carbon dioxide interacts with water—by producing a relatively high concentration of bicarbonate—*the availability of carbon seldom, if ever, limits marine primary production.*

The Availability of Nutrients

In cases where light and carbon are abundant, other molecules may limit photosynthesis. For our purposes, we refer to these compounds as *mineral nutrients*, or simply *nutrients* for short. To be precise, a nutrient is any small, inorganic molecule necessary for the growth of phytoplankton that is not itself a reactant in photosynthesis. Carbon dioxide and water are small, inorganic molecules required for growth, but as reactants in photosynthesis, they are not considered nutrients. In contrast, iron, which is needed for the molecular machinery of photosynthesis, but is not a reactant, *is* a nutrient.

[9] The pH of a solution is a measure of its hydrogen ion concentration: pH= $-\log_{10}$(H^+ concentration), where concentration is measured in moles per liter. Thus, a pH of 8.05 means that seawater has a hydrogen ion concentration of $10^{-8.05}$ moles per liter. 1 mole of H^+ ions $=6.02\times10^{23}$ ions.

The major mineral nutrients in seawater are:

- Ammonia (in the form of ammonium ion, NH_4^+, a source of nitrogen)
- Nitrate (NO_3^-, also a source of nitrogen)
- Phosphate (PO_4^{3-}, a source of phosphorus)
- Silicic acid ($Si(OH)_4$, a source of silicon)
- Iron

Ammonia, nitrate, phosphate, and silicic acid are used in large quantities by many phytoplankton and can be present in substantial concentrations in seawater. For these reasons, they are known as *macronutrients*. In contrast, iron is required in small quantities and is present in seawater at low concentrations; it is an example of a *micronutrient*.

Nitrogen: A Limiting Factor

The nutrients listed above are the ingredients one would expect to find in a marine fertilizer. However, if you have a background as a terrestrial gardener, the recipe for ocean fertilizer is a bit different from what you might expect. If you go to your neighborhood nursery to buy fertilizer for your garden, your choices are categorized by the relative concentrations of three nutrients. For example, a typical fertilizer listed as 16:16:16 has equally large amounts of nitrogen (in the form of ammonia and urea), phosphate (as calcium dihydrogen phosphate or superphosphate of lime), and potassium (as potassium carbonate—potash); all three nutrients are scarce in soil and must be provided.

If there were a corresponding system for marine fertilizer, the Redfield formula (equation 4.5) tells us that it would have a typical composition of 16:1:0; nitrogen is of utmost importance, a smidge of phosphate will do, and potassium doesn't even enter the equation. This ratio is exceedingly important. We have already seen that the availability of carbon is never a limiting factor in marine photosynthesis, and near the ocean's surface, light isn't limiting either. As a result, the Redfield ratio of 16:1 indicates that the primary factor limiting production by phytoplankton near the ocean's surface is the availability of nitrogen.

This conclusion needs some qualification. First, although potassium isn't present in the Redfield equation, it is nonetheless important: phytoplankton require potassium to adjust the ionic composition of their cytoplasm. However, seawater contains an abundance of potassium (table 3.1), so no extra is needed. Second, although phosphate is required in much smaller quantities than nitrogen, it could still (in theory) be a limiting factor. In fact, in the current ocean, this is seldom if ever the case. And finally, silicic acid (the source of silicon for diatom frustules) and iron may limit primary production in some places in the sea.

Despite these qualifications, our original generality stands, and even bears repeating: *the primary factor limiting production by phytoplankton near the ocean's surface is the availability of nitrogen*. Because nitrogen is so important, it is best if we explore its availability in some detail.

Sources of Fixed Nitrogen

As we have seen, phytoplankton require nitrogen to build proteins and genetic material. At first glance, this would not appear to be a problem. Nitrogen gas

(N_2) forms about 80% of the atmosphere and is reasonably soluble in seawater. Thus, nitrogen itself is always present in abundance. Unfortunately, very few organisms can use nitrogen in its gaseous form. Instead, for most organisms, nitrogen must first be *fixed*.

When we used the term "fixation" in regards to carbon, we defined it as the conversion of inorganic to organic carbon, in particular the conversion of carbon dioxide to glucose. In the case of nitrogen, the term is used differently. Nitrogen is fixed when it is converted to a form suitable for incorporation into organic molecules. Fixed nitrogen is typically provided in two forms: either ammonia or nitrate. Ammonia is formed biologically, whereas the initial fixation that leads to nitrate occurs abiotically.

Biological Nitrogen Fixation: *Ammonia*. The ability to convert nitrogen gas to ammonia is exceedingly rare among living organisms, but some specialized bacteria have evolved the necessary machinery. We note for future use that this machinery, like that of photosynthesis and respiration, requires iron.

On land, symbiotic nitrogen-fixing bacteria live in the roots of legumes such as beans. These bacteria take up nitrogen from the air and convert it to ammonia, which can then be used by the beans as a form of fertilizer. After the plants die, their roots remain in the soil, and as they decompose, they release ammonia and nitrogenous organic compounds into the dirt. As a result, crop rotation is an effective means of maintaining soil nitrogen. Soil depleted of fixed nitrogen (by corn, for instance) can be rejuvenated by beans planted on the plot.

In the ocean, nitrogen is fixed primarily by cyanobacteria. Some of these bacteria are benthic, living on coral reefs and in seagrass beds, where they produce roughly 10% to 15% of the nitrogen fixed in the sea even though these habitats form only about 0.25% of the overall ocean area. Virtually all the remaining fixation, 85% to 90% of the total, is accomplished by planktonic cyanobacteria.

There are several aspects of pelagic nitrogen fixation that we should note for future use. First, the per-area rate of nitrogen fixation in the open ocean is relatively small. Except in the rare, short-lived case of a concentrated bloom, the planktonic cyanobacteria found under a square meter of ocean surface do not crank out ammonia nearly as fast as the cyanobacteria in one square meter of coral reef or sea-grass bed. Second, because nitrogen is converted into ammonia by cyanobacteria, which require light for photosynthesis, production of new ammonia is confined largely to the well-lit surface layer of the ocean. And finally, we emphasize a point made earlier in this chapter: only a few species of nitrogen-fixing cyanobacteria, notably *Trichodesmium*, grow in the open ocean, and they are common only in the warm water of the tropics and subtropics.

The low rate of oceanic nitrogen fixation first becomes important when we examine the fate of the ammonia produced by cyanobacteria. In surface waters, as fixed nitrogen is slowly produced, it is quickly taken up by phytoplankton and incorporated into various nitrogen-containing organic molecules: proteins and DNA, for instance. These phytoplankton can then be eaten by herbivores, and the herbivores are eaten by predators. As these consumers respire and grow, some of their nitrogen is converted into ammonia and released into seawater as waste. This waste ammonia is again quickly taken up by phytoplankton. Similarly, when consumers die and rot, much of their nitrogen is released as ammonia, but it too is rapidly absorbed by phytoplankton. Thus, in the euphotic zone where phytoplankton are abundant, the low rate of nitrogen fixation, coupled with the continual and rapid absorption of ammonia by primary producers, ensures that ammonia's concentration in seawater is low. This point will become important when, in chapter 6, we explore the control of global primary production.

Abiotic Nitrogen Fixation: *Nitrate.* In contrast to biological fixation, which produces ammonia, abiotic fixation leads to the production of nitrate. The chief abiotic sources of fixed nitrogen are lightning, fires, agriculture, and industry, and these occur mostly on (or over) land. When a bolt of lightning arcs through the atmosphere, its immense heat causes some of the air's nitrogen to combine with oxygen, yielding a variety of compounds that can be converted into nitrate by terrestrial bacteria. Similarly, the heat of fires (forest fires, for instance) can oxidize N_2, again leading to the formation of nitrate. Some of this terrestrial nitrate dissolves in water and flows to the ocean. In the past century, human society has added significantly to the production of nitrate; about a third of the nitrate produced on earth is now produced by humankind (often as fertilizer), and much of this also makes its way into the sea. In summary, some of the ocean's nitrate is introduced into the sea by runoff from the land.

In addition, nitrate is produced by chemoautotrophic bacteria in the ocean itself, a process that takes two steps. *Nitrite bacteria* use ammonia as an energy source, oxidizing it to form nitrite (NO_2^-), which they then release into the water. *Nitrate bacteria* then take up the nitrite, use it as an energy source, and release nitrate. These nitrifying bacteria (both nitrite and nitrate bacteria) are inhibited by light, so the oxidation of ammonia to nitrate occurs only in the dark ocean interior.

This leads to two important results. First, because phytoplankton cannot grow in the dark, they are absent from the ocean interior and therefore cannot absorb nitrate from the dark waters where it is produced. As a result, the nitrate produced by nitrifying bacteria in the ocean interior tends to accumulate, leading to high concentrations at depth. Second, because nitrite bacteria take up ammonia, the ammonia concentration in the ocean interior is low. When we consolidate our understanding of ammonia's production and uptake, we come to an important general conclusion: *the concentration of ammonia is typically low everywhere in the sea*: in surface waters because it is rapidly taken up by phytoplankton, in the ocean interior because it is converted to nitrite and nitrate by bacteria.

Note that nitrate produced by nitrifying bacteria is not a source of newly fixed nitrogen. Nitrifying bacteria merely convert the previously fixed nitrogen in ammonia to a new form.

In summary, ammonia and nitrate are the primary forms in which newly fixed nitrogen is introduced into the sea: ammonia from biological sources in the sea itself, nitrate from abiotic sources on land. Ammonia is the nitrogenous nutrient most commonly used by phytoplankton in the well-lit surface waters of the ocean, where it is continuously recycled. Nitrate is the common nitrogenous nutrient in the dark ocean interior. As we will see in chapter 6, the rate at which newly fixed nitrogen is delivered to the ocean is less than the rate at which phytoplankton can use it, and this forms a basic problem in the planet's life-support system.

Fixed Nitrogen: Final Thoughts

Before we leave the subject of nitrogenous nutrients, three additional subjects deserve mention. First, although both ammonia and nitrate can act as fertilizer for phytoplankton, ammonia is the more "fundamental" of the two. After ingestion, it can be used "as is," whereas nitrate must first be reduced to ammonia within each organism before it can be fashioned into proteins or nucleic acids.

Second, in addition to nitrogen-fixing bacteria, there are *denitrifying bacteria* that convert nitrate back to nitrogen gas. As we will see in chapter 6, the oxygen concentration in seawater is low at depths of 500 to 1000 meters, inhibiting respiration in most organisms. But some bacteria in these hypoxic waters can substitute nitrate for oxygen in their metabolic chemistry, providing energy for the bacteria and converting nitrate to gaseous nitrogen. The N_2 then diffuses into the atmosphere.

We previously noted that the concentration of oxygen in earth's atmosphere depends on its production by photosynthesis. We now can see that the concentration of nitrogen in the atmosphere also depends on biology. If there were no denitrifying bacteria to replenish the atmosphere with N_2, much of the atmosphere's nitrogen would eventually be sequestered as nitrate in soil and seawater. Thus, without denitrifying bacteria, the composition of our atmosphere would be much different than it is today.

And finally, we are now in a position to plot the pathways nitrogen travels in the ocean (figure 4.13), and you would do well to familiarize yourself with this roadmap of the nitrogen cycle. As we will see, because fixed nitrogen is often the limiting factor in primary production, this map is tremendously important. However, it is also important to realize that the map shown here is not the full

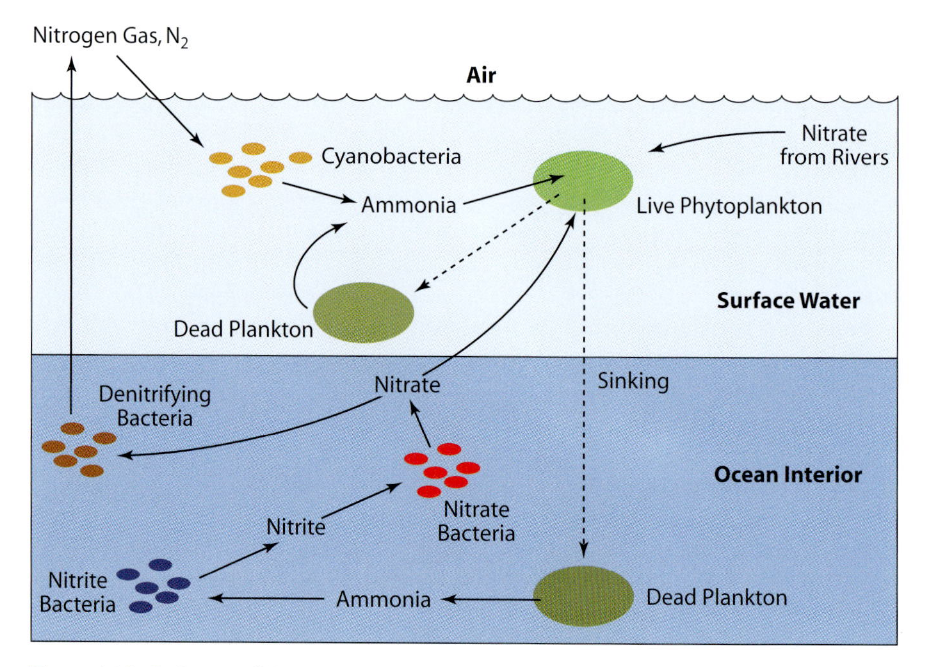

Figure 4.13. Pathways of the oceanic nitrogen cycle.

story. In addition to the paths themselves, we need to know the rates of travel along the paths and the mechanisms that control these rates. These subjects are explored in chapters 6 and 9.

Summary

The sun provides an immense source of light energy to the sea, some of which is harvested by phytoplankton. The physics of life in water make it advantageous for these organisms to be small—typically less than 100 micrometers in length—and they maintain a tenuous existence in the ocean's surface waters, photosynthesizing, growing, dying, and sinking. The net production of the oceans' phytoplankton is about 50 billion metric tons per year. This rate depends on light and nutrient availability, and one particular nutrient—fixed nitrogen—is often the limiting factor. Therefore, our task is now twofold: First, we need to track the flow of energy and organic molecules through the ocean. In essence, we ask: where does all that production go? Second, we desire to predict when and where light and nutrients are both available to phytoplankton. That is, what physical factors control the rate of production? These questions are the subject of the next several chapters.

Further Reading

Denny, M. W. (1993). *Air and Water: The Physics of Life's Media*. Princeton University Press, Princeton, NJ.

Mann, K., and J. R. Lazier (2006). *Dynamics of Marine Ecosystems* (3rd edition). Blackwell Publishing, Malden, MA.

Pilson, M.E.Q. (1998). *An Introduction to the Chemistry of the Sea*. Prentice-Hall Inc., Upper Saddle River, NJ.

Sarmiento, J. L., and N. Gruber (2006). *Ocean Biogeochemical Dynamics*. Princeton University Press, Princeton, NJ.

Appendix

Calculating Sinking Speed

"The bigger you are, the harder you fall" is common wisdom, and phytoplankton are no exception to the rule. Indeed, the speed at which an object of a given shape and density sinks increases in proportion not just to the size of the organism, but to the *square* of its size. Why should sinking speed increase so fast?

Before we tackle this specific question, let's explore a related situation that, intellectually speaking, is a bit closer to home. Imagine that you have signed up to go skydiving. You strap on a parachute, get carried aloft by a plane, and jump out into the air, a medium much less dense than your body. What happens?

The first thing you notice is that you accelerate downward. Gravity tugs on your body's mass, and the resulting force—your weight—pulls you toward the center of the earth. Within a few seconds, you attain a speed of about 50 meters per second (roughly 110 miles per hour), a velocity that would be lethal if you were to maintain it until you hit the ground.

To avoid this nasty consequence, you pull your ripcord, releasing your parachute. After the chute has deployed, it resists being pulled through the air. In other words, as your body's weight pulls down on the chute, the chute, pulling upward in the opposite direction, resists with a force called *drag*. The magnitude of drag depends on the speed at which the chute is pulled through the air.

When the chute first opens, you and it are moving rapidly through the air, and the resulting drag is larger than your weight. Because the upward force on you is larger than the downward force, your falling speed decreases. But as your speed decreases, so does the drag on the chute. Eventually, the two forces acting on your body—weight and drag—reach an equilibrium in which the upward pull of drag equals the downward pull of gravity, and you cease accelerating. This equilibrium occurs at a relatively slow speed—about 5 meters per second—sufficiently slow for a safe landing. You drift to the ground, having had both a thrill and a good lesson in physics.

The same ideas apply to phytoplankton. Because they are denser than the seawater around them, they sink. But as they pick up downward speed, their motion is resisted by the drag of the water, a force directed upward. An equilibrium speed, known as the *terminal velocity*, results, and our job here is to calculate how that velocity depends on the size of phytoplankton cells.

Let's begin by calculating the weight of a cell in water:

$$\text{weight} = \text{gravity} \times \text{effective density} \times \text{volume} = g\rho_e V \qquad \text{(Eq. 4.12)}$$

All others things being equal, the more volume, V, a cell has, the more it weighs. This should be intuitive: large people weigh more than small people. Weight also depends on the density of the material from which the cell is made. In the case of phytoplankton, it isn't the density of the cell that matters; instead, it is its *effective density* that is important. Effective density, ρ_e in the equation above, is the difference in density between the cell and its surroundings. If the cell has the same density as seawater, seawater supports it, and the cell has no tendency to sink. In this case the cell is said to be *neutrally buoyant*. If the cell's density is less than that of seawater, it is *positively buoyant*, and it floats. The density of phytoplankton cells is typically 1030 to 1100 kilograms per cubic meter, while the density of seawater is about 1028 kilograms per cubic meter. Thus, the effective density of phytoplankton is 2 to 72 kilograms per cubic meter: they are *negatively buoyant*, and they sink.

We can take this relationship one step further if we assume that phytoplankton cells are spherical, a reasonable approximation for many species. We then can easily express the cell's volume as a function of its diameter, D:

$$\text{volume of a sphere} = \frac{\pi D^3}{6} \qquad \text{(Eq. 4.13)}$$

Thus, the weight of a spherical phytoplankton cell in seawater is

$$\text{weight} = \frac{\pi \rho_e g D^3}{6} \qquad \text{(Eq. 4.14)}$$

We now turn our attention to the drag that resists the motion of this cell. In this case, we rely on a relationship worked out by George G. Stokes in 1851:

$$\text{drag on a small sphere at slow speeds} = 3\pi\mu U D \qquad \text{(Eq. 4.15)}$$

Here μ is the viscosity of the seawater: a measure of its "stickiness," its resistance to being sheared. At a typical sea-surface temperature of 20°C, viscosity is 0.001 newton-seconds per meter. The variable U is the speed of the cell through the water, and D is again the cell's diameter.

Now that we have equations for both weight and drag, we set them equal to each other to find out what happens when an equilibrium of forces is established:

$$\frac{\pi \rho_e g D^3}{6} = 3\pi\mu U D \qquad \text{(Eq. 4.16)}$$

When we solve for U, we find that the terminal sinking velocity is

$$U = \frac{\rho_e g D^2}{18\mu} \qquad \text{(Eq. 4.17)}$$

Thus, sinking speed increases with the effective density of the cell and decreases with the viscosity of the seawater. Most important here, however, is that sinking speed increases as the square of the cell's diameter. The bigger the cell, the faster it sinks.

In making this calculation, we have assumed that the cell is spherical. Although this assumption is a rough approximation for many haptophytes, cyanobacteria, and dinoflagellates, it is a lousy approximation for chains of diatoms, which are long and slender. How does our conclusion change for diatom chains and other slender objects?

This question can be answered through the same logic used above, beginning with different equations for weight and drag. Here, we assume that a diatom chain is roughly the same shape as a prolate ellipsoid, a shape something like a cigar. We also assume that the ellipsoid sinks with its long axis horizontal; that is, it moves broadside through the water. (It turns out that this is a stable orientation for sinking prolate ellipsoids.) If the length of the ellipsoid is L and its diameter is a fraction, q, of L,

$$\text{weight} = \frac{\pi \rho_e g q^2 L^3}{6} \qquad \text{(Eq. 4.18)}$$

$$\text{drag} = \frac{4\pi\mu U L}{\ln\left(\frac{2}{q}\right) + \frac{1}{2}} \qquad \text{(Eq. 4.19)}$$

Setting these forces equal to each other, and solving for sinking speed, we find that

$$U = \frac{\rho_e g q^2 L^2 \left[\ln\left(\frac{2}{q}\right) + \frac{1}{2}\right]}{24\mu} \qquad \text{(Eq. 4.20)}$$

As with the sphere, the diatom chain's sinking speed increases with effective density and decreases with viscosity. And as with the sphere, sinking speed increases with the square of length. Thus, in this important respect, diatom chains behave the same as phytoplankton of more compact shapes.

In one important respect, however, diatom chains differ from spherical cells: for a given size (length or diameter), they sink more slowly. Sinking rates of spheres and prolate ellipsoids are shown in figure 4.A1. For a given maximum dimension (the diameter of a sphere, the length of an ellipsoid), the ellipsoid sinks more slowly than the sphere.

For reasons that are not entirely understood, live phytoplankton deviate from the predictions made by these calculations. The measured sinking rates of phytoplankton (figure 4.4) are

$$U_{meas} \approx 2.82 L^{1.177} \qquad \text{(Eq. 4.21)}$$

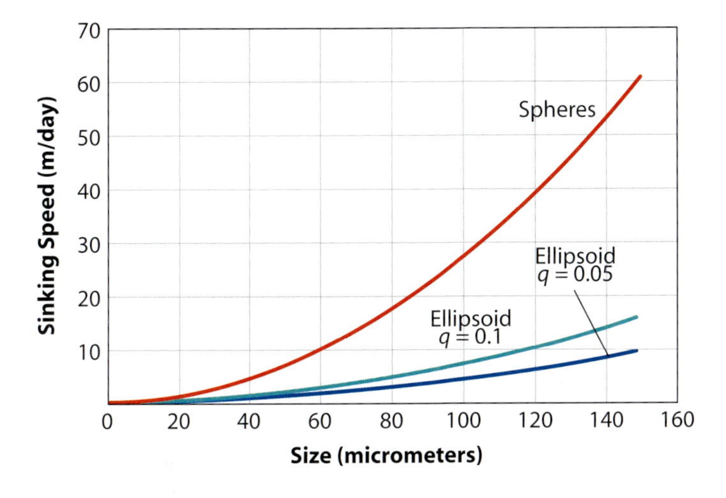

Figure 4.A1. Theoretical sinking rates for spheres and prolate ellipsoids. The ellipsoids are assumed to sink with their long axis horizontal. Here q is the ratio of diameter to length for the prolate ellipsoid.

Sinking speed increases with increasing size, but only to the power of 1.177 rather than 2. Theory and observation could be reconciled if the effective density of phytoplankton decreased in larger cells. To be exact, if $\rho_e = 0.00566L^{-0.824}$, cells will sink as measured. It may well be that the effective density of phytoplankton cells decreases with increased size due in part to the way in which their surface area varies in proportion to their mass. The mass of dense armor that coats a cell will increase roughly in proportion to L^2, whereas the volume of lower density cytoplasm increases as L^3. Thus, the larger the cell, the more nearly its density matches that of cytoplasm alone.

The Flow of Energy, Carbon, and Nutrients

In the last chapter, we explored the process by which marine phytoplankton store solar energy in organic matter. What then happens to all that organic material? Who eats whom, and how efficiently do they dine? These questions concern the *trophic structure* of the ocean,[1] and they are the subject of this chapter. While exploring these questions, we will touch on some basic principles of marine ecology.

Introducing the Consumers

Before we talk about trophic structure, we need a guide to the consumers. This task is less straightforward than our compilation of primary producers in chapter 4. In that case, all the organisms were at least photosynthetic, which lent some functional continuity to the presentation. In dealing with the consumers of the ocean, we must cope with their wildly divergent lifestyles.

How best to explore this subject? The traditional approach marches through taxonomy, presenting organisms in terms of their ancestral lineages. As a method for exploring how marine consumers function, this approach is less than ideal. It is akin to describing how the U.S. Congress works by listing the family histories of its members: interesting, but full of details that are not fully relevant. Alternatively, we could consolidate details by lumping consumers into categories based on their position in the food chain. Unfortunately, this synthetic approach can obscure important facts. For example, both heterotrophic dinoflagellates and blue whales occupy the same link in the trophic chain, but that functional similarity could obscure the fact that the two organisms differ in mass by a factor of 350,000 billion.

[1] Trophic: of or relating to nutrition.

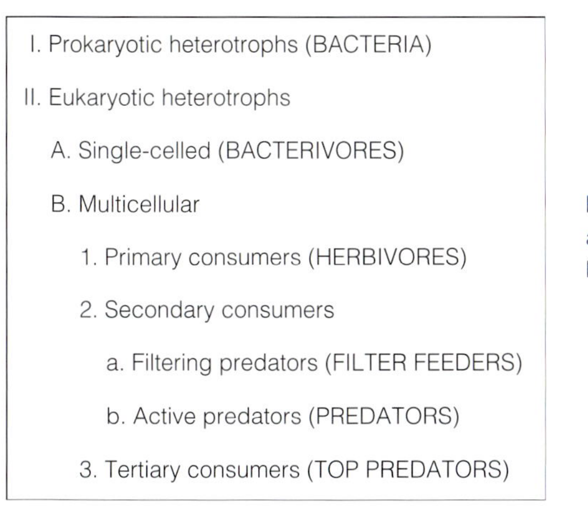

Figure 5.1. An outline of our approach to categorizing marine heterotrophs.

Here we take a different tack: we group organisms into *functional trophic groups* taking into account the complexity of each organism, what it eats, and how it captures its food. This general scheme is outlined in figure 5.1. Because small organisms are generally eaten by larger ones, this approach also tends to group organisms by size (table 5.1).

It would be an immense task to cover all known marine consumers in a single chapter, and we will not attempt it here. Instead, we impose some limits on the discussion. First, we confine ourselves to organisms found in the water column, forsaking those that live on or in the seafloor. We are not slighting benthic organisms, just recognizing their relative scarcity. As we have seen, the oceans are on average 4000 meters deep, and only the bottom meter or so, roughly 0.025% of the marine habitat, is home to benthic creatures. Thus, benthic organisms simply don't occupy enough space to have the same impact as pelagic organisms. Second, we confine ourselves to description of a few representative organisms in each trophic group. As you read about these organisms, you should keep in mind that the presentation here is at best an hors d'oeuvre to the vast smorgasbord of heterotrophic life in the sea.

For future reference, as we introduce the marine heterotrophic organisms, we note how long it takes them to reproduce. When complete, this list will enable us to draw conclusions about the relationship between size and reproduction.

Table 5.1 Marine functional trophic groups, and the typical sizes of organisms in each group

Functional Trophic Group	Size (meters)
Bacteria	0.0000005–0.000002
Bacterivores	0.00003–0.002
Herbivores	0.001–1
Filter Feeders	0.01–30
Predators	0.01–10
Top Predators	1–10

Bacteria

Let's begin with bacteria, organisms that eat dissolved molecules. We have already encountered the autotrophic bacteria of the sea—the cyanobacteria—and the heterotrophic bacteria are in many ways similar. They are small, single-celled prokaryotes (typically 0.5 to 2 micrometers in length), and they often move, using flagella. The main difference, of course, is that heterotrophic bacteria aren't photosynthetic, so they must rely on energy stored by other organisms. As we will see, much of the chemical energy utilized by heterotrophic bacteria is derived from dissolved organic material released from phytoplankton. In a sense, then, many of these bacteria could be considered herbivores, or at least herbivores after the fact. Unlike cyanobacteria, which often form chains, most marine heterotrophic bacteria are solitary, freely moving about on their own.

The concentration of dissolved food in the sea is patchy, and it is to bacteria's advantage to situate themselves where the organic soup is most concentrated. One obvious location is near dead organisms. For instance, single-celled organisms—such as bacteria themselves, phytoplankton, and the ciliates and flagellates we will encounter below—are often killed by viruses (more on this later). When an infected cell dies, it bursts, releasing new viral particles into the water, which is the point of the exercise as far as the viruses are concerned. However, bursting also releases the cell's other contents, which become food for bacteria. In another example, as the remains of small organisms (particularly gelatinous organisms) sink through the water column, they tend to sweep up other particles, forming strings and globs known as *marine snow*, one form of particulate organic matter. As they decompose, these flocculent "snowflakes" provide another source of concentrated dissolved material. And as we will see, even live, healthy phytoplankton often release carbohydrates into the water.

As bacteria attempt to locate and stay near these sources of food, motility is advantageous, and some marine bacteria are among the fastest moving bacteria on earth. A few species of marine bacteria have been clocked at over 400 micrometers per second. The bacteria in question were only about 0.6 micrometers long, so this speed is nearly 700 body-lengths per second. For comparison, a 2-meter-long human being running at the same relative speed would travel at more than 5000 kilometers per hour, faster than the fastest fighter jet.

The ability to move fast is only part of the story, however. To utilize patchy sources of food, bacteria must first be able to find them, a task seemingly beyond the behavioral capabilities of such simple organisms. However, bacteria have evolved to take advantage of the statistical properties of a process known as a random walk. Bacteria can detect the concentration of dissolved food in their immediate vicinity. When in an area of low food concentration, they frequently "tumble," changing their direction randomly. The higher the concentration of food, the less often they tumble. That is, when by chance they find themselves headed in a profitable direction, the probability that they will change direction decreases. Somewhat surprisingly, the net effect is that bacteria move toward areas of higher concentration. Thus, although their motion is not really "directed," on average it leads them where they want to go. Despite the inability of bacteria to move directly toward a source of food, speed of locomotion still counts. The faster the bacteria swim in this biased random walk, the faster they move toward areas of high concentration.

Once in an area of highly concentrated food, bacteria secrete enzymes, which break the organic material down into small molecules that can be absorbed. This digestion occurs outside the bacterial cell, which led one wit to refer to bacteria as the "ultimate swimming stomachs." Using this digestive prowess, bacteria can reproduce quickly, each cell dividing in two. Bacteria can double 1 to 10 times per day when conditions are favorable.

The study of heterotrophic marine bacteria is still in its infancy. Most of what we know about bacteria in general is derived from studies on species that can be cultured in the laboratory. For example, if one wants to know how many bacteria there are in a volume of tap water, one traditionally takes a tiny sample of the water, spreads it on a gelatin plate, and waits for colonies to grow. Each colony grows from one bacterium, so the number of colonies is equal to the initial number of bacteria in the sample. The researcher counts the number of colonies by eye, and voilà! Furthermore, it is often possible to tell from the shape and color of a colony growing on a gelatin plate what type of bacteria it contains.

Unfortunately, we do not yet know how to culture most marine bacteria, and as a consequence, much of what we know about these organisms is based on more painstaking measurements. For instance, estimates of bacterial population sizes are made by mixing samples of seawater with dyes that stain DNA. When samples are subsequently viewed under a high-power microscope and illuminated with ultraviolet light, bacteria glow and can be counted.

From such studies, it has become clear that bacteria are present in extraordinary concentrations in seawater: from 10 million to more than a billion cells per liter. At these sorts of concentrations, it might be difficult to believe there is any room for water amongst the bacteria. But these remarkably high concentrations should be kept in perspective. At a concentration of a billion cells per liter, the average spacing between bacteria is 100 micrometers, roughly 50 to 100 bacterial body lengths.[2] For human beings with a length of 2 meters, the analogous spacing would be 100 to 200 meters—one person every 1 to 2 football-field lengths—a notably sparse crowd.

The staining procedure described above is excellent for enumerating the number of bacteria present, but it paints all bacteria with the same brush: it is difficult to tell from these counts whether there is one species present, or one hundred. In contrast, the tools of molecular biology can be used to discern the number of species. DNA can be extracted from a sample of seawater, and the variety of DNA base sequences present[3] can provide a measure of the number of different types of bacteria. In this case, however, it is difficult to tell how many individuals, as opposed to species, are present, or even whether the DNA came from live or dead organisms.

[2] Imagine stacking small cubic blocks to form a larger cubic structure. For example, you could build a cube 10 blocks wide by 10 blocks deep by 10 blocks high, and it would take $10^3 = 1000$ blocks in the process. Turning this logic around, we see that the number of blocks along any edge of the large cube is equal to the cube root of the total number of blocks: $10 = \sqrt[3]{1000}$. Now, a liter is a cube 0.1 meters on a side. If there are one billion (10^9) bacteria in this cube, spaced evenly, there are $\sqrt[3]{10^9} = 1000$ bacteria along each 0.1 m edge. Thus, these bacteria are spaced $0.1\text{ m} \div 1000 = 100$ micrometers apart.

[3] The sequence of nitrogenous bases in a DNA molecule determines the sequence of amino acids produced when the DNA is "read." In other words, the base sequence specifies the genetic code of an individual.

Despite the technical difficulties inherent in the study of marine bacteria, rapid progress is being made, and you would be well served to keep your eye on the scientific news as the story unfolds.

The Bacterivores

A wide variety of organisms eat bacteria, and among the major players are ciliates, flagellates, and amoebas.

Ciliates. Ciliates are perhaps the most widespread and important predators of bacteria. They are also the largest and most complex singled-celled animals. As their name implies, the bodies of these eukaryotic protozoa are covered (at least in part) with cilia, which propel the organisms through the water. They have a gullet and mouth for ingesting small prey such as bacteria, and in addition to providing locomotion, cilia help bring food particles into the mouth. Like bacteria, ciliates grow rapidly, dividing every 0.5 to 2 days when conditions are good. Ciliates are typically 10 to 100 micrometers in length, but can be as small as 2 micrometers and as large as 3 millimeters.

Anyone who has read an introductory biology textbook has encountered at least one ciliate, *Paramecium*, a genus commonly found in freshwater ponds. Naked ciliates similar to *Paramecium* are found in the ocean. Other marine ciliates possess a barrel- or vase-shaped covering called a *lorica*, made primarily from protein. When a lorica is present, cilia are confined to the portion of the cell that extends out of the covering. One such group of loricate ciliates, the tintinnids, is especially common in the sea (figure 5.2).

Ciliates are very effective at eating bacteria and other small organisms. Experiments have shown that a single ciliate can filter all the bacteria from 10 to 50 microliters of seawater per hour. At a low concentration of bacteria, 10 million per liter, this rate amounts to 100 to 500 bacteria per hour per ciliate, an impressive meal. But at a high concentration of bacteria, 1 billion per liter, this filtration rate yields 10 to 50 *thousand* bacteria per hour per individual ciliate.

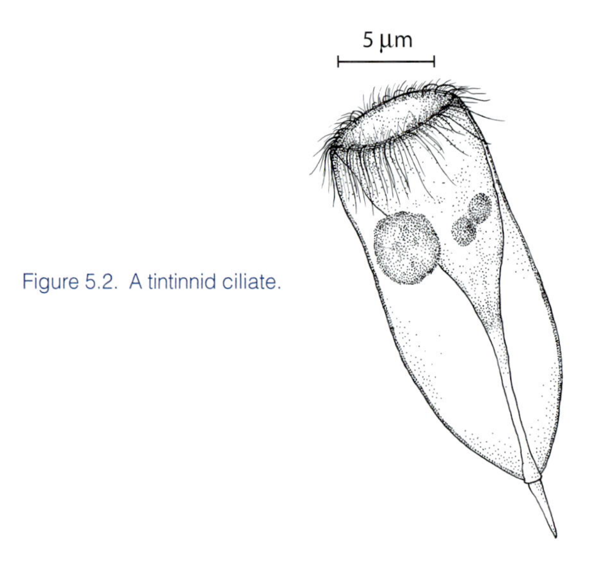

Figure 5.2. A tintinnid ciliate.

Not surprisingly, ciliates can, at times, exert heavy influence on bacterial population size.

Typical concentrations of ciliates in surface seawater vary from 1000 to 350,000 per liter. We can put this number into two perspectives. First, if each ciliate can filter 50 microliters of water per hour and there are 1000 ciliates in a liter, it would take about 20 hours for the ciliates to filter the entire volume. On the other hand, if there were 350,000 ciliates per liter, the entire volume would be filtered every 3.4 minutes. Again, these numbers suggest the potentially substantial effect ciliates can have on the bacterial populations they graze.[4]

The second perspective concerns the spacing of individuals. At the seemingly huge concentration of 350,000 ciliates per liter, the average spacing between individuals is still about 1400 micrometers, fourteen body lengths for a ciliate 100 micrometers long. Imagine yourself in a crowd where each person is separated from the next by 30 meters, and you will have an idea of the relative spacing of ciliates in the sea.

All ciliates are relatively closely related—they are members of the phylum Ciliophora—so the term "ciliate," which we have used so far to describe these organisms' morphologies, also serves to describe their evolutionary heritage.

Flagellates. In contrast, the term "flagellate" is used as a catchall descriptor for a disparate variety of single-celled eukaryotic organisms that use flagella for locomotion and feeding. Generally these organisms are small, 2 to 100 micrometers in length, exclusive of the flagellum. Members of this group are not closely related. For example, heterotrophic dinoflagellates are related to the phytoplankton we discussed in chapter 4, while choanoflagellates are more closely related to the sponges. But having flagella, both are flagellates.[5] Without the taxonomic coherence exhibited by the ciliates, we are left to describe the flagellates primarily in terms of how they function. They come in two general sorts: autotrophic flagellates (which have the ability to photosynthesize) and heterotrophic flagellates (which do not). At times, the distinction is blurred. Some autotrophic flagellates ingest bacteria, and these switch hitters are sometimes referred to as "mixotrophic." Heterotrophic flagellates include the choanoflagellates, the silicoflagellates, and those dinoflagellates that lack chloroplasts.

Individual flagellates can consume from as few as 4 to more than 200 bacteria per hour. While not as individually voracious as the ciliates, flagellates are nonetheless important predators of bacteria because they occur in very high

[4] The calculations here assume that bacteria neither move nor are mixed about in the water. If bacteria were thusly stationary, ciliates could filter each small volume sequentially, with no volume filtered twice and no bacteria missed. In reality, bacteria are more likely to be continuously and randomly redistributed. In this case, ciliates consume a given fraction of the bacteria in a given period, and the concentration of bacteria decreases exponentially. The concentration at time t is $C(0)e^{-jnt}$, where $C(0)$ is the initial concentration, j is the volume filtered by one individual in one unit of time, and n is the number of individuals present in a given volume. With a little algebra, one finds that the time it takes to reduce the bacterial concentration by half, $t_{1/2}$, is $0.69/(jn)$. This half-life is 13.8 hours when 1000 ciliates are present in a liter, 2.4 minutes when 350,000 are present.

[5] Bacterial flagella (which are stiff structures, rotated by an apparatus at their base) are fundamentally different from eukaryote flagella (which are flexible structures, driven by sliding filaments within the flagella themselves). Because of this difference, and because they are prokaryotic, bacteria are not counted as flagellates.

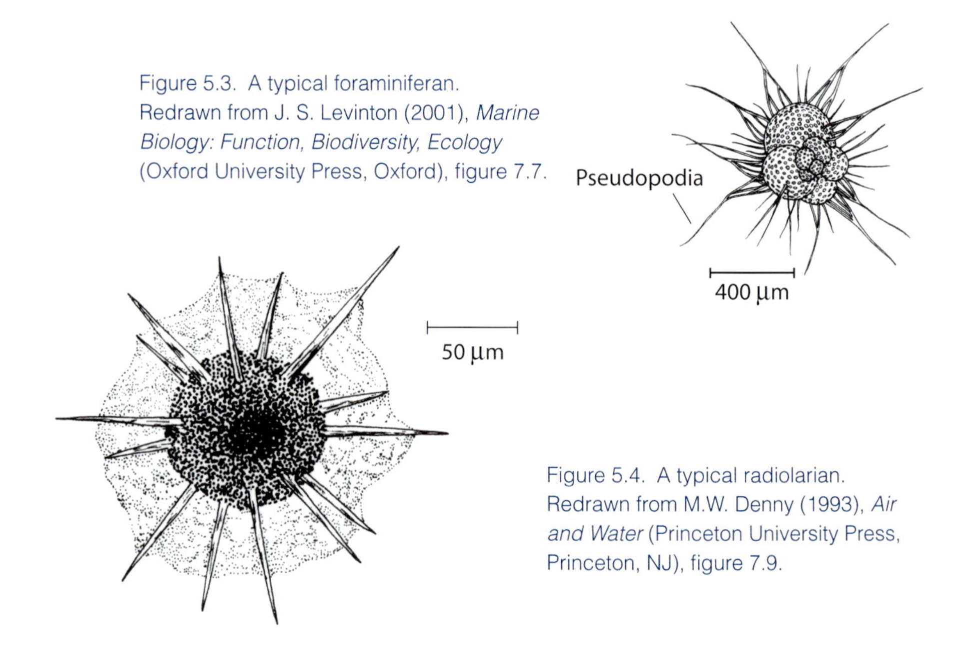

Figure 5.3. A typical foraminiferan. Redrawn from J. S. Levinton (2001), *Marine Biology: Function, Biodiversity, Ecology* (Oxford University Press, Oxford), figure 7.7.

Figure 5.4. A typical radiolarian. Redrawn from M.W. Denny (1993), *Air and Water* (Princeton University Press, Princeton, NJ), figure 7.9.

densities, as high as 10 million per liter, thirty times the maximum reported for ciliates. At this concentration, flagellates eating 200 bacteria per hour from a bacterial population of 1 billion per liter could consume the entire bacterial population in less than five hours. Flagellates can reproduce as often as nine times per day, although once per day is perhaps more typical.

Amoebas. There are a variety of marine amoebas, some of them naked and others with shells. Here we introduce two of the more common shelled varieties: the *foraminifera* and *radiolaria*.

Foraminifera, commonly known as "forams," are a morphologically diverse group of single-celled eukaryotes characterized by their calcium carbonate (or occasionally siliceous) shells, which often resemble miniature snail shells. Foraminifera are small—typically 500 to 1000 micrometers in maximum dimension—but larger than ciliates and flagellates. The forams are characterized by the amoeboid processes they extend into the water around them (figure 5.3). These slender, finger-like projections (also known as pseudopodia) are used to capture prey, which include both bacteria and phytoplankton. In addition, some foraminifera house symbiotic algae, usually dinoflagellates. In other words, they are mixotrophic: as heterotrophs they eat bacteria and algae, and as symbiotic autotrophs they eat sunlight. The symbionts may be carried to the ends of the pseudopodia during the day and retrieved at night. Foraminifera can reproduce once every 3 to 10 days, a substantially longer doubling time than that of ciliates and flagellates.

Forams are abundant throughout the ocean. After they die, their shells sink, forming one main type of marine sediments, commonly known as *globigerina ooze*, named for a common genus of these creatures. Forams can be present at concentrations up to about 100 per liter.

Radiolaria are another form of single-celled, eukaryotic amoeba, but with a different type of skeleton. In this case, the material is silicon dioxide, and the skeleton often forms a beautiful framework of perforated spheres and radial spines, some contained within the cell and others extending beyond it (figure 5.4). Radiolaria come in a greater variety of sizes than the foraminifera, ranging from 50 micrometers to a few millimeters. In other ways, radiolaria are very similar to forams: they extend pseudopodia into the water to capture bacteria, diatoms, and even larger prey such as copepods (see below); they can contain symbiotic algae, typically dinoflagellates; and their doubling time is similar to that of forams, 3–10 days.

Radiolaria are most common in tropical and semitropical waters, where their skeletons may form distinctive sediments known as *radiolarian oozes*. Radiolaria are present at concentrations up to about 20 per liter of seawater.

Herbivores

To this point, we have introduced only single-celled organisms. In contrast, all the organisms from here on out are multicellular. This increase in complexity requires us to adjust how we measure reproduction. Single-celled organisms typically reproduce by splitting in two, so it was reasonable to track their reproduction in terms of doubling time, the period from one split to the next. In contrast, multicellular organisms develop specialized gonads, and each individual can produce many young, either sexually or asexually. In this case, we track reproduction in terms of *generation time*, the time from when an organism is born to when it reproduces.

Copepods. The most abundant herbivores in the ocean are the copepods. In fact, copepods have more mass collectively than any group of animals on earth, a fact that should not be too surprising given that they eat phytoplankton, the largest mass of primary producers on earth.

Copepods are crustaceans, cousins to shrimps, crabs, and lobsters. They have a skeleton constructed from chitin, a carbohydrate similar to cellulose, and have jointed appendages. The body is typically teardrop-shaped, with a head, thorax, and abdomen (figure 5.5). There is an eyespot on the head that senses light, but it is not capable of forming an image. The most common copepods in the open ocean are the calanoid copepods, which swim using a series of five pairs of legs attached to their thorax and catch particles using fan-like motions of their more anterior appendages. When startled or in danger, copepods can escape by making quick strokes with their long antennae. When the antennae are not being used for locomotion, they are held extended into the water, where they act as sensory structures, warning the animal of the approach of both predators and prey. Copepods often undergo a vertical migration each day, rising to near-surface waters during the night and sinking to depths of several hundred meters during the day.

Calanoid copepods are typically 1 to 2 millimeters long, substantially larger than many of the single-celled organisms we have examined so far. They eat primarily phytoplankton but ingest other small organisms as well. The fecal pellets produced by copepods are an important component of particulate organic matter in the sea. As we will see in chapter 6, fecal pellets in general are important because they sink, and copepod fecal pellets sink from 12 to 225 meters per day, about 10 times as fast as phytoplankton.

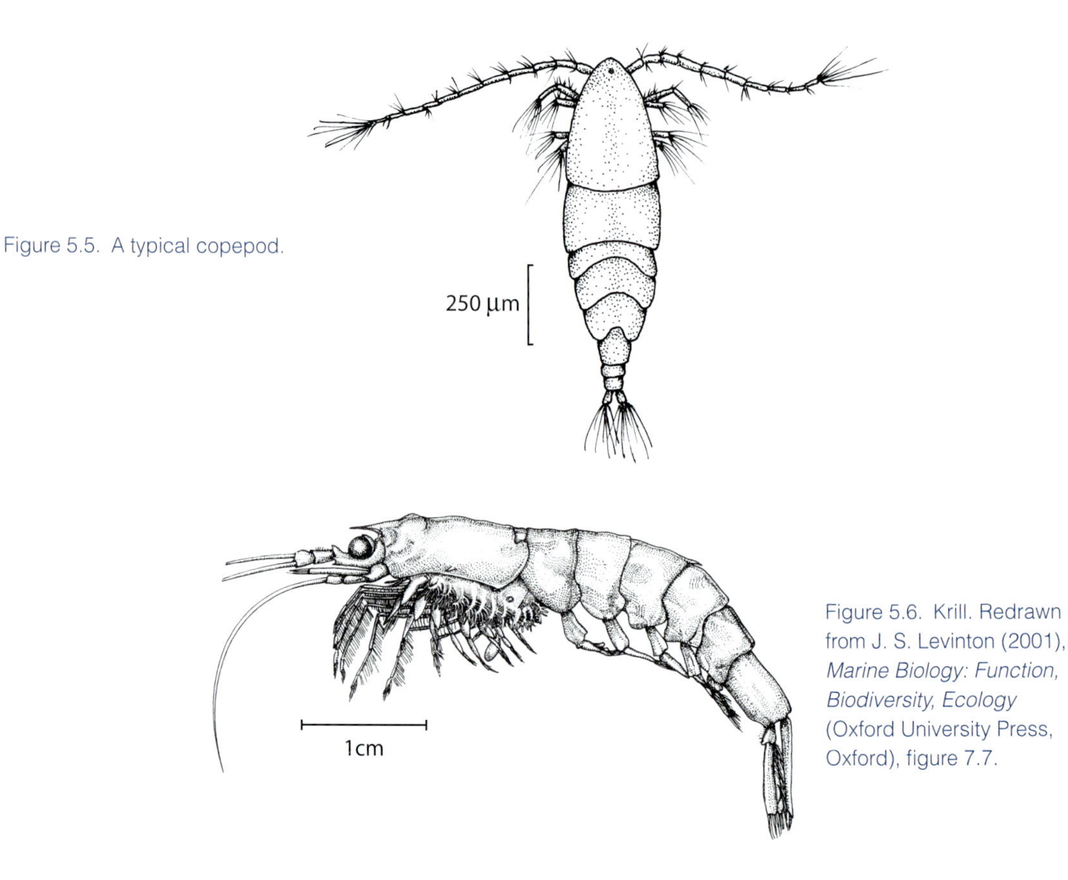

Figure 5.5. A typical copepod.

Figure 5.6. Krill. Redrawn from J. S. Levinton (2001), *Marine Biology: Function, Biodiversity, Ecology* (Oxford University Press, Oxford), figure 7.7.

The generation time of copepods is substantially longer than that of the organisms we have examined to this point: it takes from 12 to 28 days from the time of hatching until a copepod can reproduce.

Krill. Like copepods, krill are crustaceans, having a chitinous exoskeleton and jointed appendages. They resemble small shrimps (figure 5.6) and use a set of five posterior pairs of legs, the swimmerets, to swim. Feeding is accomplished by using other more anterior pairs of legs to form a filter. Krill have image-forming eyes, and are substantially larger than any of the organisms we have discussed so far, typically ranging from 1 to 5 centimeters in length. The generation time of krill is longer than that of copepods: from 30 to 120 days.

Despite their relatively large size, krill dine mostly on phytoplankton, only occasionally eating other small zooplankton such as flagellates, ciliates, and copepods. Fecal pellets produced by krill sink even faster than those excreted by copepods: up to 850 meters per day.

Like copepods, krill often undergo a daily vertical migration, up at night, down during the day. As a means of defending themselves against predators, krill often travel in swarms, in which densities can reach 30 individuals per liter. This concentration may not sound impressive relative to the concentrations we have noted for flagellates and ciliates, but again we should put it into perspective. At a concentration of 30 per liter, the average spacing between krill is about 3 centimeters. Given that the animals themselves are 1 to 5 centimeters long, at this concentration they are packed together cheek to jowl.

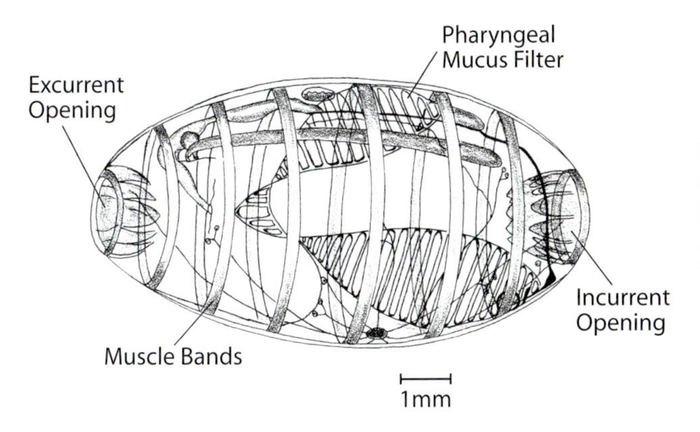

Figure 5.7. A typical solitary thaliacean, showing the internal filter. Redrawn from P. A. Meglitsch (1967), *Invertebrate Zoology* (Oxford University Press, New York), figure 12.9.

The idea behind swarms seems to be the same as that behind schooling in fish: presented with such a large group, predators can't focus on any one individual. The strategy may backfire at times. For example, baleen whales (see below) often eat entire swarms of krill, engulfing them by the ton rather than by the individual.

Krill are the dominant zooplankton of the Southern Ocean but are common members of seas worldwide, especially in highly productive waters. They rival copepods in terms of global biomass.

Thaliaceans: Salps and Doliolids. Thaliaceans are pelagic tunicates, cousins to the sea squirts found on rocky shores and among our closest invertebrate relatives. Each individual has a barrel-shaped body that encloses a conical mucus sieve (figure 5.7). As water is drawn into the barrel, food is filtered by the sieve, and the sieve—along with its trapped food—is drawn into the mouth by cilia. The current of filtered water exits out the back of the barrel. In salps, water is forced through the sieve by rhythmic contraction of muscles in the body wall, and the filtered water acts as a jet that propels the animal. In other words, to eat, salps must swim. Doliolids also use muscles to power jet propulsion, but flow of water through the sieve can be effected by cilia alone, so doliolids can feed while staying stationary.

Thaliaceans have a complex life cycle: individuals are alternately solitary and aggregated. In aggregations, individuals lie side by side to form the walls of a larger cylinder, with their excurrent siphons combining to make one large jet (figure 5.8). Salps and doliolids are capable of rapid reproduction and growth. Their generation times range from three days to about three weeks, and they can increase their body mass by 10% per hour. This high rate of growth is possible at least in part because of the gelatinous nature of the animal's body. Because the body is more than 95% water, the mass of a salp or doliolid can increase without much expenditure of organic material. The high water content of the body also renders these animals nearly neutrally buoyant.

Salps and doliolids can capture particles as small as 0.7 micrometers in diameter, but they consume primarily phytoplankton. Solitary individuals are large relative to most of the animals we have encountered so far—they can be ten or more centimeters in length. Aggregations are larger still, attaining lengths of several meters.

Salps and doliolids are found throughout the ocean, but are particularly common in tropical and semitropical waters and in the Southern Ocean. In productive

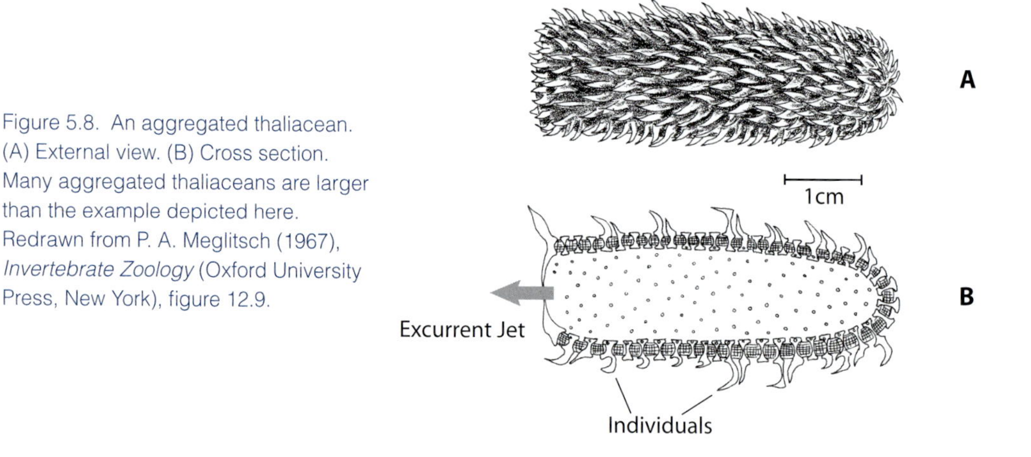

Figure 5.8. An aggregated thaliacean. (A) External view. (B) Cross section. Many aggregated thaliaceans are larger than the example depicted here. Redrawn from P. A. Meglitsch (1967), *Invertebrate Zoology* (Oxford University Press, New York), figure 12.9.

coastal waters, they have been observed in swarms covering more than 9000 square kilometers, with densities as high as seven individuals per liter.

Larvaceans. Also known as *appendicularians*, larvaceans are among the strangest organisms in the ocean. Like salps, they are pelagic tunicates, and the animal itself looks superficially like a tadpole—a small body similar to that of a solitary salp, with the addition of an undulatory tail. But the animal survives through a unique strategy: it builds a hollow, gelatinous house that it then inhabits, beating its tail to power a stream of water through its domicile (figure 5.9). Larvaceans are typically small, a few millimeters or less in maximum dimension, but their houses are substantially larger, up to a meter in diameter. Water moves into the house through coarse mucus sieves (which remove any particles too large for the animal to eat, including diatoms), and then through a much finer sieve on which particles down to 0.1 micrometer length are trapped. Only then are particles directed to the mouth. When the sieves become clogged, which may happen several times per day, the animal abandons its house and builds another. Abandoned larvacean houses sink, taking the trapped particles with them. Because the houses are much larger than untrapped particles, particles trapped in larvacean houses sink faster than they otherwise would, about 50 meters per day for small houses, and up to 800 meters per day for some giant larvacean houses.

Large swarms of larvaceans have been observed, with densities as high as 26 individuals per liter. Like thaliaceans, larvaceans can reproduce and grow rapidly. A young appendicularian can build its first house a mere 12 to 20 hours after it is conceived and can be sexually mature within 5 to 21 days.

Thaliaceans and larvaceans are very efficient at filtering seawater. For example, larvaceans filter 5 to 60 milliliters of water per hour per individual. Thus the dense swarm of 26 individuals per liter noted above could filter all the water around it in 0.6 to 7 hours. Salps are even more voracious than larvaceans: some solitary individuals can filter as much as 5 liters per hour. At this feeding rate and the density found in swarms (7 individuals per liter), the water could be completely filtered in a little less than two minutes. Large swarms of salps can single-handedly graze down an entire phytoplankton bloom.

Some of the organic matter eaten by thaliaceans is excreted as large fecal pellets, which then sink at speeds of 300 to 2700 meters per day.

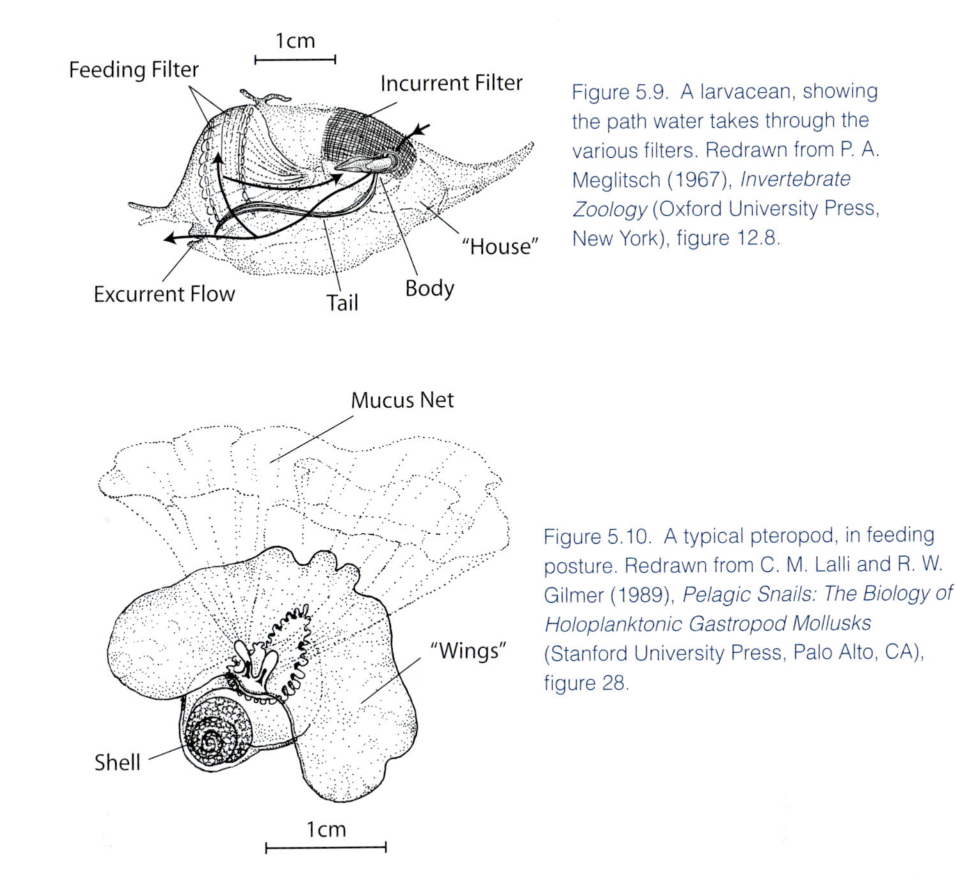

Figure 5.9. A larvacean, showing the path water takes through the various filters. Redrawn from P. A. Meglitsch (1967), *Invertebrate Zoology* (Oxford University Press, New York), figure 12.8.

Figure 5.10. A typical pteropod, in feeding posture. Redrawn from C. M. Lalli and R. W. Gilmer (1989), *Pelagic Snails: The Biology of Holoplanktonic Gastropod Mollusks* (Stanford University Press, Palo Alto, CA), figure 28.

Pteropods. Pteropods—commonly known as sea butterflies—are small snails whose foot has evolved into a pair of wing-like lobes that flap rhythmically, propelling the animal through the water. Some pteropods (the gymnosomes) have lost their shells, but others (the thecosomes) retain a fragile version of the typical snail shell.

Gymnosomes are predators, but thecosomes are herbivores, feeding passively as they slowly sink through the water, dragging behind them a mucus web that entangles phytoplankton (figure 5.10). The web can be 5 to 10 times the diameter of the shell. Once coated with food, the web is retrieved and eaten. The snail then swims up to begin the cycle again.

Pteropods are small. Their shells are typically less than a centimeter in diameter, and their wingspan may be five times the diameter of the shell. Despite their small size, pteropods have a relatively long time to first reproduction, 210 to 240 days.

In areas where pteropods are common, shells of their dead comprise a major component of shallow sediments, which are then known as *pteropod oozes*.

Fish. Fish need no introduction; their general body form and life style are familiar to anyone who has ever visited an aquarium or eaten at a seafood restaurant. They come in a wide variety of sizes and shapes, and some of them filter phytoplankton from the water.

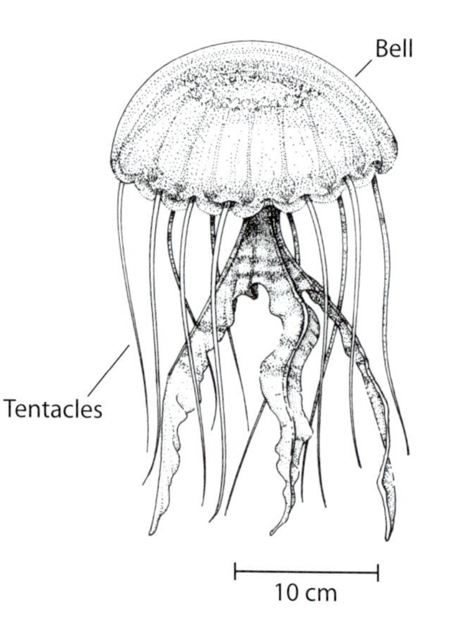

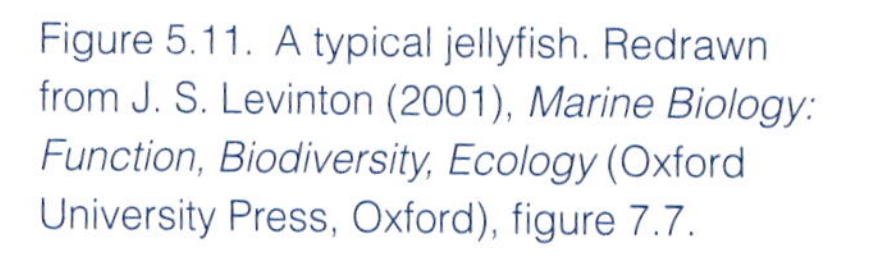
Figure 5.11. A typical jellyfish. Redrawn from J. S. Levinton (2001), *Marine Biology: Function, Biodiversity, Ecology* (Oxford University Press, Oxford), figure 7.7.

Fish are vertebrates; that is, they have vertebral columns. In the chimeras, sharks, skates, and rays, the skeleton is made of cartilage, the material that stiffens the bridge of your nose. In contrast, bony fishes have calcified skeletons. As large, active swimmers, fish require specialized structures (gills) to rapidly absorb oxygen from the water around them and to release carbon dioxide waste. In both cartilaginous and bony fishes, filter feeding is accomplished by forcing water through gill rakers, sieve-like structures situated between the throat and gills. Once separated from the water, food is passed down the gullet. The gill rakers of herrings, menhaden, and similar fishes are finely spaced, allowing these species to sieve phytoplankton from the sea.

Fish typically reproduce once per year, but in many species females do not reach reproductive age until they are several years old. Fish can live for a surprisingly long time: orange roughy can live 100 to 150 years, and a rockfish found off the coast of California was recently estimated to be 210 years old.

Other Herbivores. We earlier categorized ciliates, flagellates, and amoebas as bacterivores, but they also eat phytoplankton. Thus, this list of major herbivore players should include ciliates, flagellates, and amoebas as well.

Filter Feeders

Our next category is a loosely defined group of organisms who make their living by filtering animals, largely herbivores, from the water. The boundaries of this group are necessarily vague. For example, copepods and krill also filter food from the water around them, but we leave them out of this group because they eat primarily phytoplankton rather than animals.

Jellyfish. Jellyfish are members of the Cnidaria, a phylum that includes corals and sea anemones. Their bodies consist of a gelatinous, bowl-shaped bell that encloses both a volume of water and the organism's innards (figure 5.11). During locomotion, muscles in the bell contract, expelling a jet of water and

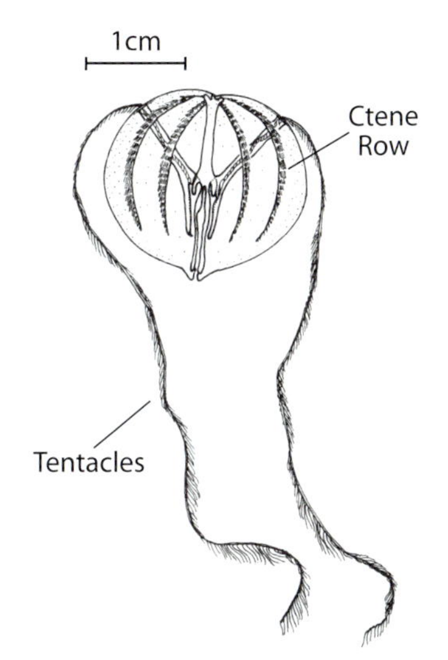

Figure 5.12. A typical ctenophore. Redrawn from J. S. Levinton (2001), *Marine Biology: Function, Biodiversity, Ecology* (Oxford University Press, Oxford), figure 7.7.

propelling the jellyfish forward. Tentacles attached to the bell's rim trail behind; as the tentacles sweep through the water, they can contact prey (often copepods), which are then captured using stinging cells known as *nematocysts*.

Nematocysts are a marvel of biological engineering. The type commonly found on jellyfish tentacles consist of a barb attached to the end of a long hollow thread. This miniature harpoon and line are stored inside out and tightly coiled inside a specialized cell called a *cnidocyte*. When the cell is triggered by contact with prey, water rushes into the cnidocyte, forced in by osmotic pressure, causing the hollow thread to rapidly turn itself right-side out. If you have ever blown air into a rubber glove to get the fingers right-side out, you have the idea of how this works. The entire process happens in a matter of mere milliseconds, and the barb can move with an acceleration 40,000 times that of gravity. As the thread everts, the barb at its end penetrates the prey, and in some nematocysts a potent toxin is then injected. The poison from just a few nematocysts is sufficient to subdue a copepod, for instance, and the paralyzed animal can then be transferred to the jellyfish's mouth in the middle of the bell. Other types of nematocysts are less lethal; they simply entangle or stick onto prey.

Most nematocysts are too weak to penetrate human skin, and when you touch a jellyfish tentacle that has these relatively benign nematocysts, you are likely to notice only that the tentacle feels "sticky" as the outer layer of your skin is harpooned. However, nematocysts of a few jellyfish are capable of penetrating skin, and they inflict painful and even fatal stings. Cnidocytes can be engulfed intact by other animals (such as ctenophores, see below) and the captured nematocysts are used for protection in the same fashion they are employed by jellyfish.

Jellyfish come in a wide variety of sizes, from less than a centimeter across the bell to nearly a meter. Some of the larger jellyfish are capable of capturing substantial prey, such as krill and small fish.

Comb Jellies. The comb jellies, or *ctenophores*, are egg-shaped gelatinous organisms that move using eight rows of ciliated plates (figure 5.12). Each plate is known as a ctene, from the Greek word for comb. The individual plates in each row stroke in sequence, forming waves that pass down the body and propel the animal forward. In the process, prey stick to the body itself, to sticky lobes that extend from the body, or to two tentacles trailing behind. The length of a typical ctenophore body can be up to 20 centimeters, but most are smaller: 1 to 10 centimeters long. One unusually flattened species is nearly a meter in length. The generation time for comb jellies is 20 to 30 days under optimal conditions.

Comb jellies are among the most beautiful organisms in the sea. The arrangement of individual cilia (whose bases are fused to form each ctene) is such that they act like a diffraction grating, separating white light into its spectral components. As a result, a swimming ctenophore looks like a scintillating, locomotory rainbow. As beautiful as this effect is, it likely has little function. In the ocean, white light is found only in the shallowest surface waters (chapter 4). At the depths where most ctenophores live, light is dim and bluish, and comb jellies appear relatively drab. Or they do until they turn on their own light shows. Many ctenophores are intensely luminescent. Blue-green light is produced in rows of organs inside the body, which appear to backlight the ctenes. The function of this luminescence is uncertain.

Unlike jellyfish, only one rare genus of ctenophores (*Haeckelia*) has nematocysts, and these are obtained from the jellyfish the comb jelly eats.

Fish. Filter-feeding fish come in two general categories. First, the young stages of many species filter food until they become large enough to catch prey by other means. Second, the adults of some species continue to feed in this fashion. Anchovy, sardines, mackerels, shad, pilchards, and alewives are among the common filter feeders. The largest fish that filter feeds is the whale shark, *Rhincodon typus*, which can grow to a length of 12 meters. A close second is the basking shark, *Cetorhinus maximus*, at 6 to 8 meters long.

Whales. And then there are whales. They are mammals, like you and I, which means that they are warm-blooded and breathe air. Thus, unlike fish, whales are tied to the surface of the ocean, where they must return every few minutes to breathe. Despite this constraint, whales can dive to great depths, a fact discovered when whale carcasses were found entangled in telegraph cables retrieved from the ocean floor.

Many whales—the toothed whales—are active predators, and we discuss them below. However, others—the baleen whales—are the preeminent filter feeders of the ocean. These whales, which include humpback, minke, right, sei, fin, gray, Bryde's, and blue whales, feed by opening their mouths and engulfing huge quantities of water along with any organisms therein. This mouthful is then strained through stiff plates of *baleen* that hang from the roof of the whale's mouth, their hair-like edges forming a sieve. In this fashion, water exits, but food (commonly krill) stays behind to be scraped off by the tongue and swallowed. A single whale can eat 2000 to 9000 kilograms per day.

Whales become sexually mature at 3 to 10 years of age, and the gestation time is another nine months to more than a year. Thus, the generation time for whales is 4 to 11 years.

Whales are truly huge. An average humpback whale is 14 meters long and weighs about 80,000 kilograms. Blue whales, the largest organisms that have ever lived on earth, can reach a length of 33 meters and weigh an astounding 190,000 kilograms. Baleen whales are found in all the world's oceans, but they are particularly common in temperate and polar seas.

The great whales were heavily hunted by human beings in the nineteenth and twentieth centuries, some nearly to extinction. Restrictions on whaling were imposed in the late twentieth century, and a few populations are slowly recovering. The International Whaling Commission is charged with keeping track of whale population numbers; their estimates suggest that there are currently about 2 million whales in the ocean. Now, the ocean contains 1.4 billion cubic kilometers of water (1.4×10^{21} liters), so this works out to approximately 1.4×10^{-15} whales (1.4 femtowhales) per liter, a concentration twenty orders of magnitude lower than that of ciliates. Whales are relatively scarce.

This comparison might be misleading. Although they are present at much lower concentration than ciliates, each whale is immensely larger than an individual ciliate. For example, it would take roughly 1.5×10^{14} (150 million million) ciliates to fill the volume of one humpback whale. Thus, 2 million whales is approximately the volumetric equivalent of 3×10^{20} ciliates. However, even if we were to slice all earth's baleen whales into ciliate-sized chunks and spread the chunks throughout the 100-meter-deep euphotic zone, there would be only 8 chunks per liter, far less than the 1000 to 350,000 individuals per liter typical of ciliates in surface waters.

Predators

The principle predators of the sea are chaetognaths, fish, squids, birds, and various mammals.

Chaetognaths. Commonly known as arrow worms, chaetognaths ("spiny jaws") are the principle predators of copepods. The mass of chaetognaths in the oceans may be 10% to 30% that of the copepods they eat, making the arrow worms a major player in the flow of energy through the marine habitat.

Arrow worms have a simple body, consisting of a head with the apparatus needed to capture prey—jaws, mouth, and sensory organs—and an elongated body that provides propulsion (figure 5.13). The tail end of the body sports several pairs of fins resembling the fletching of an arrow, hence chaetognaths' common name. The animals swim by oscillating the body, and they typically move with a darting gait, flicking their tail to converge on their prey. Each prey item, typically a copepod, is grasped with the spines of the jaws, which in many species inject a dose of tetrodotoxin, paralyzing the prey. The tetrodotoxin is produced by bacteria rather than by the arrow worm itself, although it is currently unclear how the toxin is accumulated and injected. Chaetognaths typically undergo a daily vertical migration similar to that of their copepod prey: down in the day and up at night.

Chaetognaths vary in size from 2 to 120 millimeters, although most fall in the range from 3 to 25 millimeters. One to seven generations are produced per year.

Fish. We have already described the general characteristics of fish, and predatory fish are no different except that, in general, they are bigger and faster than filter-feeding fish. Fish are commonly visual predators, using their eyes to find and track prey. Typical predators include salmon, cod, sharks, tuna, marlin, and swordfish.

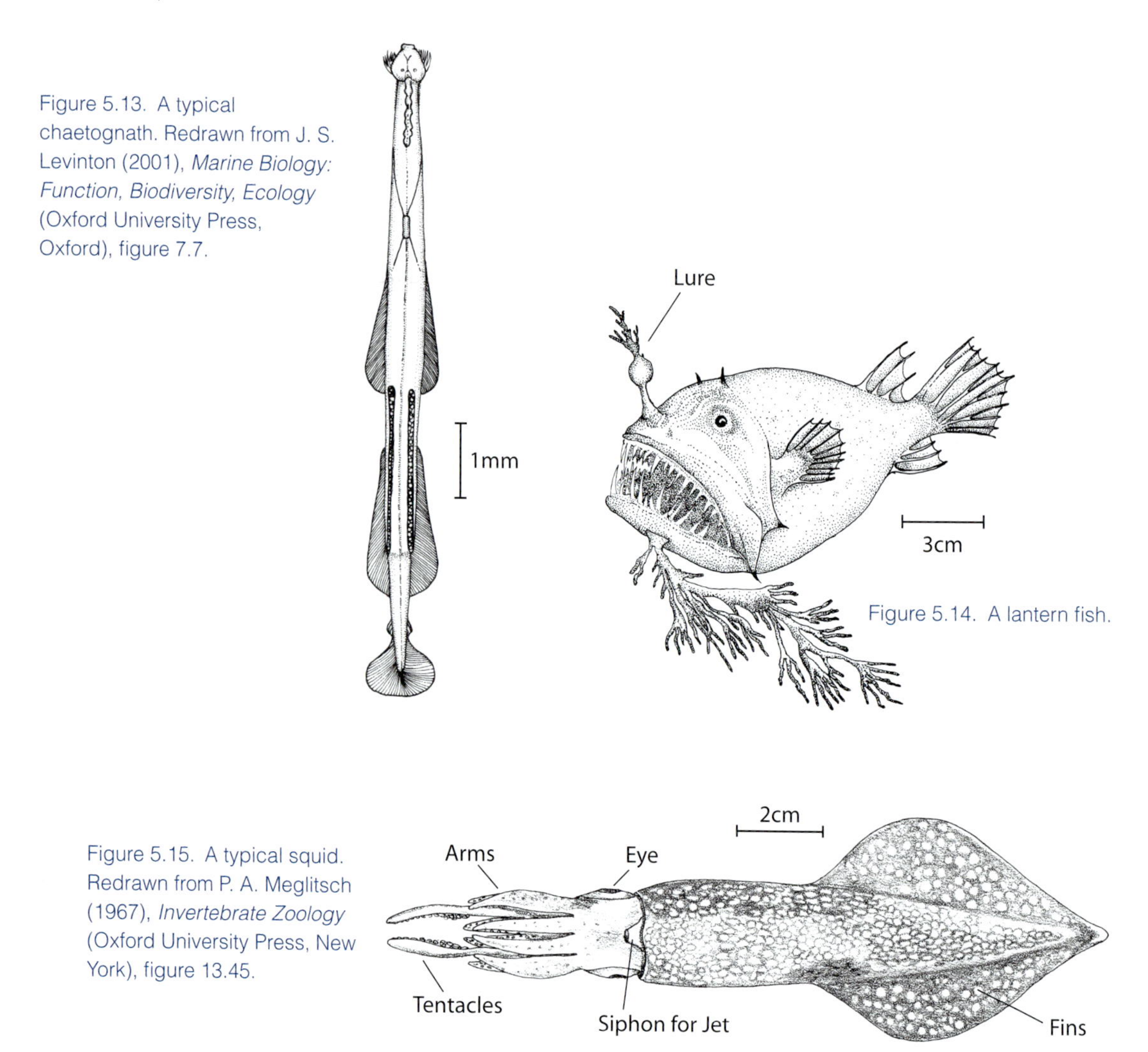

Figure 5.13. A typical chaetognath. Redrawn from J. S. Levinton (2001), *Marine Biology: Function, Biodiversity, Ecology* (Oxford University Press, Oxford), figure 7.7.

Figure 5.14. A lantern fish.

Figure 5.15. A typical squid. Redrawn from P. A. Meglitsch (1967), *Invertebrate Zoology* (Oxford University Press, New York), figure 13.45.

Other fish species are sit-and-wait predators. These predators include the deep-sea angler fish, which uses a small luminescent "lure" attached to its forehead to attract prey, which are then quickly engulfed (figure 5.14).

Squid. Squid are mollusks, distant cousins to snails and clams and close cousins to octopuses and cuttlefish. A squid's body consists of a hollow, muscular mantle (which encloses the guts), a head that supports a pair of eyes, a biting mouth that resembles a bird's beak, and a set of eight arms and two tentacles (figure 5.15). Squid move either by flapping a pair of fins attached to the sides of the mantle or by squirting water from the mantle cavity. This jet propulsion can be very rapid, and is used by squid to escape predators.

Prey capture often relies on the rapid extension of the pair of tentacles. When transverse muscles in the tentacles contract, the tentacles become longer and skinnier. This extension can shoot the sucker-laden ends of the tentacles toward prey, a process that takes only a few hundredths of a second.

Squid occur in a wide variety of sizes, with bodies sometimes only a few centimeters long, and sometimes meters in length, as with the giant *Architeuthis*, whose mantle may be 6 meters long. Squid typically eat fish and other squid. Like fish, they commonly reproduce annually, and like some fish, most squid die immediately after reproducing. Squid are found in all oceans and at all depths.

Birds. There are a wide variety of marine predatory birds: penguins, albatrosses, terns, gulls, cormorants, pelicans, puffins, murres, frigate birds, and others. They are typically visual predators. Many (such as gulls, terns, and pelicans) spot their prey from the air, then dive a few meters into the water to capture individual fish. Others, such as cormorants, are adept swimmers and dive tens of meters to chase down prey. Penguins, found only in the polar and subpolar waters of the southern hemisphere, have taken this tactic to an extreme. Their wings have evolved into flippers, giving these birds the distinctive ability to "fly" underwater, where they are fast and maneuverable. As a result of the change in wing morphology, however, they can no longer fly in air. Birds commonly reproduce once per year, and females typically reach reproductive age after one year.

Mammals. Predatory mammals include small toothed whales, seals, and sea lions. As mammals, all these organisms are tied to the ocean's surface, to which they must periodically return to breathe. But like the baleen whales discussed above, they cope amazingly well with this constraint. For example, elephant seals spend months at a time at sea, diving to depths of 1500 meters, where they hunt and even sleep, returning to the surface only long enough to take a few breaths before diving again.

Toothed whales are similar in overall body form to the baleen whales discussed above, but they lack baleen plates and instead have teeth. Whales that fall into this category include belugas, narwhals, and various beaked whales, as well as porpoises and dolphins. Toothed whales are found throughout the ocean, but seals and sea lions are typically confined to coastal waters, in part because the richest fish populations are close to shore (chapters 9), and in part because these mammals give birth on land. Seals, sea lions, and toothed whales typically become sexually mature at 3 to 5 years of age.

Top Predators

Lastly, there are the top predators, predators so fierce that they themselves are immune to predation. As one might intuitively suspect, relatively few species claim this distinction: common top predators are the large toothed whales (sperm whales and killer whales) and large sharks. Note that these top predators are an exception to the rule that size increases with trophic level. Just as lions and wolves are smaller than wildebeests and caribou, sharks and orcas are smaller than the baleen whales they sometimes eat.

And then there are humans, the ultimate predators on the planet. It is traditional for us to think of ourselves as "fishermen," a term that allows us to make a distinction between our actions and those of red-toothed nature, but in reality we are predators. Human society's increasingly pervasive effects on the marine trophic system are discussed at length in chapter 11.

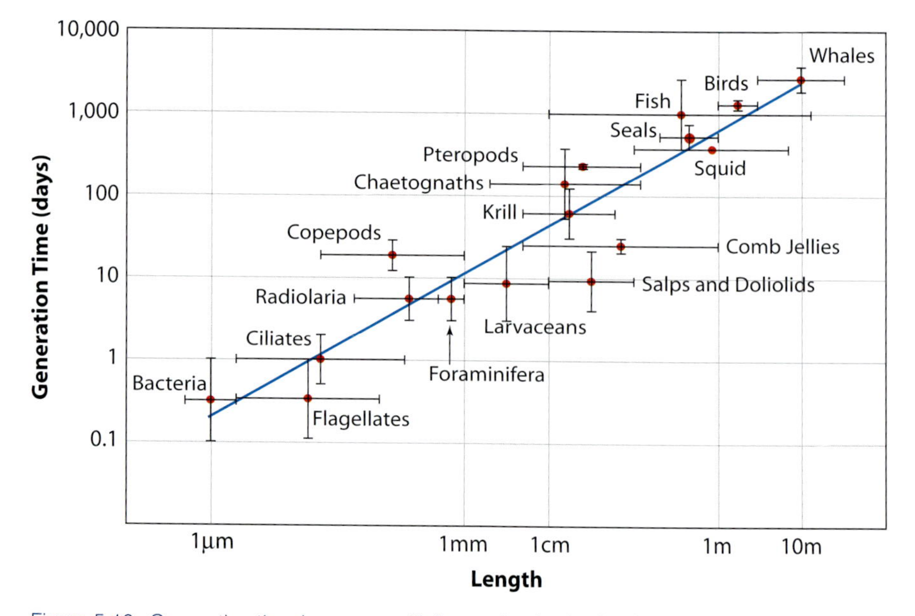

Figure 5.16. Generation time increases with increasing body size for heterotrophic marine organisms. For the values shown here, $\log_{10}(\text{generation time}) = [0.579 \times \log_{10}(\text{length})] + 2.77$, where generation time is measured in days and length is measured in meters. In other words, generation time increases roughly as the square root of body length. Pelagic jellyfish have not been included in this figure; there are insufficient data on their reproductive habits.

Synthesis

There are several themes that emerge from this brief overview of pelagic organisms. First, the size of individuals and their concentrations in the water column are related to what they eat: herbivores are small (often single-celled) and incredibly abundant, while top predators are large and scarce. We will explore the reasons for this correlation later in this chapter.

Second, it should be apparent that the division of animals into distinct trophic groups is an idealization that only loosely matches reality. For example, we have encountered animals that eat both phytoplankton and other animals, making them both herbivores and predators. In particular, the group I have labeled "filter feeders" is artificial, a category based more on how these animals feed than on the type of food they ingest. Despite its shortcomings, categorization of animals into trophic groups provides a simple and useful mechanism to cope with the diversity of marine consumers. As long as you realize that this categorization is intended as a guide rather than "the truth," you will not be led astray.

Third, generation time—the time from birth to reproduction—varies as a function of size. The larger the organism, the longer its generation time (figure 5.16). This trend will become important when we consider the relationship between trophic groups. For example, because they are small, bacteria presented with a new source of food can rapidly reproduce to take advantage of the newfound bounty. In contrast, it would take a population of whales years to respond to an increase in krill concentration.

Note that the axes on this graph are logarithmic, a transformation necessary if we are to see the details across a range of sizes that spans more than seven orders of magnitude. Note also the position of gelatinous organisms such as comb jellies, salps, doliolids, and larvaceans. They lie below the general trend line, indicating that, for the rate at which they reproduce, they are relatively large. This is due in part to the fact that these organisms' bodies are highly dilute. Whereas the body of a fish is 70% to 80% water, the body of a comb jelly might be more than 95% water. Thus, in these organisms, a small amount of living tissue translates into an unusually large body. In contrast, the crustaceans (copepods and krill) lie above the general trend line; for their size, they reproduce relatively slowly.

And finally, I must emphasize that in this discussion we have dealt only with the *known* animals of the sea: animals that live near the surface where they can be readily seen or hooked on a fishing line, animals slow enough to be caught in nets and sturdy enough still to be identified. Given the vast size of the oceanic habitat, it is certain that there are many species we know nothing about, and hardly a month goes by when scientists don't discover some new deep-sea fish or unknown comb jelly. Perhaps the best example of our current inability to thoroughly census the ocean is the giant squid, *Architeuthis*. Dead specimens have occasionally been recovered for more than a hundred years, but until October, 2004, a live *Architeuthis* had never been observed in its natural habitat. Then, Japanese researchers managed to photograph a giant squid as it attacked a bait. In December, 2006, the same research team managed to capture a young giant squid. If it has taken science this long to photograph and capture a six-meter-long squid, what else is out there waiting to be discovered?

Energy, Carbon, and Nutrients

Having described the players, we now turn our attention to how they contribute to, and control, the flow of energy and carbon in the ocean, flow that provides a useful perspective on how the ocean works. First, we will track the flow of energy through the food chain. Energy enters as sunlight and is stored in phytoplankton. It is subsequently passed along to the herbivores that eat the phytoplankton and to the predators that eat the herbivores and each other. As we will see, the details of these energy transfers allow us to calculate the relative population size of the various links in the trophic chain.

It is important to note, however, that the flow of energy is a one-way process. Energy enters as sunlight, flows along the food chain, and exits as heat, a state in which it is relatively useless to living things. The system continues to work only because there is a continuous influx of solar energy.

In contrast, there is a fixed amount of carbon on earth. If production were the only process operating in the oceans, all the available carbon would long ago have been fixed, and no carbon dioxide would remain to serve as the raw material for photosynthesis. The fact that carbon dioxide is still present in seawater, and photosynthesis is still active in the ocean today, is evidence that, unlike energy, carbon is somehow recycled. Similarly, nutrients are recycled, and this recycling is interesting, important, and the subject of much current research. In this chapter we will begin to understand the paths along which carbon and nutrients flow, a foundation on which we will build in subsequent chapters.

Let us first turn our attention to the flow of energy.

Energy Flow

Consider one possible fate of a diatom. Having diligently absorbed sunlight and CO_2, the cell has grown to mature size and stored considerable energy in its organic material. If left alone, it would soon divide, reproducing itself to start anew. But fate has different plans for this particular diatom: it is eaten by a copepod. After its consumption, the diatom is digested, and its organic material is broken down into small constituent molecules. The digested diatom is then used to build new copepod tissue, allowing the copepod to grow a little. In this fashion, some of the solar energy stored in the diatom (primary production) is passed on to—and stored in—the copepod (secondary production).

The process is surprisingly inefficient, however. The copepod uses energy to find, capture, and eat the diatom. More energy is used to digest this prey into small bits, and still more energy is needed to excrete waste products and to build the digested material into new copepod tissue. In sum, only about 10% of the energy stored in the diatom remains as energy stored in the copepod. This factor of 10% is known as the *trophic efficiency*, *E*. The rest of the energy (that needed for capture, digestion, excretion, and synthesis) ends up as heat and is of no further use to the copepod or any other organism.

A similar picture emerges when we consider the fate of the copepod. Having fattened itself on diatoms, the copepod is eaten by a small fish. But in the process of capturing and digesting the copepod and using the remains to build new fish tissue, only about 10% of the copepod's stored energy ends up stored in the fish. The other 90% is converted to useless heat. This process is repeated for the squid that eats the small fish, the salmon that eats the squid, and the human being that eats the salmon for Saturday night dinner. With each act of ingestion, roughly 10% of the remaining stored solar energy is assimilated into new tissue and the rest is converted to waste heat.

I should note that there is nothing magical about a value of 10% for trophic efficiency. In fact, the precise value varies considerably from one species to the next and even from time to time within a species. Values as low as 5% are not uncommon. Arrow worms have a trophic efficiency of about 30%, some ciliates and flagellates have efficiencies as high as 70%, and some bacteria have efficiencies approaching 90%. Nonetheless, when averaged over members of a trophic group, the value of 10% is representative, and for the sake of simplicity, we use it here.

The inefficiency of energy transfer among organisms is traditionally conveyed using the *trophic pyramid of energy*, as shown in figure 5.17. The large base of the pyramid depicts the first *trophic level*: the energy stored in primary producers. On the next level of the pyramid are herbivores—including bacteria that eat primary production—and the energy stored at this level is less than that of the primary producers. Next up are the predators—including, for trophic-level purposes, the bacterivores. And at the peak of the pyramid are the top predators. With each step up the pyramid—each transition from one *trophic level* to the next—stored energy is reduced.

There are several caveats I should note regarding this trophic pyramid. First, as it is presented here, the pyramid is incomplete. Some predators eat other predators before they themselves are eaten. Thus, there may be several levels of *intermediate predators* that, for simplicity, I have lumped together

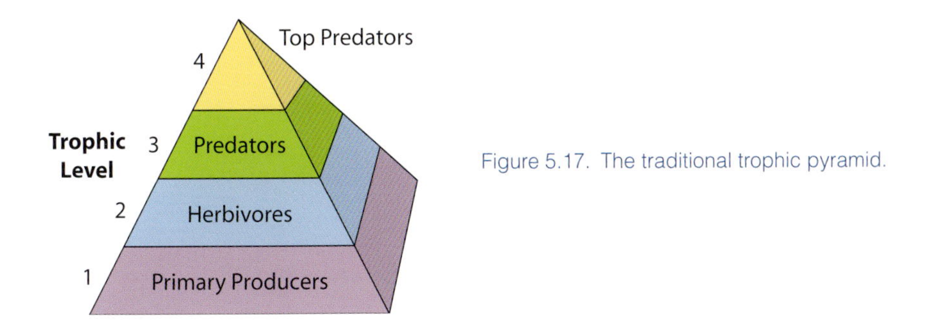

Figure 5.17. The traditional trophic pyramid.

here. Note also that the trophic *groups* we used to introduce the consumers do not directly correspond to trophic *levels*. Filter feeders, for instance, may eat both phytoplankton and zooplankton, and therefore straddle trophic levels 2 (the herbivores) and 3 (the predators). Lastly, bacteria are difficult to assign to a trophic level. Here, we assign them to trophic level 2 based on the fact that they primarily consume dissolved organic material, much of which is primary production. If bacteria lie on trophic level 2, bacterivores must lie on level 3.

As a visual heuristic tool, the trophic pyramid is effective, but it can also be misleading. By depicting only energy that remains, the trophic pyramid deemphasizes energy that is lost. A better analogy might be that of a river (figure 5.18). The main stream entering at the left is the flow of solar energy into the ocean, only a minor fraction of which is absorbed and stored by phytoplankton. All the energy that enters phytoplankton eventually exits, but only about 10% flows as stored energy into the next trophic level, as depicted by the small stream leading from producers to herbivores. The remaining energy flow exits into the "sea" of heat. Similar splitting of streams occurs for predators and top predators, by which point the flow of stored energy is a mere trickle. If 90% of the useful energy is lost at each split, only 0.1% of the energy originally stored in primary producers is present in the top predators. To put this another way, for each joule of energy stored in the tissue of a top predator, 1000 joules must initially be stored by phytoplankton.

The notion of a trophic river serves as the basis for an important calculation. Consider, for instance, the flow of energy from herbivores to predators. Let's suppose that each herbivore stores energy at a rate S_1 and that there are N_1 herbivores in the local population. In this case, the total rate at which newly stored energy is available to predators is S_1N_1. As we have seen, however, only a fraction E of this available energy can actually be converted into new tissue by predators. Thus, the rate at which energy is available to be converted into growth by predators is $E\,S_1N_1$.

Let us now suppose that each predator stores energy (grows) at rate S_2 and that there are N_2 predators in the local population. In that case, the total rate at which energy is used by the predators for growth is S_2N_2. At equilibrium—that is, when neither herbivore nor predator population is changing in size—the rate at which energy is provided to predators for their growth is just equal to the rate at which they use it:

$$ES_1N_1 = S_2N_2 \qquad \text{(Eq. 5.1)}$$

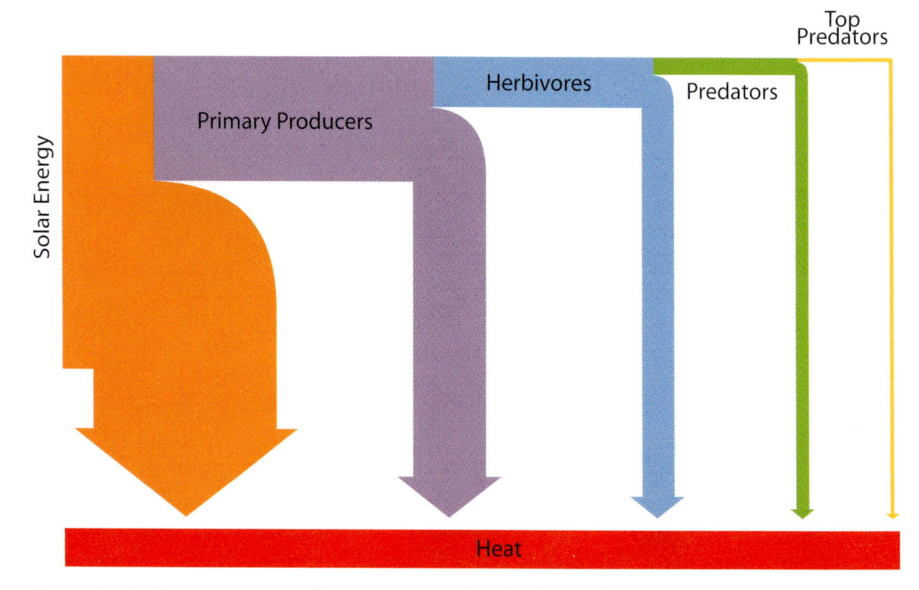

Figure 5.18. The trophic river. For ease in drawing the figure, I have used a trophic efficiency of 0.3 rather than the more typical value of 0.1.

Rearranging this equation, we find that

$$\frac{N_2}{N_1} = E\frac{S_1}{S_2} \qquad \text{(Eq. 5.2)}$$

This relationship provides a recipe for calculating the size of the predator population relative to that of the herbivores. For example, we noted above that trophic efficiency E is typically about 10%. In this case

$$\frac{N_2}{N_1} \approx \frac{1}{10}\frac{S_1}{S_2} \qquad \text{(Eq. 5.3)}$$

Thus, if the rate at which individual herbivores store energy were similar to that for individual predators (that is, $S_1 \approx S_2$), the size of the predator population, N_2, would be only about a tenth that of the herbivore population, N_1. In reality, predators are likely to be larger than herbivores (see table 5.1), and due to their larger mass, each individual predator will add new tissue at a faster rate than each individual herbivore. In that case, $S_2 > S_1$, and the number of predators that can be supported will be less than a tenth the size of the herbivore population. For instance, if the mass of a predator is ten times that of an herbivore, we would expect[6] that $S_2 \approx 5.6\ S_1$. In this hypothetical case,

[6] As a general rule, the metabolic rate of organisms, an index of the rate at which they can store energy, increases as mass$^{3/4}$. Thus, an organism 10 times the mass of another would store energy at approximately $10^{3/4} = 5.62$ times the rate of the other. An organism 100 times as large would store energy 31.6 times as fast. There is considerable variation around this rule, however, so these calculations should be taken with a grain of salt.

$$\frac{N_1}{N_2} \approx \frac{1}{10}\frac{S_1}{5.6S_1} = 0.018 \qquad \text{(Eq. 5.4)}$$

and the number of predators is only 1.8% that of herbivores.

This same calculation can be applied to any two adjacent trophic levels: the lower level has size N_1 and growth rate S_1, and the higher level has size N_2 and growth rate S_2.

This calculation provides an explanation for one of the important general rules of marine biology: the population size of organisms present in the ocean decreases drastically with increasing trophic level. As we saw in our survey of the animals, herbivores far outnumber predators, and predators far outnumber top predators.

It is important to note that this calculation has been couched in terms of the number of individuals in each trophic level. If we desire to compare biomass between trophic levels, we need to take into account the mass of each individual. For instance, in the example given above, the number of predators is only 1.8% that of herbivores, but each predator weighs ten times as much as each herbivore. In this case, the total mass of predators is 18% that of herbivores. If the trophic efficiency is high and the rates of growth are similar between trophic levels (as happens for some phytoplankton and herbivores, for instance), it is possible (although unusual) that the mass of organisms at a higher trophic level can be greater than that at a lower trophic level.

The Microbial Loop

The outline of the trophic river presented above could apply as easily to terrestrial ecosystems as it does to life in the oceans. For example, grasses capture solar energy and grow, but they are eaten by insects. The insects are in turn eaten by songbirds, which are lunch for hawks. However, in the ocean, there is another pathway for energy, one that has limited analogy in terrestrial systems.

Consider an alternative fate for the organic material in a diatom. If, instead of being eaten by a copepod, the cell dies as the result of a viral infection, its contents are spilled into the ocean as dissolved organic material (often referred to by the acronym DOM). Phytoplankton contribute to the DOM even when alive and healthy. Depending on nutrient availability, phytoplankton may release a substantial fraction of their photosynthetic production as dissolved organic material. This is particularly prevalent under conditions where light and carbon dioxide are abundant but nitrate is scarce. In this situation, phytoplankton can readily manufacture carbohydrates, but lacking the nitrogen needed to make proteins and nucleic acids, they release the carbohydrates into the water rather than store them. As much as 50% of carbon fixed by phytoplankton is released as DOM. As long as this spilled organic matter is in solution, it cannot be passed along the trophic river, in which case up to 50% of the solar energy trapped by phytoplankton is wasted from the start.

Indeed, some of the spilled organic material remains dissolved, but a surprisingly large fraction is instead ingested by bacteria. This ingestion is important because it converts energy-rich organic material from photosynthesis back into a particulate form that is readily available to other consumers. For instance, as we have seen, bacteria are voraciously hunted by ciliates and flagellates, which in turn can be eaten by copepods and other small animals. In this fashion, the ability of bacteria to absorb dissolved organic material salvages much of the energy released by phytoplankton, returning it to the trophic stream. The recovery of energy by bacteria and their predators is known as the *microbial loop*, and its importance in marine ecology has been recognized only in recent years.

The trophic role of the microbial loop is augmented by the efficiency of the organisms involved. In contrast to the low trophic efficiencies of multicellular animals (typically 10%), the trophic efficiencies of bacteria can be remarkably high (typically about 50%, but in a few cases up to 90%), and the trophic efficiencies of some of their predators (ciliates and flagellates) can be nearly as high (up to 70%). As a result, the population size of members of the microbial loop can be surprisingly large (see equation 5.2). For example, if the trophic efficiency is 0.5 in the transfer of energy from cyanobacteria to heterotrophic bacteria, and the growth rates of the two types of organism are similar (that is, $S_1 = S_2$), the number of bacteria at trophic level 2 could be 50% the number of cyanobacteria at trophic level 1.

At this point, an important question might occur to you. If up to half the carbon fixed by phytoplankton is spilled into the sea as DOM, why haven't more organisms evolved mechanisms to take advantage of this soup? Why do marine ecosystems rely on bacteria to salvage this energy? In fact, many organisms have the capability to absorb dissolved organic matter. For example, the larval stages of many marine invertebrate animals can absorb DOM. However, bacteria have an inherent advantage: they are very small.

Dissolved organic material is absorbed by the surface of an organism, so the rate at which this material can be supplied to the organism is proportional to its surface area. In contrast, the demand for organic material is set by the organism's volume: the larger the volume of an organism, the greater the rate at which material must be delivered to maintain its metabolism. Now, for organisms of a given shape, surface area is proportional to the square of its length, L, whereas the volume within is proportional to the cube of length. Thus,

$$\frac{\text{surface area}}{\text{volume}} \propto \frac{L^2}{L^3} = \frac{1}{L} \qquad \text{(Eq. 5.5)}$$

As a result, the ratio of surface area to volume gets progressively larger as organisms decrease in size. For example, a bacterium 2 micrometers long has roughly 50 times as much surface area per volume as a 100-micrometer-long ciliate. Thus, by being small, bacteria have an advantage in the uptake of dissolved organic material. The details of this relationship need to be modified if shape changes with size, but for marine organisms, observed changes in shape are not sufficient to offset the basic argument: because they are small, bacteria have more surface area per volume, and so have an advantage in absorbing dissolved organic material.

Bacteria have other advantages as well. They have evolved extraordinarily effective means of absorbing DOM, so that even when compared surface area to surface area, they are more effective than other organisms. And, as we have seen, bacterial trophic efficiencies are much higher than those in most animals: they make better use of the material they absorb.

In summary, the ability of bacteria to scavenge dissolved organic material salvages much of the energy that would otherwise be lost to the trophic flow when phytoplankton die or otherwise release carbohydrates. As a result, the microbial loop forms an important pathway in the marine trophic river.

Viruses: A Loop within the Microbial Loop

Before leaving the microbial loop, I should note one other important trophic interaction: marine bacteria are commonly infected with viruses. Typically, 50% of bacteria in the sea are infected by bacteriophages, viruses that specialize in preying upon bacteria, and 20% to 40% of these infected bacteria die each day. Thus, bacteriophages may be as important as—and possibly more important than—ciliates and flagellates in controlling bacterial population dynamics.

Viral control of bacterial populations can have important trophic consequences. We have just learned that by converting dissolved organic matter into particulate form, bacteria can reinject into the trophic stream energy that would otherwise be lost. But when a bacterium dies from a viral infection, the cell bursts (a process known as *lysis*), and its contents are released into the water as dissolved organic material. This includes the newly formed viruses that triggered the bacterium's rupture: they are too small to be effectively filtered from the water (0.002 to 0.2 micrometers in length), and therefore are effectively "dissolved." As a result, the trophic effect of bacteriophage infections is to short-circuit the microbial loop. Lysis releases the dissolved organic material incorporated into bacteria before the bacteria can be consumed by animals higher on the food chain.

The importance and abundance of viruses in the sea have only been recently recognized. Because of their small size, viruses are difficult to detect, but recent advances in fluorescent-stain technology now allow oceanographers to efficiently count viruses in seawater. Current estimates show that there are approximately 3 billion viruses per liter in the deep sea and an incredible 100 billion viruses per liter in productive coastal waters. Summed over the entire ocean (1.4 billion cubic kilometers of seawater), these values suggest that there are approximately 4×10^{30} viruses in the sea.

This is another of those numbers too huge to easily grasp. As we did for cyanobacteria in chapter 4, we can attempt to fathom the magnitude of viral population size by placing individuals end to end. Multiplying 4×10^{30} viruses by 0.1 micrometers per individual, we come to the mind-boggling conclusion that if all marine viruses were lined up they would extend 4×10^{20} kilometers, a distance of 42 million light-years. This is roughly a hundred times the distance across the Milky Way galaxy, another number too large to grasp. Perhaps another viewpoint will help. Each virus contains about 0.2×10^{-15} grams of carbon, so the 4×10^{30} viruses in the ocean have a mass of approximately 800 million metric tons. This is equal to the amount of carbon fixed by the entire ocean in 20 days.

Upstream vs. Downstream Control

As we have seen, the inevitable loss of energy from one trophic level to the next provides a means by which we can predict the relative population size of organisms at each trophic level. That is, if we know how much energy enters the trophic process via a given population of phytoplankton, we can (in theory) follow the energy downstream to calculate the maximum potential population size of herbivores, filter feeders, and predators. To put it yet another way, in this view of the world, the population size at each trophic level is controlled by processes upstream in the trophic river.

This approach works fine for the herbivores, predators, and top predators in the system, but what about the primary producers? Phytoplankton are situated at the source of the trophic river; there is no upstream population on which to base our calculation. So, what controls the standing crop of phytoplankton?

Part of the answer is already in hand. As discussed in chapter 4, the availability of raw materials—light, carbon, and nutrients—potentially limits the population size of phytoplankton in a given volume of ocean water. In this sense, the population size of primary producers can indeed be set by factors upstream in the trophic process.

However, there is one important factor that we have not yet discussed: control by herbivores. As herbivores acquire energy from phytoplankton, they inevitably reduce the population size of these primary producers. Thus, it is possible that the population size of phytoplankton is limited not only by factors upstream, but by factors downstream as well, in this case by herbivores. As we will see, the same logic also applies to other trophic levels: the population size of herbivores can be controlled by predators, the population of predators by top predators, and so forth.

To see how this works, we first explore the manner in which a phytoplankton population would grow in the absence of any interaction with downstream trophic levels. A hypothetical history is shown in figure 5.19. The graph begins with a small population. Because there are initially few organisms vying for available resources, the population can grow rapidly. In fact, if resources are unlimited, the population grows ever more rapidly, increasing exponentially through time as it doubles in size with each phytoplankton division. At some point, however, the population reaches a size at which resources begin to be limiting, and the rate of population growth decreases. Eventually, the population of phytoplankton reaches a size at which resources are utilized at the same rate they are provided. At this size, the population has reached its *carrying capacity*, and no further growth is possible. (For the mathematically inclined, the

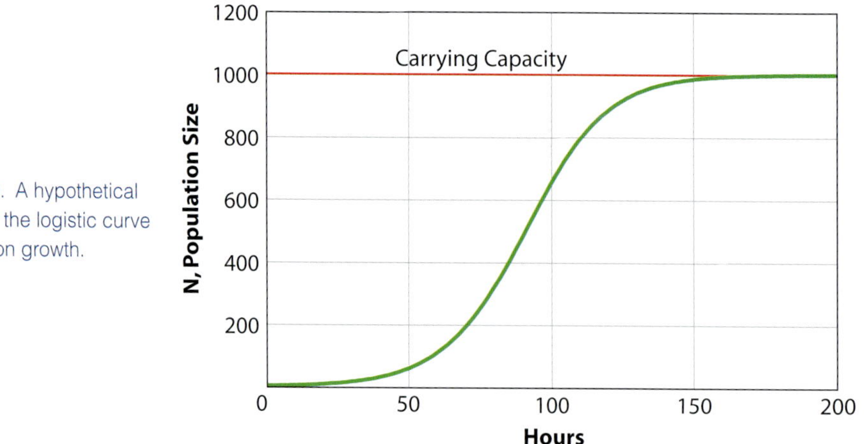

Figure 5.19. A hypothetical example of the logistic curve of population growth.

appendix to this chapter provides a discussion of an equation—the *logistic equation*—used to describe this pattern of population growth.)

Now we'll start the process all over again, but this time let's introduce a population of prudent herbivores into the system. (Real herbivores are seldom prudent, but for present purposes let's assume that our hypothetical herbivores act in their own best interest.) To survive, the herbivores need to eat some of the phytoplankton, but being prudent, they know that they shouldn't eat them all. If they did, there wouldn't be any left to provide their next meal. With a bit of thought, the hypothetical herbivores figure out that if, in a given period, they eat an amount just equal to the growth in the phytoplankton population size, they can dine forever. That is, if at the end of a day herbivores eat only the new phytoplankton produced since morning, the system returns to its starting point, and each day is just like the next—a trophic analogue to the goose that lays golden eggs.

We can depict this type of prudent grazing on a graph (figure 5.20). In this first experiment, we introduce our herbivores at the beginning of Day 2. They wait until the end of the day and then crop the phytoplankton population back to the level at which they first encountered it. The number of phytoplankton consumed is shown by the length of the vertical interval at the end of Day 2, a measure of the *sustainable yield*. In Day 3, the phytoplankton population grows by the same amount it grew in Day 2, the herbivores crop it back, receiving the same benefit as the day before, and the process can be repeated *ad infinitum*.

But what if instead of introducing our herbivores on Day 2, we wait until Day 4? By this time, the phytoplankton population has grown to a greater size and is increasing at a greater rate. Again the herbivores wait until the end of the day and crop the surplus population—a process that can be repeated indefinitely—but in this case, the herbivores get a bigger meal (figure 5.20, Experiment 2). In this second example, because herbivores maintain the phytoplankton population at a size where it grows rapidly, the sustainable yield of energy is large.

And finally, we could wait until Day 6 to introduce our herbivores to the system (figure 5.20, Experiment 3). In this case, the phytoplankton population has grown very large, but because it is near its carrying capacity, further increase

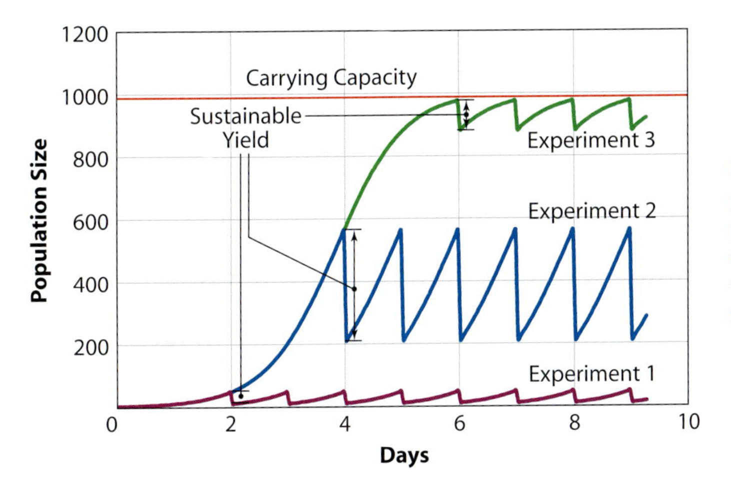

Figure 5.20. Prudent predators can control population size. Note that sustainable yield is maximized when population size is near the middle of its range.

happens slowly. The prudent herbivores take this into account, again maintaining the phytoplankton population at a constant size, but the amount the herbivores can eat in a day is small because population growth is slow.

Two important facts should be evident from these thought experiments. First, the presence of herbivores can control phytoplankton population size. As shown in figure 5.19, in the absence of herbivores, the phytoplankton population size can be quite large. However, herbivores can limit and maintain population size—the standing crop—at any size less than the carrying capacity. In this case, it is not upstream factors (the availability of light, carbon, and nutrients) that limit the phytoplankton population; instead it is the effect of herbivores downstream that exerts control.

Second, the rate of energy flowing through the system depends on the size of the phytoplankton population maintained by the herbivores. If herbivores maintain the population at either small or large size, they cannot sustainably eat very much each day, and the rate of energy flowing to the second trophic level is relatively small. In contrast, if herbivores maintain the phytoplankton population at medium size, the rate of energy flow is maximized.

In these experiments, we have assumed an unrealistic prudence on the part of the herbivore population. In reality, herbivores are likely to be greedy: presented with a large population of phytoplankton, they will eat as much as they can. If they consistently eat more phytoplankton than the population produces in a day, there are fewer phytoplankton left to provide their next meal, and the phytoplankton population decreases through time. Eventually, there will be too few phytoplankton to feed the herbivores, and herbivores will die. With the herbivore population thus reduced by starvation, the small remaining phytoplankton population can increase, starting the whole process over again. In this fashion, when downstream control is present, it is possible for the population sizes of both phytoplankton and herbivores to oscillate through time, and this two-trophic-level system can exhibit complex dynamics.

But why stop with just two trophic levels? Introduction of predators into this system, a third trophic level, can have a variety of effects. On one hand, if the population of predators is particularly smart, it can maintain the herbivore population at a size sufficient to maintain the phytoplankton population at medium size, thereby maximizing energy flow. In this fashion, by keeping herbivores in check, predators could stabilize and optimize the system. On the other hand, indiscriminate predators could amplify the oscillations among trophic levels, potentially destabilizing the system. Adding top predators only increases the potential for complex interactions.

These sorts of complexities provide ecologists with endless fodder for speculation and research, but alas, we will not be able to delve into the details here. There are, however, two messages you should take to heart. First, while upstream controls can, in large part, determine the *relative* population size at different trophic levels, the *absolute* sizes are likely set by downstream controls. Second, in the real-world absence of intelligent, prudent herbivores and predators, it is unlikely that populations are ever at equilibrium.

The upstream and downstream controls we have described here are often given other names in the ecological literature. For example, herbivores are higher on the trophic pyramid than primary producers. Thus, what we call downstream control—herbivores controlling primary producers—could equally well be termed "top-down control"—a higher trophic level controlling a lower trophic level. By the same logic, upstream control corresponds to "bottom-up" control.

Doubling Time

The example used above presents the pattern of population growth in the context of phytoplankton. Because their doubling time is short, in the absence of grazing, the population can approach its carrying capacity in just a few days. Other organisms may have longer doubling times, and will therefore approach their carrying capacities more slowly. An example is shown in figure 5.21, which contrasts the response of phytoplankton and foraminifera populations to their introduction into a new habitat. The phytoplankton, which double twice per day, respond to the bounty of this new habitat much more quickly than the foraminifera, which double once every five days. We will briefly explore the importance of the different response times of phytoplankton and their grazers in chapter 10.

The Flow of Carbon and Nutrients

So far, we have focused our attention on the flow of energy through marine ecosystems. Let us now switch gears and track the flow of matter, specifically the flow of carbon and nutrients.

In part, the flow of matter follows the flow of energy. Carbon and nutrients are taken up by primary producers and packaged into organic material, some of which is transferred to herbivores when phytoplankton are grazed. Carbon and nutrients are then passed to the various levels of predators. As with energy, some matter is lost from living organisms as one moves along the trophic river. In contrast to energy, however, this "lost" matter does not exit the system altogether. Instead, it resides for a time in one of seven interconnected "pools" before it begins the process all over again. (I use the term "pool" here in an abstract sense, intending to identify categories of matter rather than to label any particular volume of seawater.) For the moment, our task is simply to identify

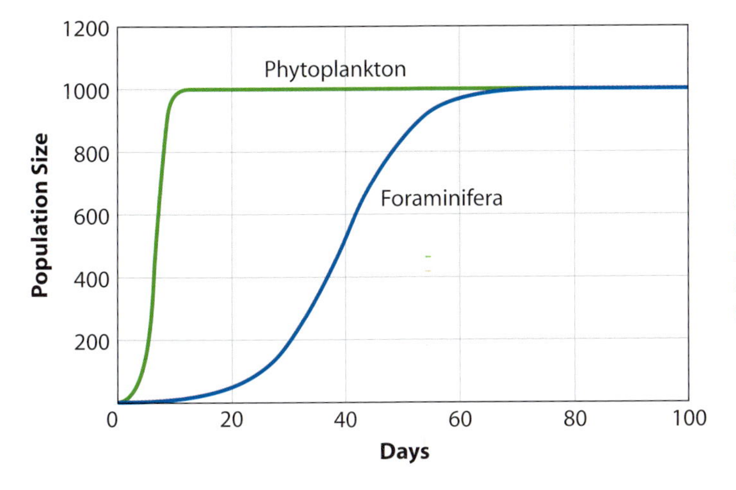

Figure 5.21. The rate of population growth varies among species with different doubling times; phytoplankton reproduce much more rapidly than do foraminifera.

and describe these pools and their connections; we will deal with their dynamics in chapters 9 and 10.

Pool 1: Living Matter. The first pool is composed of living organisms: all the carbon and nutrients contained in all trophic levels in the sea.

Pools 2 and 3: Particulate Matter. When plants and animals die, often they do not immediately disintegrate; for a while they remain as discrete particles. Some of these particles are organic (particulate organic matter, the remains of living tissue), others are inorganic (the remains of shells and skeletons, for example).

Pool 4: Dissolved Organic Matter. We have already encountered the fourth pool, the pool of dissolved organic matter.

Pool 5: Dissolved Inorganic Matter. Dissolved inorganic matter comes in two general varieties: inorganic carbon (carbon dioxide, bicarbonate, and carbonate) and mineral nutrients.

Pool 6: Sediments. Shells, bones, and dead bodies that sink to the bottom of the ocean accumulate, forming mud and ooze that can potentially be compacted into solid sediment.

Pool 7: The Atmosphere. And lastly, there is the atmosphere, a reservoir of nitrogen and carbon dioxide.

The major connections among these pools are shown in figure 5.22, information best digested in parts. First, note that, in most cases, the connections among pools are one-way. For example, through excretion and death, herbivores and predators contribute nitrogenous nutrients to the pools of dissolved inorganic matter. But except in unusual cases, they do not take nutrients back out of this pool. Phytoplankton often contribute to the pool of DOM by releasing carbohydrates, but they cannot readily reabsorb these sugars. As carbon-rich sediments are subducted, carbon can be injected into the mantle, later to be released into the atmosphere as carbon dioxide. There is no direct route by which atmospheric carbon dioxide enters the sediment, however.

In contrast, two important connections are two-way; both involve dissolved organic material. Primary producers contribute to the pool of dissolved inor-

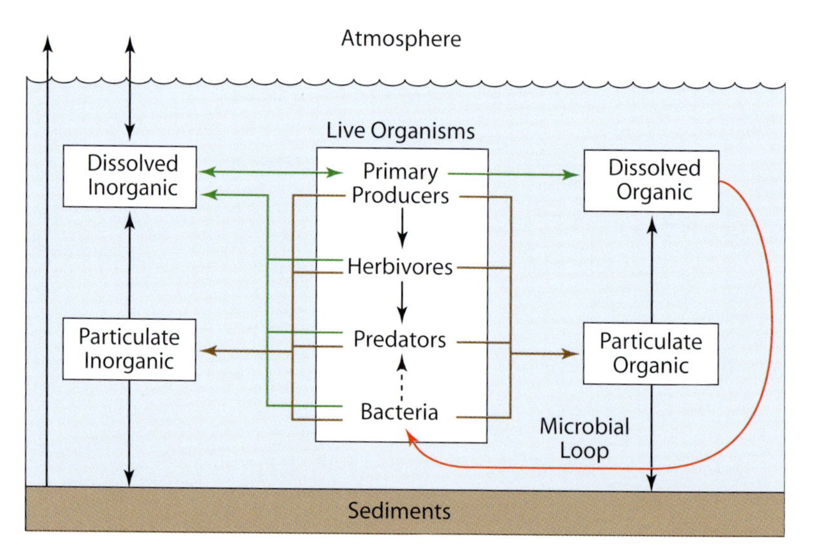

Figure 5.22. Paths of oceanic carbon and nutrient flow. In this diagram, I have lumped the filter feeders with other predators.

ganic material by releasing carbon dioxide during respiration and nutrients after death. But as we have seen, phytoplankton can also take up carbon dioxide and nutrients from this pool. In the other important two-way connection, carbon dioxide moves back and forth between the atmosphere and the pool of dissolved inorganic material.

Next, compare the two pools of dissolved material. The inorganic pool is well connected. Material can move into it by a variety of paths, and material can move out to both the atmosphere and primary producers. In contrast, the pool of dissolved organic material is relatively isolated. Material can move in as phytoplankton release carbohydrates and dead organisms rot, but the only major route out is through bacteria. This path, shown by the dashed lines, is the microbial loop.[7]

One final tidbit of information should be noted when considering figure 5.22: different pools respond to gravity in different ways. Under gravity's acceleration, living organisms, as well as particulate organic and inorganic material, tend to sink, and we will explore the consequences in detail in chapters 6 and 10. The atmosphere and sediments are also acted upon by gravity, but they have already sunk as far as they can go. In contrast, the pools of dissolved material do not sink. A glass of iced tea provides an analogy. Sugar crystals are denser than tea, and they initially sink to the bottom of the glass. Once dissolved, however, they lose their tendency to sink. No matter how long you wait, sugar dissolved in tea will not settle back down to the bottom of the glass. The same thing applies to any small molecule dissolved in seawater, including mineral nutrients. Just as the propensity of particles to sink has ramifications, the ability of dissolved material to stay in solution is important for how the ocean works, and we deal with these consequences in chapters 6 and 9.

Although figure 5.22 summarizes the paths of carbon and nitrogenous nutrient flow, we need to note carefully that the figure itself does not give us any information about the *rates* of flow along these paths. We have discussed some of the rates in relative terms: the flow of matter from primary producers to herbivores is approximately ten times the flow from herbivores to predators, for instance. But our discussion has not yet developed to the point where we can calculate the absolute magnitudes of these and other flows.

Summary

One can easily be overwhelmed by the staggering variety of heterotrophic organisms that live in the sea. To make sense of how they function and interact, one needs different perspectives, and in this chapter we have explored three:

- *Trophic Levels.* Order among organisms can be perceived by keeping track of who eats whom, and we have categorized heterotrophs into trophic levels: herbivores (including bacteria), predators (including bacterivores), and top predators. In general, this categorization also lumps animals by size; most animals are bigger than their prey.

[7] As we have noted, dissolved organic matter can be taken up by other organisms as well. For example, the planktonic larval stages of many mollusks can absorb dissolved organic molecules. Nonetheless, bacteria account for the vast bulk of reabsorbed DOM.

- *Energy Flow*. The flow of useful energy from one trophic level to the next is governed by trophic efficiency, which is typically about 10%. As a result, population size generally decreases as one travels down the trophic stream. In this respect, the initial input of solar energy into primary producers exerts upstream control over all heterotrophic life in the sea. But ecological interactions also play a major role. By controlling the size of the phytoplankton population, herbivores and their predators impose downstream control on the flow of energy.
- *Carbon and Nutrient Flow*. The flow of energy through the marine ecosystem is accompanied by flows of carbon and nutrients. However, unlike energy, which ultimately exits the system as heat, carbon and nutrients are potentially recycled, moving among seven pools.

These three perspectives provide the basis for understanding how biology interacts with the physics of the ocean, and this interaction is the subject of the next five chapters.

And a Warning

As a brief introduction to the animals of the sea and the flow of carbon, nutrients, and energy among them, this chapter has inevitably neglected some important topics. For instance, we have not tracked the flow of phosphate and silicic acid, and we have given extremely short shrift to the process of sedimentation. We touched only briefly on the roles of viruses in the ocean, and we have totally ignored benthic organisms, the animals that live on or in the seafloor. Perhaps most egregiously, we have not even mentioned one of the major domains of life, the Archaea. These tiny organisms superficially resemble bacteria, but differ in some important aspects of their membrane chemistry and the mechanism by which they produce proteins. They have recently been found to occur in great abundance throughout the oceans, and some can even fix nitrogen. But their role in the overall trophic dynamics of the sea is currently uncertain. Clearly there are details of heterotrophic organisms that we have ignored; if God is in the details, this chapter has been a most unholy treatment of life in the sea.

But we must move on. I hope that this brief overview has piqued your interest in marine biology and marine ecology. Please take some time on your own to delve into the details by consulting any of the multitudinous specialized texts on these subjects.

Further Reading

Azam, F., et al. (1983). The ecological role of water-column microbes in the sea. *Marine Ecology Progress Series* 10: 257–263.

Alldredge, A. L., and L. P. Madin (1982). Pelagic tunicates: Unique herbivores in the marine plankton. *BioSciences* 32: 655–663.

Bond, C. E. (1996). *The Biology of Fishes* (2nd edition). Saunders College Publishing, New York.

Bone, Q. (ed.) (1998). *The Biology of Pelagic Tunicates*. Oxford University Press, New York.

Bone, Q., H. Kapp, and A. C. Pierrot-Bults (eds.) (1991). *The Biology of Chaetognaths*. Oxford University Press, New York.

Caprulio, G. M. (1990). *Ecology of Marine Protozoa*. Oxford University Press, New York.

Denny, M. W., and S. Gaines (1999). *Chance in Biology*. Princeton University Press, Princeton, NJ.

Ellis, R. (1991). *Men and Whales*. Alfred Knopf, New York.

Fuhrman, J. A. (1999). Marine viruses and their biogeochemical and ecological effects. *Nature* 399: 541–548.

Lalli, C. M., and R. W. Gilmer (1989). *Pelagic Snails: The Biology of Holoplanktonic Gastropod Mollusks*. Stanford University Press, Stanford, CA.

Leviton, J. S. (2001). *Marine Biology: Function, Biodiversity, Ecology* (2nd edition). Oxford University Press, New York.

Mann, K., and J. R. Lazier (2006). *Dynamics of Marine Ecosystems* (3d edition). Blackwell Publishing, Malden, MA.

Sarmiento, J. L., and N. Gruber (2006). *Ocean Biogeochemical Dynamics*. Princeton University Press, Princeton, NJ.

Appendix

The Logistic Equation

In this appendix we explore the mathematics commonly used to describe how population size changes through time. We begin with a simple concept: the more individuals there are in a population, the more young they are likely to produce in a given period. In other words, the larger the population is, the faster it can potentially grow. The essence of this simple idea is easily expressed as a differential equation:

$$\frac{dN(t)}{dt} = rN(t) \qquad \text{(Eq. 5.6)}$$

Here, $N(t)$ is the number of individuals in the population at time t, and $dN(t)/dt$ is the rate of change of population size with respect to time. The proportionality between rate of change and population size is set by the constant, r, the *intrinsic rate of increase*. The larger r is, the faster the population grows.

Equation 5.6 does an excellent job of describing the rate of growth at a particular time, but often we desire something more. What about absolute population size rather than its rate of change? How large is the population at a particular time? To answer these questions, we need to solve equation 5.6 as a function of t. The first step is to separate variables, rearranging the equation to group all terms containing $N(t)$ on one side and all terms containing t on the other:

$$\frac{1}{N(t)}\,dN(t) = r\,dt \qquad \text{(Eq. 5.7)}$$

Next, we integrate, essentially summing up the infinitesimal changes in population size and time:

$$\int \frac{1}{N(t)} dN(t) = r \int dt \qquad \text{(Eq. 5.8)}$$

We now recall two basic facts from calculus:

$$\int \frac{1}{x} dx = \ln x + \text{constant} \qquad \text{(Eq. 5.9)}$$
$$\int dx = x + \text{constant}$$

Here, x is a generic variable.

Applying these concepts to equation 5.8, we conclude that

$$\ln N(t) = rt + C \qquad \text{(Eq. 5.10)}$$

where C is the constant of integration. Next, we exponentiate both sides of the equation:

$$e^{\ln N(t)} = e^{(rt + C)}$$
$$N(t) = e^{C} e^{rt} \qquad \text{(Eq. 5.11)}$$

We are almost there; we need only to establish the value of e^C by applying an appropriate boundary condition. This is most easily done by specifying the initial population size, $N(0)$. At $t = 0$:

$$N(0) = e^{C} e^{r0}$$
$$N(0) = e^{C} \qquad \text{(Eq. 5.12)}$$

Substituting this value into equation 5.11 brings us to our answer:

$$N(t) = N(0) e^{rt} \qquad \text{(Eq. 5.13)}$$

In other words, if the rate of change of population size is proportional to population size (equation 5.6), population size grows exponentially through time (equation 5.13). The absolute size of the population at any time depends on the initial size of the population, $N(0)$.

Equation 5.13 is a simple and elegant expression, but there is a problem with this kind of exponential growth. Because the rate of population growth increases as population size expands, the number of individuals is predicted to head rapidly toward infinity. However, in reality, any population must bump into practical limits: limited space available in its habitat, limited food for growth, etc. To arrive at a realistic description of population growth, we must go back to our basic assumptions and adjust equation 5.6 to incorporate limits on population size.

There are several ways in which limits can be manifested. One of the simplest is to expand equation 5.6 to include a term, shown here in square brackets, that gets smaller as the population grows:

$$\frac{dN(t)}{dt} = rN(t)\left[\frac{K - N(t)}{K}\right] \qquad \text{(Eq. 5.14)}$$

In this new term, K is the *carrying capacity* for the population, the maximum size it can reach. Without specifying what sets K, we can nonetheless see that it does its job. When $N(t)$ is small relative to K, the expression in brackets is very nearly 1, and equation 5.14 is approximately the same as equation 5.6. In other words, when $N(t)$ is small, population size increases exponentially. But as population size approaches K, the expression in brackets approaches 0, forcing the rate of growth to decrease. Equation 5.14 is the *logistic equation.* You will find it cited in any text on ecological theory, and it was used here as the basis for figure 5.19.

The form of equation 5.14 makes it evident how limits are applied to population growth—the closer population size gets to carrying capacity, the slower it expands—but it does not explicitly relate population size to time. To draw a graph such as figure 5.19, we need to solve equation 5.14 in the same sense that we solved equation 5.6. Because derivation of this relationship is a bit more complicated than that of equation 5.13, we jump straight to the final result rather than dwell on the mathematical details:

$$N(t) = \frac{1}{\left[\frac{1}{N(0)} - \frac{1}{K}\right] e^{-rt} + \frac{1}{K}} \qquad \text{(Eq. 5.15)}$$

$N(t)$ is again population size at time t, $N(0)$ is population size at $t=0$, and as before, r and K are the intrinsic rate of increase and carrying capacity, respectively.

The logistic equation (expressed as either equation 5.14 or equation 5.15) is a useful tool for modeling population growth, but we must be careful to note that it contains an important implicit assumption. Time, t, is a continuous variable: it can take on any value. Therefore, in using equation 5.15, a function of this continuous variable, we assume that population size can take on any value and increase smoothly through time. There are many cases, however, when this assumption is unrealistic. For instance, consider a population in which individuals reproduce only at set intervals; phytoplankton provide a hypothetical example. If we were to start with one phytoplankton cell, we could watch it grow until it split into two. These two cells might then grow in unison and split simultaneously, increasing population size to four. At the next splitting, the population size would be eight, and so on. Thus, if cells divide at set intervals, population size increases in discrete jumps: we cannot have a population of three individuals, or seven, or thirteen. How can we describe this kind of interval-by-interval growth?

Let's begin by returning to equation 5.14, the differential form of the logistic equation:

$$\frac{dN(t)}{dt} = rN(t)\left[\frac{K - N(t)}{K}\right] \qquad \text{(Eq. 5.16)}$$

Multiplying out the right-hand side of the equation, we see that

$$\frac{dN(t)}{dt} = \frac{rKN(t) - rN^2(t)}{K}$$
$$\frac{dN(t)}{dt} = rN(t) - \left(\frac{r}{K}\right)N^2(t) \qquad \text{(Eq. 5.17)}$$

In essence, this equation again says that the infinitesimal change in population size, $dN(t)$, in an infinitesimal interval of time, dt, is determined by the constants r and K and the instantaneous population size $N(t)$. If, however, reproduction happens at set intervals, we must express the rate of growth in terms of the discrete jump in population size, ΔN, which happens at intervals Δt:

$$\frac{\Delta N}{\Delta t} = rN_n - \left(\frac{r}{K}\right)N_n^2 \quad \text{(Eq. 5.18)}$$

Note that I have changed the nomenclature for population size. Here, N_n is the number of individuals present at the beginning of interval n. For example, N_0 is initial population size, size at the start of interval 0, N_1 is the population size at the start of interval 1, and so forth.

We now make an important shift in our presentation. If $\Delta N/\Delta t$ is the change in population size in an interval, it is equal to the difference in size between interval n and interval $n+1$. Thus,

$$\frac{\Delta N}{\Delta t} = N_{n+1} - N_n \quad \text{(Eq. 5.19)}$$

Substituting this new expression into equation 5.18, we see that

$$N_{n+1} - N_n = rN_n - \left(\frac{r}{K}\right)N_n^2 \quad \text{(Eq. 5.20)}$$

Now for a bit of mathematical legerdemain. First, we manipulate previous constants to create two new variables:

$$a \equiv r + 1$$

$$b \equiv \frac{r}{K} \quad \text{(Eq. 5.21)}$$

Substituting these variables in equation 5.19 (and noting that $r = a - 1$) allows us to make the following maneuvers:

$$N_{n+1} - N_n = (a-1)N_n - bN_n^2 \quad \text{(Eq. 5.22)}$$

$$N_{n+1} - N_n = aN_n - N_n - bN_n^2 \quad \text{(Eq. 5.23)}$$

$$N_{n+1} = aN_n - bN_n^2 \quad \text{(Eq. 5.24)}$$

This relation is the answer we seek. Equation 5.24 is a recipe for calculating logistic growth of a population, interval by interval. Given values for a, b, and the population size at interval n, we can calculate the population size at the end of interval $n+1$. From that value, we can calculate the population size at interval $n+2$, and so forth.

The logistic equation, in the form of equation 5.24, has been the subject of a great deal of interest and study. Some has been directed toward the kinds of usage to which we have put the logistic equation in this chapter, and we will discuss a practical application of this math in chapter 11 when we look at the dynamics

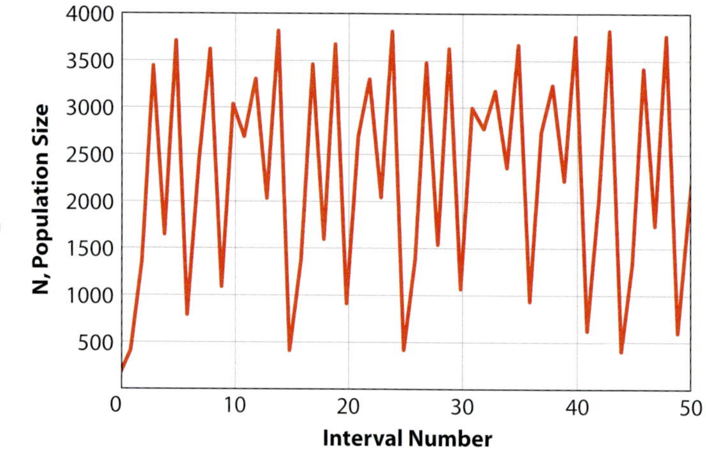

Figure 5.A1. For certain parameter values, Eq. 5.24 predicts chaotic fluctuations in population size. Here $a=3.9$ and $b=0.001$.

of fisheries. In recent years, however, equation 5.24 has been studied from another (and very surprising) angle. For certain combinations of a and b, population size does not follow a predictable path through time; instead, it jumps around chaotically. In fact, the quirky behavior of this form of the logistic equation is one of the classic examples of *chaotic dynamics*, a new field of mathematics that has had far-reaching impact on our understanding of both the physical and natural worlds. An example is shown in figure 5.A1, which depicts the results of equation 5.24 with $a=3.9$, $b=0.001$, and starting population size of 100 individuals. Much fun is to be had by playing with equation 5.24 using different values of a and b.

For a readable introduction to the theory of chaos, you should consult F. C. Moon (1992), *Chaotic and Fractal Dynamics* (John Wiley and Sons, New York).

The Dilemma of the Two-Layered Ocean

Now that we have sufficient background information about the size, shape, composition, and biology of the oceans, we can begin to put the pieces together and examine the whole. This will be a story in four parts. In the current chapter we will see how the ocean is composed of two distinct layers, and will arrive at the disturbing conclusion that because of this two-layered nature the ocean is a really lousy system for supporting life. In chapter 7 we will digress to explore the physics of fluids as they move on our rotating earth. The strange effects that result (Coriolis acceleration and geostrophic flow) provide a physical basis for exploring the large-scale wind patterns and currents in the surface ocean. These motions, which form a heat engine powered by the sun, are discussed in chapter 8. Finally, in chapter 9, we tie all of these physical factors together to explain the processes by which the dilemma of the two-layered ocean is at least partially overcome in the sea. At the end of chapter 9, you will be amazed at how much we can explain about life on our planet.

The Two-Layered Ocean: A Description

We begin this chapter by giving away its punch line in figure 6.1. In a very stylized and general sense, the essence of the open ocean is captured within this set of graphs, and it is best if we take time to consider them in some detail.

The axis running from the top of the page toward the bottom is an index of depth. The ocean's surface is at the top, and the farther down the page we look, the deeper into the ocean we travel. Note that this axis is not drawn to scale; to see details near the surface I have expanded the axis there, and have I compressed the axis for much of the deep ocean.

Each of the axes running across the page is an index of the magnitude of some characteristic of the ocean. Values increase to the right. For example, axis

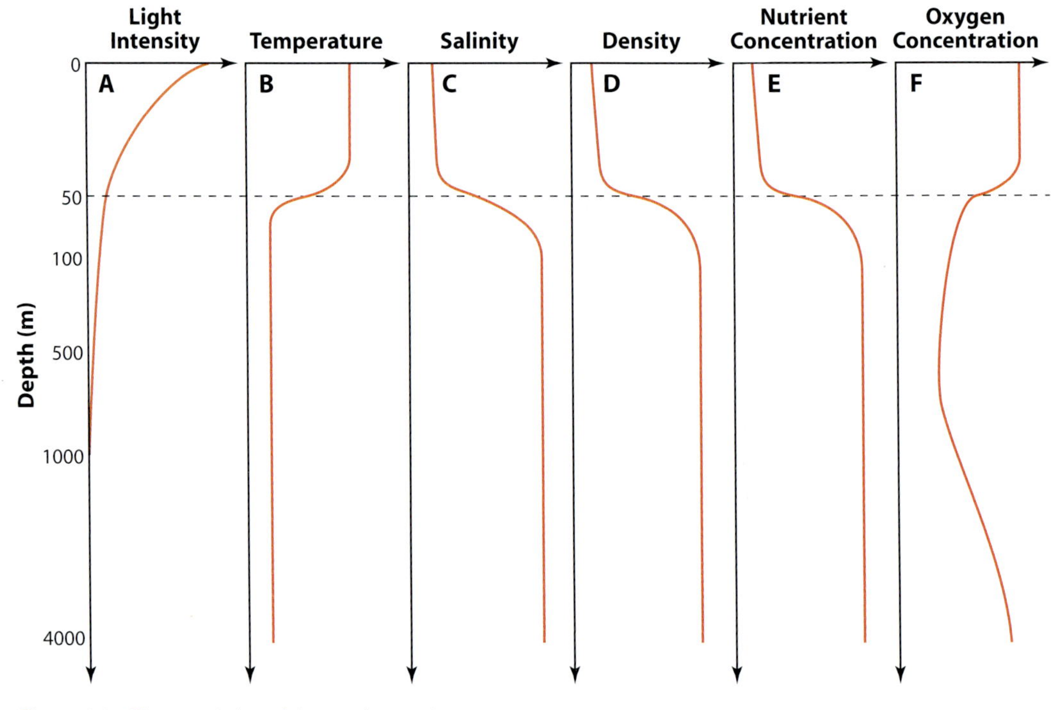

Figure 6.1. Characteristics of the two-layered ocean.

A reiterates the information we learned in chapter 4 about how light behaves in the sea. At the ocean's surface, light is typically bright, and the intensity decreases exponentially as we move deeper in the ocean. Graphs B–F display similar depth-dependent information for the temperature, salinity, and density of ocean water, and for the concentration of nutrients and oxygen. The nutrients I refer to here include the nitrogenous nutrients discussed in chapter 4, but also include others such as silicic acid and phosphate.

The Thermocline

Note that graphs B–F share a common attribute: at a depth of about 50 meters something drastic happens. For instance, in graph B we see that seawater is uniformly warm near the ocean's surface, but that at a depth of about 50 meters the temperature abruptly decreases. This steep gradient in temperature is the *seasonal thermocline* (or simply, the *thermocline*, for short). In graphs C–E we find that the salinity of the water, its density, and the concentration of nutrients are all uniformly low near the surface, but below the critical depth of 50 meters they are higher. Each of these gradients (in these three cases, a shift from low to high with increasing depth) has a technical name. The abrupt change in salinity is the *halocline,* the steep gradient in density is the *pycnocline*, and the shift in nutrient concentration is the *nutricline*.

The concentration of dissolved oxygen (graph F) shows a pattern that is somewhat similar to temperature: it is high at the surface, and decreases rapidly as we descend below 50 meters. In this case, however, there is an added fillip.

The concentration of oxygen reaches a minimum at a depth of 500 to 1000 meters, and then increases at greater depth.

Of these characteristics—light, temperature, salinity, density, nutrients, and oxygen—temperature has historically been the easiest to measure. Oceanographers have been submerging thermometers in the oceans for more than 200 years, and as a result, information about the thermal structure of the sea is widespread and familiar. Although easy with modern technology, direct measurements of the other characteristics do not have the same lengthy history. As a result, it has become traditional in oceanography to use the term "thermocline" as shorthand for all of the physical changes that occur together at some certain depth in the ocean. Throughout the rest of this book, when we casually mention the thermocline, it may help to remind yourself that this term might refer to more than just temperature.

Salient characteristics of the thermocline are commonly measured using a device that continuously records the conductivity of seawater (an index of salinity), the water's temperature, and its pressure (a measure of the instrument's depth). This conductivity-temperature-depth (CTD) instrument is lowered into the ocean, recording as it descends. From the combined data regarding salinity, temperature, and pressure, the density at each depth can be calculated. In this fashion, the CTD instrument allows oceanographers to measure the data needed to plot graphs 6.1B, C, and D. A dissolved oxygen meter can also be included in the instrument package, providing information for graph 6.1F. Nutrient concentrations are not as readily measured *in situ*; typically, discrete water samples are collected at selected depths and returned to the laboratory for analysis.

The Dilemma

The combined changes in the character of seawater at a depth of about 50 meters allow us to characterize the ocean as consisting of two strata. There is a thin upper layer above the thermocline that is light and well oxygenated, two characteristics conducive to life. The light is sufficient to allow phytoplankton to effectively harvest energy from the sun and store it as glucose, and the oxygen can support heterotrophic animals that eat the phytoplankton. There is a catch, though. As we learned in chapter 4, photosynthesis requires both light and nutrients, and because the water in this upper layer has a low concentration of nutrients (see graph E of figure 6.1), phytoplankton have difficulty growing. Without an actively growing phytoplankton population, the flow of energy down the trophic river is stifled, and animals have difficulty surviving. Thus, the upper layer is not well structured to support life.

What about the second, lower layer? Below the thermocline, nutrients are abundant and, at least in the top portion of the lower layer, oxygen is present in sufficient quantity to support animals. But again there is a catch. Because light is attenuated by seawater, this thick bottom layer of the ocean is dark. No light, no phytoplankton (no matter how concentrated the nutrients); and, again, no phytoplankton, no animals. Thus, the lower layer is no better structured to support life than is the top layer.

And that about covers it. The ocean has two distinct layers, neither of which can readily support life. Overall, a lousy way to run a planet.

The Two Layers Are Stable

At this point you might be asking yourself an obvious question. The only reason the top layer is problematic is a lack of nutrients, and (as shown in figure 6.1E) there are plenty of nutrients available close by, down in the lower layer. Why can't some of the nutrient-rich water below be mixed up into the top layer? It's a nice thought, but as a practical matter it is difficult to bring about.

Recall that water in the upper layer has both a high temperature and a low salinity. As a result, the water in the top layer has a lower density than the cold, saline water below (figure 6.1D). In essence, the low-density fluid of the top layer floats on the high-density fluid of the lower layer like oil on water. As a result, the system is very stable. Think of how hard you have to shake a bottle of salad dressing to mix the oil and vinegar, and you will have an idea of what is required to overcome the stability of a layer of low-density fluid floating on a layer of higher-density fluid.

Through the application of some simple physics, we can be more quantitative about this stability. Lifting dense water from below the thermocline up into the lower-density surface layer is analogous to lifting a bucket of water from a swimming pool up into the air above. In both cases, a force must be applied to move a dense medium upward into a less dense medium. Recall that mechanical energy is the product of an applied force and the distance that force moves its point of application. Thus, raising a bucket out of the pool—or water from below the thermocline—requires that energy be expended.

How much energy are we talking about? Let's take a representative example in which the surface layer is 50 meters thick and water above and below the abrupt thermocline differs in density by two kilograms per cubic meter (a difference of only about 0.2%). Let's lift water from below the thermocline up into the surface layer, where it is then uniformly mixed. Increasing the volume of the surface layer in this fashion has the effect of pushing the thermocline deeper. In this case, for each square meter of ocean, 491 joules are required to mix enough dense water up from below to move the thermocline down by just 1 meter. (See the appendix to this chapter for the details of this calculation.) There are a million square meters in a square kilometer, so the energy required to push a square kilometer of thermocline down a meter is 491 million joules. And there are 361 million square kilometers in the ocean, for a grand total of 1.8×10^{17} joules required to displace the thermocline of the earth's seas a mere meter, energy equal to the total used by human society in two weeks. If the difference in density above and below the thermocline is greater, or the thermocline is displaced farther, even more energy is required.

Thus, without the imposition of some very energetic mixing process, nutrient-laden water in the lower layer of the ocean is prohibited from moving up in bulk. And without upward transport of nutrients, there can be little photosynthesis. Because of the stability of the system, there is no easy solution to the basic dilemma of the two-layered ocean.

A Reality Check

This conclusion may seem at odds with reality. You may well have the impression from chapters 4 and 5 that there is plenty of life in the sea—billions of metric tons of carbon fixed each year! swarms of salps hundreds of kilometers long!—and on that basis you could conclude that an abundance of plants and animals have found some way around our dilemma. How can we reconcile these two views: the grim picture we have painted here and the fact that you can buy cheap canned tuna at your local grocery?

First, we note that the abundance of life in the oceans is more apparent than real. As human beings, we tend to encounter the oceans primarily at their edges. For reasons that we will explore in chapter 9, these coastal waters are among the few places in the ocean that can support plentiful plant and animal life. As exceptions to the rule, these productive areas skew our impression of the ocean writ large.

In fact, most of the ocean supports only a low concentration of living things. Bring to mind a clear blue tropical sea, filled with the kind of water one finds in the central Pacific Ocean, for instance. The very clarity of the water is evidence of the poverty of life within it. If that water were full of phytoplankton, for instance, it would be a muddy green, not crystal blue.

To put this in more concrete terms, consider these facts. Using modern fishing technology, we are currently taking from the oceans about as much fish as they can sustainably produce, about 85 million metric tons per year. If we try to fish any harder, there will simply be fewer fish to catch (see chapter 11). Now 85 million tons may sound like a lot, but it is only about 10% of the amount of meat and milk produced on land each year. And this relatively small amount of fish is produced in an area more than twice that of the land. When compared, area for area, to a cattle ranch in Texas or a corn farm in Iowa, the oceans are an ineffective farm.

Second, we must note that the two-layer rule is general, but not universal. Go to the center of the Pacific, Atlantic, or Indian Oceans and there will be two layers as advertised. If, however, you pick a shallow part of the ocean—the continental shelf, or the Grand Banks east of Canada, for example—you will see a different story. There, for reasons that we will soon understand, the two layers as we have described them never really exist. Similarly, there are areas of the ocean—the temperate seas, for instance—where there are two layers in summer, but the layers merge in winter. We will eventually deal with these exceptions, but their existence should not overshadow the fact that the vast bulk of the ocean does indeed consist of two layers, and is therefore subject to our dilemma.

And finally, as we have noted above, even in the parts of the ocean that do have two distinct layers, there are partial solutions to the basic dilemma of the two-layered ocean. Here and there physics conspires to move nutrient-rich water up into the light (chapter 9). However, before we can understand these solutions, we must first understand the details of the central problem, and that is the subject of this chapter.

Why Is the Surface Warm?

Figure 6.1 is a statement of fact, a description of what the open ocean typically looks like. But that alone is not enough. To understand how life can survive by circumventing the dilemma of a two-layered ocean, we first need to understand the mechanisms that create the layers.

Let's start by examining the pattern of temperature in the ocean (figure 6.1B). Recall that the surface layer of the sea is relatively warm, shifting abruptly to colder water at a depth of about 50 meters. In light of this graph, there are two facts we need to explain: first, why is the ocean surface warm, and second, why are the ocean depths cold?

Absorption of Solar Heat

The warmth of surface water is easily explained. As we learned in chapter 4, sunlight is absorbed by seawater; this is why the deep ocean is dark. And as light energy from the sun is absorbed, it is effectively converted to heat.

Most of the sunlight impinging on the ocean is absorbed in the first few meters below the surface (see figure 4.10). For example, if 6% of light is absorbed in each meter of water the light passes through, as would happen even in relatively clear ocean water, 50% of all light is absorbed in just the top 12 meters of the ocean. If the seawater is a bit more murky, so that 10% of light is absorbed in each meter, 50% of the light is absorbed in just the top 7 meters.

If most of the light is absorbed in the top 7–12 meters of the upper layer, wouldn't that mean that this thin skin of the water column would heat up more than the rest of the water above the thermocline? And, if so, why is all 50 meters of the upper layer at more or less the same temperature (figure 6.1B)?

Turbulent Mixing

Indeed, if the surface water of the ocean were not stirred, the thermocline would be much nearer to the surface than it is. Furthermore, instead of having a uniform temperature in the surface layer and the abrupt shift in temperature that we typically see at the thermocline, there would be a gradual decrease in temperature from the surface downward.

In fact, however, the surface of the ocean *is* constantly being stirred. As wind blows over the water—a near-universal state for the ocean—friction between rapidly moving air and the sluggishly moving sea has the effect of stirring the water. If the wind is slow, the stirring can be gentle. If the wind is sufficiently fast, it can produce waves on the ocean surface, and as these waves break, forming whitecaps, the consequent stirring can be energetic. This stirring, intense or not, is the turbulent mixing we briefly introduced in chapter 4.

The mixing that wind engenders in water has the form of eddies that swirl, stretch, and move away from the surface, mixing up the water around them. These are the same sort of turbulent eddies evident when one stirs cream into a cup of coffee, and just as the stirred-up eddies in a cup tend to disperse the cream evenly, the wind-stirred eddies in the ocean's upper layer tend to evenly disperse throughout the layer any heat absorbed near the surface. The net result is an upper layer, commonly known as the "*mixed layer,*" of nearly uniform, warm temperature.

The turbulence that explains the uniformity of temperature in the surface mixed layer is also capable of explaining in part why, in shallow parts of the ocean—over continental shelves, for instance—two layers do not usually exist. If the sea is sufficiently shallow and the wind is sufficiently strong, the effects of wind-driven turbulent mixing can extend all the way to the seafloor. In this case, mixing erases any tendency for a cold bottom layer to form, and the entire water column has only one top-to-bottom layer with a well-mixed composition. We will return to this effect later.

Seasonal and Permanent Thermoclines

At this point, I have to admit to having foisted a generalization upon you. In figure 6.1, I have shown an abrupt transition at a depth of 50 meters between the surface layer of warm water and the deep layer of cold water. In the real ocean, the structure of the thermocline can deviate from this simple picture in several respects. In fact, there are two thermoclines in the ocean. The first is the seasonal thermocline we have dealt with so far; and its depth may vary through the year. In addition, there is a *permanent thermocline*, with a depth that is greater and more constant. Let's pause for a moment to explore the characteristics of these two thermoclines.

The Seasonal Thermocline.

The seasonal thermocline depicted in figure 6.1 is a representative example, but at any given location the depth of the actual thermocline may differ. In temperate seas, the ocean surface is heated primarily in summer, and the seasonal thermocline has a maximum depth of about 50 meters as shown. In winter its depth may increase as the surface layer cools and mixing extends deeper (figure 6.2). In contrast, in the tropics, where there is less variation in heating with the season, the seasonal thermocline may be 200 meters deep or deeper, and its depth varies little through the year. In polar seas,

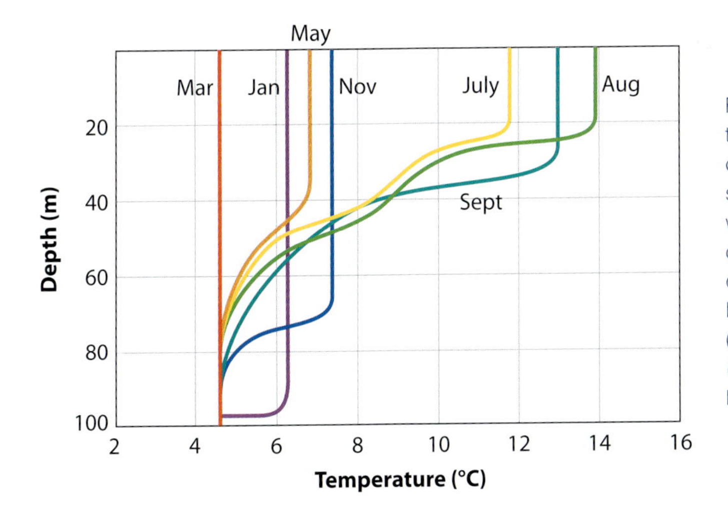

Figure 6.2. In temperate seas, the depth of the thermocline changes with the season; shallow in summer and deep in winter. In the tropics, the depth of the thermocline is more constant through the year. Redrawn from G. L. Pickard (1963), *Descriptive Physical Oceanography* (Pergamon Press, London), figure 9.

where mixing is intense and light is dim for much of the year, there may be no thermocline at all. Because of this variability, you should regard the figure of 50 meters as a very rough rule of thumb for the depth of the seasonal thermocline.

The Permanent Thermocline. The presence of the time-varying seasonal thermocline results in a larger-scale gradient of temperature that extends deeper into the ocean and varies little from one year to the next. In this *permanent thermocline*, temperature gradually decreases from 6 to 10°C at the base of the seasonal thermocline to 0 to 4°C at a depth of about 1500 meters. As with the seasonal thermocline, the permanent thermocline inhibits mixing between the surface and deep layers of the ocean. But, because the gradient in temperature—and therefore in density—is so gradual in the permanent thermocline, the inhibition of mixing is not as stringent as that imposed by the seasonal thermocline. For our purposes, we need not concern ourselves with the permanent thermocline, and through the remainder of the book when we talk about "the thermocline," it is the seasonal thermocline we have in mind.

Why Is the Bottom Layer Cold?

At this point, we can explain why the surface layer of the ocean is uniformly warm: it is heated by the sun and mixed by the wind. But why is the deep layer cold? This low temperature turns out to be more difficult to explain.

First, we note that the depths of the ocean are not cold simply as a matter of physical necessity. At least three times in the earth's history the bottom waters of the ocean have been considerably warmer than they are today. The most recent episode was about 100 million years ago when the deep ocean reached a temperature of 13 to 14°C, about 10°C warmer than it is now. In fact, given that the oceans have been basking in the sun for more than 3 billion years, one would think that they should have heated through. Thus, it simply won't do to assert that deep waters of the ocean are cold either because they have to be or because they always have been. Instead, we need a mechanism that explains their low temperature.

The Flux of Sunlight

As we saw in chapter 4, the sun is a source of light energy. To answer the question of why the depths are cold, we begin by exploring the pattern in which this energy arrives at earth's atmosphere. Imagine yourself traveling to the top of the atmosphere and taking with you a square board one meter on a side. The board is painted flat black so that it absorbs all light that hits it. Once in space, if you were to hold the board broadside into the sunlight, it would absorb, and turn into heat, about 1370 joules of solar energy every second (figure 6.3A). This rate, 1370 watts per square meter, is known as the solar constant. To put this in perspective, this is sufficient energy to keep thirteen 100-watt light bulbs lit continuously, with a few watts left over to run a computer or two, all from one square meter of absorbing surface.

Note that this is the energy the board would absorb if it were held broadside to the sunlight. If, instead, the board is held at an angle to the sun, it intercepts

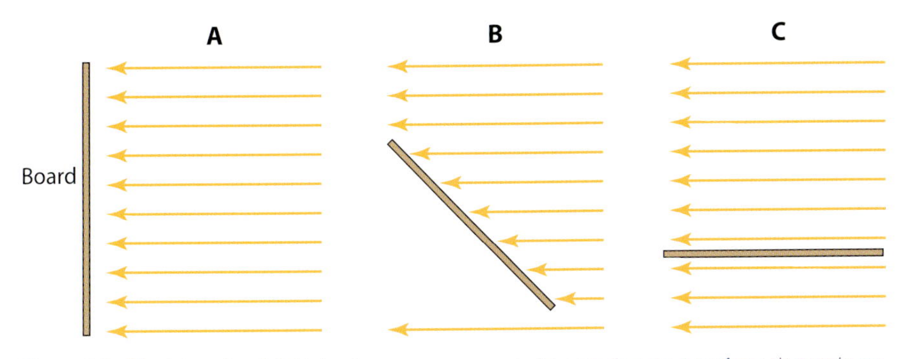

Figure 6.3. The intensity of light (watts per square meter) impinging on a surface depends on the angle between light and surface. Light is most intense when the surface is perpendicular to the direction in which the light approaches (A) and least intense when the surface is parallel to the direction in which light approaches (C).

a smaller patch of light, and absorbs less heat (figure 6.3B). Taken to an extreme, if the board were held edge on to the sunlight, it would absorb virtually no heat at all (figure 6.3C).

The same ideas apply to earth's surface. The projected area the earth as a whole exposes to the sun is equal to π times the square of the earth's radius, approximately 1.3×10^{14} square meters. When this area is multiplied by a factor of 1370 watts per square meter we find that the sunny side of the earth intercepts 1.8×10^{17} joules of sunlight every second, the value noted in chapter 4. That's 180 *billion* megawatts of solar energy. As sunlight strikes the planet, about 40% of it is reflected back into space (by clouds and snow, for instance), but roughly 1.0×10^{17} watts, still a staggeringly large number, is absorbed by the atmosphere, land, and ocean.

Thermal Equilibrium

Because the earth absorbs this amount of heat every second, it is reasonable to think that it might be heating up. If one were to start with a cold earth and instantly "turn on" the sun, this would indeed be true. Such a scenario would in a sense be analogous to turning on an incandescent electric light bulb. The instant you flip the switch, energy is delivered to the bulb (in this case in the form of electrical current rather than light), and the bulb rapidly heats up. But at the same time that it heats up, the bulb radiates energy out. For the bulb, some of this radiated energy is in the form of visible light (after all, it is a light bulb), but a similar process is true of any object that is heated. The hotter the object, the more energy it radiates. (The rate at which energy is radiated is proportional to the fourth power of the object's temperature. Thus, a small increase in temperature results in a large increase in radiated energy.) Soon after the switch is turned on, the bulb reaches an equilibrium in which the rate at which energy radiates out is just equal to the rate at which electrical energy is supplied. Once at equilibrium, the temperature of the bulb stays constant.

The same physics applies to the earth taken as a whole. In the course of its 4.6-billion-year history, earth's temperature has risen until the energy it radiates out into space is just equal to the light energy it absorbs from sunlight. This equilibrium results in an average surface temperature of about 15°C. Whereas the temperature of a light bulb filament is several thousand degrees Celsius, resulting in radiation of visible light, at the relatively low surface temperature of the earth, the ground radiates infrared light.

In summary, the earth has the average temperature it does because it has reached an equilibrium with the energy delivered to it by sunlight.

The Effect of Latitude

That is far from the complete story, however. Recall that the amount of heat energy absorbed by our board in space is affected by the angle of the board to the sunlight. When broadside, the board absorbs the most energy; when tilted, less. The same is true for the surface of the earth. In the tropics, where the sun is high overhead at noon, the surface of the ocean is nearly broadside to the sun and each square meter of it absorbs the maximal amount of heat. As one travels toward the poles, however, the noontime sun is lower in the sky (figure 6.4). As a result, a typical square meter of sea surface at high latitudes is substantially tilted to incoming sunlight, and it absorbs less heat. Thus, the rate at which heat energy is delivered to the ocean depends upon latitude: the influx of heat is maximal in the tropics and minimal at the poles.

This simple picture is complicated a bit by the fact that earth's axis, the line through the earth connecting the north and south geographic poles, is tilted with respect to the plane in which the earth moves around the sun (figure 6.5). In June—the beginning of summer in the northern hemisphere—earth's north pole is tilted toward the sun (figure 6.5A), and northern latitudes receive more sunlight than average. Six months later, in the northern winter, earth's axis is tilted away from the sun (figure 6.5B), and less solar heat is absorbed. The same process applies in the southern hemisphere, but the timing of the seasons is reversed.

If we were to average over the entire year the amount of heat absorbed by the ocean at different latitudes, the graph would look like the red line shown in figure 6.6. As you should expect, average heat influx is greatest near the equator, the most tropical place on earth, and is least (but not zero) at the poles.

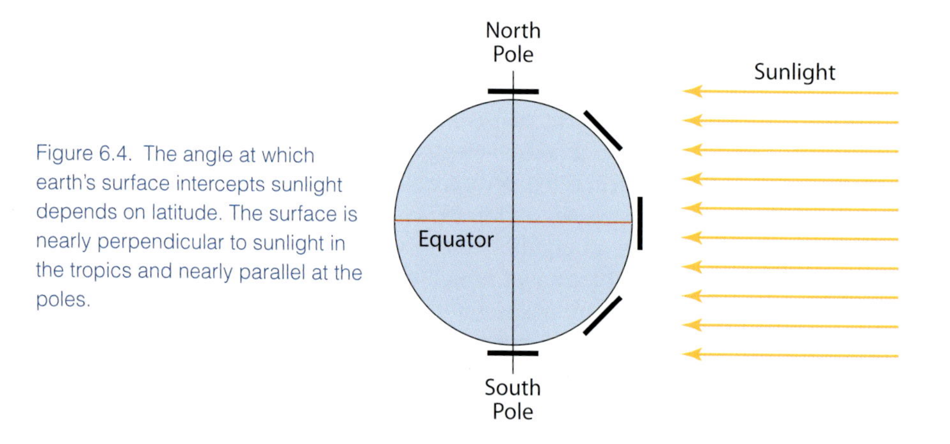

Figure 6.4. The angle at which earth's surface intercepts sunlight depends on latitude. The surface is nearly perpendicular to sunlight in the tropics and nearly parallel at the poles.

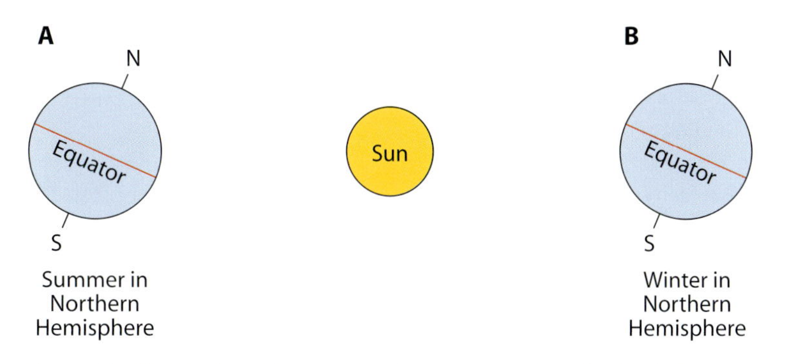

Figure 6.5. The tilt of earth's axis drives the seasons. When the North Pole is tilted toward the sun, it is summer in the northern hemisphere and winter in the southern hemisphere. Six months later, when the North Pole is tilted away from the sun, it is winter in the northern hemisphere and summer in the southern hemisphere.

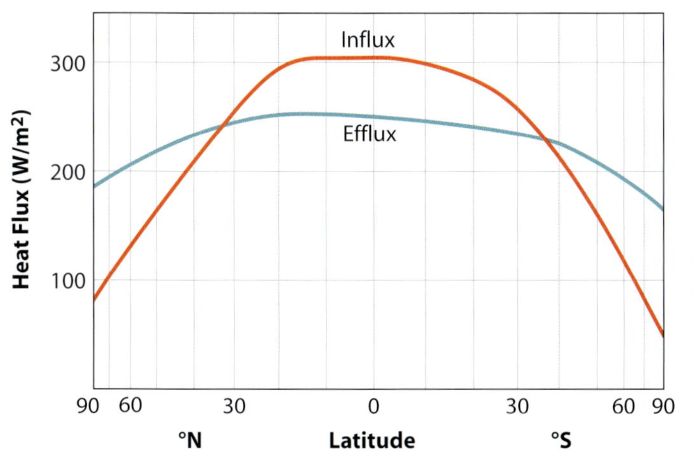

Figure 6.6. Yearly averaged heat influx and efflux vary with latitude. Near the equator, influx exceeds efflux; near the poles, efflux exceeds influx.

Influx vs. Efflux

You probably have encountered this explanation before. It is, however, only half the story, and the second half is likely to be less familiar. Consider heat efflux, the rate at which the earth radiates heat out into space. We noted above that the rate at which a body radiates heat depends on the temperature of the object. So, if we measure the temperature of the ground at different latitudes, we can estimate the instantaneous rate at which heat is radiated out at any particular time. Averaging this radiation over the year, we can then draw on figure 6.6 a green line showing the rate of heat loss, the heat efflux, again as a function of latitude.

Even a quick glance at figure 6.6 tells us that something strange is going on. For areas in the tropics, there is more heat arriving each year than there is heat leaving. In this case, by the logic we have used, we would expect the tropics to be getting continually hotter. At the same time, there is more heat leaving polar areas than arrives, and these areas should be getting continually colder.

Only at a latitude of 30° to 40° North or South does the yearly heat influx equal the yearly heat efflux. According to this graph, these latitudes are the only places on earth where the temperature should stay the same from one year to the next.

Again we have reached a conclusion that deviates from reality. Over the history of the earth, the tropics have not been getting hotter and hotter. Nor have Arctic and Antarctic areas been getting colder and colder. Instead, aside from a few relatively minor fluctuations, such as ice ages, the temperature at any place on earth has stayed about the same.

Global Warming

Given the current ballyhoo over global warming, this statement may seem naïve, if not just plain wrong. The Intergovernmental Panel on Climate Change has concluded (and at least some governments have admitted) that humankind's collective effect on the earth's atmosphere is leading to an increase in average temperature, an increase that is likely to amount to a few degrees Celsius over the next century. How, then, can I assert that temperature stays about the same?

The answer is that we are talking about processes with vastly different rates. On the one hand, yes, the earth is currently (and distressingly) heating up, but the rate is relatively slow, a few degrees per century as noted above. On the other hand, given the disparity between the heat influx and the heat efflux apparent in figure 6.6, we would expect equatorial areas to be heating up, and polar areas to be cooling down, at perhaps 10°C *per year*. This is clearly not happening.

Thus, for the purposes of the present argument, it is correct to say that the temperature of any spot on earth stays more or less the same from one year to the next, and that this fact is at odds with the unbalanced influx and efflux of light energy. This is not to deny the reality of global warming, but its relatively slow and slight effect can be ignored for the purposes of the present discussion. We will return to the subject in chapter 10.

The Transport of Heat

Back to the disparity between the imbalance of heat flux and the constancy of temperature. The tropics have a surplus of heat available to them, while the polar areas have too little. Perhaps we can reconcile the situation by somehow taking heat from the tropics and delivering it to the poles. In fact this could work out really nicely—as a quantitative matter, the excess heat influx at latitudes between 30° North and 40° South is almost exactly equal to the heat flux deficit of latitudes more poleward. All we need is a mechanism for transporting heat in the appropriate directions.

As you probably suspect, a mechanism is available. Actually, there are three major mechanisms that transport heat energy from the tropics toward the poles. Two of these—winds and the currents they drive in the surface ocean—we will deal with in detail in chapter 8. In the meantime, we explore the third mechanism: *thermohaline circulation.*

Thermohaline Circulation

Let's return to the pattern of heat influx and efflux shown in figure 6.6, and imagine what the effects would be on the ocean, beginning near the poles. Given their current temperature, the surface waters in the Arctic and Antarctic seas are, on average, radiating heat out into space faster than they are absorbing heat from the sun. We would thus expect this surface water to be getting colder through time. What happens if it does?

Brine Formation and the Deep Ocean

Let's take the process to an extreme: when the ocean's surface water reaches a temperature of –1.89°C, it begins to freeze. Recall from chapter 3 that the freezing of seawater is a complicated process. As water molecules try to form crystalline ice, they expel the dissolved salts. The resulting sea ice is nearly fresh, and as more and more seawater freezes, a continuous stream of very cold (–1.89°C or colder), very saline water oozes down from the underside of the ice. Because this oozing brine is both cold and saline, it is denser than the "regular" seawater around it, and being denser, the brine sinks. So, wherever sea ice is being actively formed, very cold, very saline seawater is produced, and this dense water flows downward under the pull of gravity.

Where does the dense brine go? Well, it continues downward until it hits something denser than itself. On the way down, the cold, saline brine produced at the surface mixes a bit with the water around it, getting warmer and less saline as it goes, but by the time it reaches the seafloor under the ice, it is still the coldest, saltiest water around. If the seafloor were everywhere flat, that would be the end of the story, but as we have seen (chapter 2), the seafloor has a rolling topography of basins and ridges, and the brine flows over this topography. For example, the Artic Ocean is contained in a group of deep basins separated by shallow sills from the North Atlantic and Pacific Oceans. As a result, after the cold, saline water produced by ice has filled up the deep arctic basins, it begins to slop over into the adjacent seas like water from an overflowing bathtub.

Well, it would slop over into the Pacific, except that the connection between the Arctic and Pacific Oceans, the Bering Strait, is currently so shallow—about 45 meters—and so narrow that it effectively acts like a dam. Thus, virtually all of the cold, dense water produced in the Artic Ocean flows instead into the Atlantic, emerging through gaps between Greenland and Iceland and between Iceland and Norway.

In the process of flowing out of the Arctic basin, the dense brine mixes with the local surface water. Although it has not yet had a chance to freeze, this surface water has nonetheless been substantially cooled by the net loss of heat characteristic of the polar regions, and it is cold and dense. This mixture of brine and surface water emerges into the Atlantic, where it takes on the title of *North Atlantic Deep Water.*

In the Antarctic, the story is identical in principle, but different in detail. Where Antarctic ice is formed on seawater, primarily in the Weddell and Ross Seas, cold, dense seawater is formed, and this brine is mixed with cold surface water and flows downhill, just as water does in the Arctic. In the case of Antarctic water, however, the flow is not impeded by any shallows. As a result, the dense, cold water produced in Antarctic seas flows out into all the major oceans (Pacific, Atlantic, and Indian), where it is known as the *Antarctic Bottom Water*.

North Atlantic Deep Water. Antarctic Bottom Water. The names are telling us something. It turns out that, on average, the water that flows north out of Antarctica is slightly denser than the water flowing south from the Arctic. As a result, when the two flows meet, the North Atlantic Deep Water flows over the Antarctic Bottom Water. A simplified version of the overall pattern of flow is shown in figure 6.7. The message of this figure is that no matter where you look in the major oceans, the water at the seafloor got there because it was cold and saline, having been produced by the cooling of seawater near the poles.

And that is the answer to the question of why the deep oceans are cold.

Oxygen in the Deep Ocean

Note that the flow of water from the poles to the deep oceans can also explain why water in the ocean depths has a high concentration of dissolved oxygen (figure 6.1F). As the North Atlantic Deep Water and Antarctic Bottom Water are formed, they are at the surface of the ocean, where they become saturated with oxygen. Furthermore, because this water is cold, the saturation oxygen concentration is initially high as the water sinks into the depths. (More gas can be dissolved in cold water than in warm water.) Unless this oxygen is somehow used up by bacteria or animals before it gets to the seafloor, it is still well oxygenated when it arrives. In fact, the rate at which oxygen is respired in these cold deep waters is quite slow, and their high oxygen content is therefore not surprising. In short, the deep oceans are well oxygenated because they are ventilated by water that originated at the surface near the poles.

Heat from the Tropics

To this point, we have answered the second of our major questions—why the deep oceans are cold—but we are still left to explain how this transport of water can act to transport heat away from the tropics.

In one respect, it is quite simple. If water is pumped away from the polar seas in the flow of the North Atlantic Deep Water and Antarctic Bottom Water, other water must flow into the polar seas to take its place. Where does this imported water come from? It would be difficult for it to come in as rain. After all, we are talking about the polar seas and conditions sufficient to freeze seawater. Any precipitation is likely to come in the form of snow, which will then accumulate on top of the ice. Some water enters polar oceans from rivers. But most of the water pumped out of the polar seas into the ocean depths is replaced by an inward flow of surface seawater originally from the tropics. Because this water has basked in the tropical sun, it is warm. Thus, as cold water flows away from the polar seas, warm water flows toward them, carrying heat energy along with it. It is this influx of heat that helps keep the poles from getting colder and colder.

The pattern of heat flux in this circulation is exactly what we require to begin to balance the heat budgets for both the polar areas and the tropics. The tropics have a surplus of heat, and some of it is transported away by the bulk movement of surface water toward the poles. The polar seas have a surplus of "cold," if you will, and some of this is transported out by frigid, saline water moving toward the equator. In this fashion, the unusual manner in which seawater freezes leads to a large-scale mechanism for transport of solar heat energy.

Upwelling and Tidal Mixing

There is one final connection we must make before this picture is complete. We have suggested that surface water moves toward the poles from the tropics. What replaces the tropical water? The obvious answer is that just as the cold, dense water produced at the poles is replaced by water from the tropics, the warm tropical surface water is replaced by the cold, dense water that was pumped out at the poles. The two transports in opposite directions can be connected to form a cycle (figure 6.7). This is indeed what happens, but the mechanism by which the cycle is completed is somewhat surprising.

As the picture stands now, cold, dense water is delivered to the seafloor in the tropics, and warm water moves poleward at the surface. Somehow, we need to raise the deep dense water up to replace the poleward flow. In one sense, there really isn't a problem. Having been produced at the surface in polar areas, dense water has gravitational potential energy that can be expended as the water descends, and this energy can be used to force deep water upward in the tropics. (See the appendix to this chapter for a brief introduction to gravitational potential energy.) This effect undoubtedly accounts for some of the upward flow that completes the cycle. It turns out, however, that two other mechanisms are more important.

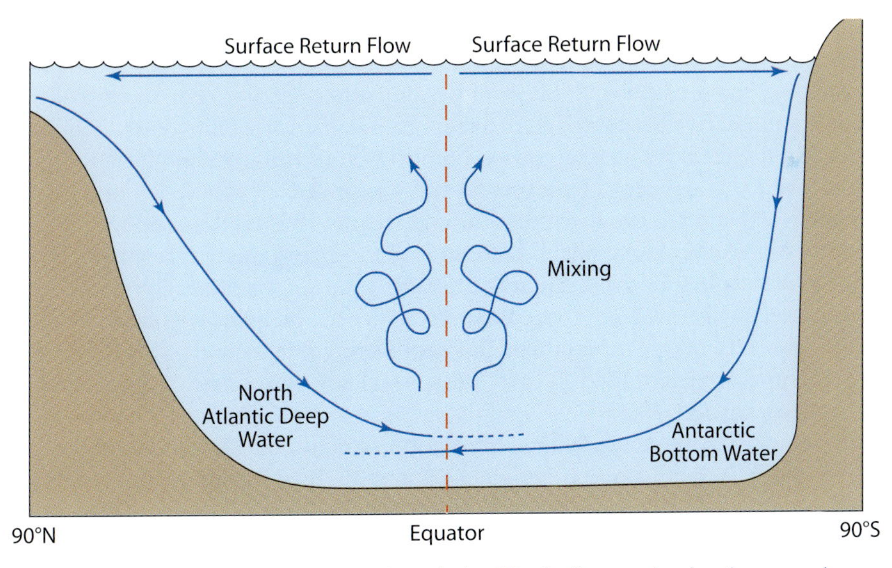

Figure 6.7. A north-south cross section through the Atlantic Ocean, showing the general pattern of thermohaline circulation. For a more detailed picture, see figure 9.9.

First, deep water is brought to the surface in specific areas of the ocean subjected to *upwelling*. Upwelling is most common along the eastern boundaries of the oceans, in the Southern Ocean, and along the equator. The mechanism of upwelling, which accounts for this odd spatial distribution, is the subject of chapter 8, and it is best if we postpone our discussion of the process until then.

The second mechanism responsible for the upward return of deep water involves the tides. As ocean tides ebb and flow, the friction of water's motion across the seabed creates turbulent mixing, which stirs the sea. Most of this mixing occurs in shallow parts of the ocean (on continental shelves, for instance), but there is also mixing associated with the interaction of tides with mid-ocean ridges and sea mounts. As the tides stir the ocean, deep, cold water is gradually mixed back to the surface, helping to complete the cycle of circulation.

In summary, cold water sinks near the poles and moves equatorward through the deep ocean. Tropical water is then lifted, mixed, or welled up to the surface, where it is heated and moves toward the poles. This grand cycle is known as *thermohaline circulation*. *Thermo* expresses the fact that the circulation involves the transport of heat; *haline* that the circulation also involves the salinity of seawater (saline, haline, same thing). In casual reference to the cyclical manner in which it transports both water and heat, the thermohaline circulation is sometimes referred to as the "ocean conveyor."

Although figure 6.7 accurately portrays the fundamental concepts of the thermohaline circulation, the actual path taken by water in a thermohaline cycle is somewhat more convoluted. These additional details will be dealt with in chapter 9.

Detecting the Thermohaline Circulation

How do we know that the thermohaline circulation is really there, and how fast does it move? Some of the most direct evidence comes as a serendipitous result of the atmospheric nuclear tests that the former Soviet Union performed in the 1950s and '60s. When a thermonuclear device is exploded above ground, large amounts of exotic, radioactive materials are injected into the atmosphere. Some of these decay rapidly, but others persist. For example, tritium, a radioactive isotope of hydrogen, has a half-life of 12.3 years. Much of the tritium produced in the Soviet bomb tests eventually dissolved in rain, which then was delivered by rivers to the Arctic Ocean, where it mixed with seawater and entered the thermohaline circulation. By sampling the deep water flowing out of the Arctic Sea, oceanographers tracked the bomb-spawned tritium as it traveled south. In the first 30 years after the bomb tests, the tritium-marked deep water traveled from about 60° N to 45° N, a rate of approximately 100 kilometers per year, or about 0.3 millimeters per second. More recently, the thermohaline circulation has been tracked using naturally occurring radioactive carbon, ^{14}C.

In summary, the thermohaline circulation is real, but the water motion involved is really quite slow. If we were to start with an ocean of uniform temperature, it would take thousands of years for the thermohaline circulation to fill the deep oceans with cold water. On the other hand, because the thermohaline circulation has in fact been flowing more or less continuously for thousands of years, we now have a deep ocean that is cold.

Thermohaline Circulation and the Thermocline

The circulatory nature of the thermohaline circulation raises another question. As cold water rises up to complete the cycle of flow, how does this vertical motion affect the two-layered ocean? If cold water is continuously being stirred up to replace warm surface water, wouldn't that erase the thermocline? The answer is both "yes" and "no." Yes, in the rare areas where water wells up energetically, the thermocline can be lifted or even eradicated by the upward transport of cold water. But in the absence of upwelling—that is, in most of the ocean—the upward mixing of water in the thermohaline circulation is far too slow to affect the thermocline appreciably.

A Brief Review

Let's review what we know at this point. The surface of the ocean is warm because it is heated by the sun. The warm water at the surface has a lower density than the cold, saline water below, and therefore the surface layer is stable, even in the presence of moderate stirring from wind. The waters of the deep ocean are rich in nutrients, but because of the stable separation of the top and bottom layers, these nutrients cannot be mixed up to the surface. That leads to the problem noted at the beginning of this chapter. Although it is light enough in the top layer to power photosynthesis, the lack of nutrients prevents plants from growing. In the deep layer, nutrients are not a problem, but the lack of light prevents effective photosynthesis.

Biological Pumps

Given that lack of nutrients in the top layer of the ocean is the central dilemma faced by oceanic life, it behooves us to understand why the surface concentration of nutrients is low. Let's begin with a hypothetical ocean that has the same two layers we have been dealing with, but two important differences: (1) it has no phytoplankton and (2) there is a high concentration of nutrients in the surface layer. In other words, the top layer of this hypothetical ocean is warm and light, and has a relatively low salinity, a high concentration of oxygen, and (unlike the real ocean) a high concentration of dissolved nutrients. In the absence of phytoplankton, the high concentration of nutrients would persist indefinitely because, when nutrients are dissolved in seawater, they do not sink.

Uptake and Sinking

Now let's introduce into the top layer of our hypothetical ocean some phytoplankton and the consumers that eat them. Three things happen. First, having plenty of light and nutrients available, the phytoplankton grow like mad. Second, as they grow, the phytoplankton take nutrients out of the water, concentrating them in cells where the nutrients are used to build new organic material.

And finally, because they are denser than seawater and too large to be dissolved, the phytoplankton sink.

Before they sink out of the top layer, some of these phytoplankton are eaten and incorporated into consumers' bodies, but ingestion only delays the inevitable. Much of the food animals eat is subsequently evacuated as fecal pellets, which, being large, rapidly sink. And eventually each animal dies, at which point it too begins to sink

Although the water below the thermocline is denser than the water above, it is seldom as dense as a phytoplankton cell, fecal particle, or dead animal. Thus, when a sinking object reaches the interface between the upper and lower layers of the ocean, it is likely to keep on sinking. In other words, although the difference in density between the surface and deep layers forms an effective barrier to the mixing of water between layers, it forms at best a highly leaky barrier to sinking phytoplankton, fecal pellets, and animals.

Decomposition

What happens to these organisms and fecal matter after they have sunk into the deep layer of the ocean? As phytoplankton sink into the lower layer, light intensity diminishes and they die. The dead animals, dead phytoplankton, and fecal pellets are then digested by bacteria, releasing nutrients back into the water column. For example, as cells rot, ammonia is released, which is then converted to nitrate (see chapter 4). The problem is that by the time they are back in solution, these nutrients are below the thermocline, and there is then no easy way for them to get back up into the surface layer.

The net result of this cycle of grow-sink-die is that plants and animals act as part of a *biological pump* that transports nutrients out of the ocean's surface layer, down into the ocean's interior. This is the basic mechanism that accounts for the low concentration of nutrients in the surface layer.

Nutrients are not the only molecules pumped from surface waters. As we have seen, when phytoplankton grow, they absorb carbon compounds from the water around them. Typically, carbon is taken up in the form of bicarbonate ions (HCO_3^-), which are then converted to CO_2 and used in photosynthesis. When the phytoplankton cell dies and sinks, it takes with it the carbon absorbed from seawater. If the phytoplankton cell is eaten, its carbon ends up in animal tissue or fecal pellets, both of which eventually sink. Thus, carbon is pumped out of the surface layer in the same fashion as nutrients. In this chapter, we examine the biological pump of carbon as it relates to primary production, and we will return to this pump in chapter 10 when we explore global warming.

Pumping of both nutrients and carbon can be augmented by the daily vertical migrations undertaken by some organisms. At night, copepods swim to shallow depths where they banquet on phytoplankton. However, by the time that food has been digested and fecal pellets are formed, the animals have descended hundreds of meters, well below the thermocline, where they hide during the day. In this fashion, the carbon and nitrogen in copepod fecal pellets is actively pumped out of the ocean's surface layer. Other vertical migrators, such as salps and fish, contribute to this augmented pumping.

Recycling

The full story of the biological pump is a bit more complicated, of course. For example, as we discovered in chapter 4, the rate at which an object sinks depends on its size. Small cells (bacteria and cyanobacteria, for instance) sink more slowly than large cells (such as diatoms and dinoflagellates), and these small cells often die and decompose before they sink out of the surface layer. In this case, the nutrients and carbon contained in these cells are released back into the surface waters, where they can again be taken up by phytoplankton and used in growth. Larger cells, which would otherwise sink out of the surface layer, sometimes die from viral infections and release their contents, and animals excrete ammonia and urea, returning these nutrients to the water. In sum, there are several mechanisms that can short-circuit the biological pump by locally recycling nutrients and carbon in the surface layer. As we will see, as much as 90% of carbon fixed in the surface ocean is recycled rather than being exported by the biological pump.

This recycling necessitates a second consideration of how we measure primary production. As we progress with our examination of the two-layered ocean, we need to be careful to distinguish between *recycled* and *new* production.[1]

To define these terms, consider a hapless watchmaker. In his shop, he has one watch, which he continuously takes apart and reassembles. If we were to observe him for a day, we could keep track of the number of times he takes the parts available and builds them into a functioning watch, and this number is his productivity: X watches per day. But this assessment of productivity is somewhat misleading. No matter how many times the watchmaker assembles it, he still has only one watch. In this case, all his productivity is tied to recycling the limited set of parts at hand, and if he wants to stay productive, he can't sell the one watch he has. This is *recycled production.*

In contrast, we could provide the watchmaker with a new set of watch parts at the beginning of the day and again every time he finishes assembling a watch. At the end of the day, he has again assembled X watches, but in this case all X watches are constructed from new parts. He could sell those X watches and still have the same watch he started with. In this case, his production is the same (X watches), but his *new production*, the production he can sell, is much higher. Note that an outside observer would not have to see the watchmaker in action to tally up his new production. The observer could measure new production simply by keeping track of the number of watches offered for sale.

Phytoplankton are analogous to watches: they can be constructed from either recycled or new materials. And we need to keep track of their materials in the same way we would for watches: phytoplankton production can be recycled or new.

How can recycled production in the ocean be distinguished from new production? To understand the practical measurement of new production, we return to our watchmaker and attend to some details. In this scenario, let's assume that the watchmaker has in his shop an abundance of gears, hands, and cases, all the ingredients needed to make watches, with one exception: the only mainspring he has is the beat-up bronze mainspring in his one intact watch. He

[1] Continuing the practice from chapter 4, I use the term "production," as shorthand for "net primary production."

can take that watch apart and build a new one, perhaps using new gears and hands, but every time he builds a watch he must use the bronze mainspring. In other words, mainsprings are his limiting resource; without new mainsprings, he can only recycle his production.

Now, let us supply the watchmaker with new parts: an abundance of gears, hands, cases, and, one by one, new mainsprings. But these new mainsprings are made of steel, rather than bronze, and can thereby be distinguished from the mainspring the watchmaker was previously recycling. For each new mainspring we give the watchmaker, one new watch can be built and sold. While waiting for new mainsprings to arrive, the watchmaker can still recycle the parts he has, so his overall production—the sum of both new and recycled production—may be higher than his new production.

In this scenario, an outside observer could keep track of the watchmaker's new production in two ways. First, as before, he could count the number of watches sold. One of those watches might contain the old, bronze mainspring, but that shouldn't matter. As long as the watchmaker sells a watch only after a new mainspring has been delivered, the number of watches that leaves the shop is still equal to new production.

Alternatively, the observer could count the number of steel mainsprings delivered. For each mainspring, the watchmaker produces one watch for sale, so the number of mainsprings delivered is likewise a measure of new production.

Again, phytoplankton are analogous to watches. As we noted in chapter 4, primary production in the surface ocean is typically limited by the availability of nitrogenous nutrients: ammonia and nitrate. Ammonia is roughly analogous to the bronze mainspring in our hypothetical watches; it is continuously recycled in the surface ocean as organisms die and excrete wastes. Yes, newly fixed ammonia is produced by cyanobacteria, but the rate at which ammonia is delivered from this source is small compared to the rate at which ammonia is recycled. Thus, production based on ammonia uptake is largely recycled production. On the other hand, in the open ocean, nitrate is formed by bacteria only in the dark ocean interior, and it is therefore not available to be recycled in the surface layer. Delivery of nitrate to the surface ocean is therefore analogous to the delivery of steel mainsprings. Carbon fixed in association with uptake of nitrate is new production.

This analogy suggests two practical methods for measuring new production in the sea. First, one could measure the rate at which carbon is exported to the ocean interior. This method is analogous to counting the number of watches sold, and it is usually accomplished using sediment traps moored below the thermocline. Each trap is essentially a bottle with its open mouth directed upward to catch the rain of particles sinking down from the surface layer. Knowing the area of the bottle's mouth, the time for which it was deployed, and the weight of sediment captured, one can estimate the rate at which carbon is exported.

Alternatively, in analogy to measuring the delivery of steel mainsprings, one could measure the rate at which nitrate is delivered to the surface layer. This is done indirectly. First, one measures the concentration of nitrate in surface water. One then injects a small amount of labeled nitrate[2] into a sample of surface

[2] Nitrate is typically labeled using a stable isotope of nitrogen, ^{15}N. This isotope contains one more neutron than is present in typical nitrogen, ^{14}N, and it is very scarce in nature. ^{15}N is not radioactive, but it can be differentiated from ^{14}N using a mass spectrometer, a device that measures the mass of individual molecules.

water such that it forms about 1% of the total nitrate present. Phytoplankton in the sample are allowed to photosynthesize, and one then measures the fraction of labeled nitrogen taken up relative to all other forms of nitrogen. From these measurements, one can calculate the fraction of overall production that is new.

For example, let's assume that labeled nitrate forms 1% of the total nitrate available to the experimental phytoplankton. In other words, for every labeled nitrogen molecule taken up, we can deduce that a total of 100 nitrate molecules were absorbed. Let's further assume that labeled nitrogen forms 0.1% of all nitrogen taken up during the experiment. (The actual value of this fraction is what one would measure in a real-world experiment.) This implies that $100 \times 0.1\% = 10\%$ of nitrogen absorbed was nitrate, and consequently, 10% of total production by phytoplankton was new.

Neither of these measurements of new production is easy to make, and estimates of new productivity are much more scarce than estimates of net productivity. From the few measurements that have been made, it appears that the fraction of total production that is new depends on total net productivity (figure 6.8). In areas of low net productivity—less than 50 grams of carbon fixed per square meter per year—the dominant phytoplankton are small cyanobacteria that sink very slowly. As a result, 90% or more of their production is recycled, and only about 10% is new production exported into the ocean interior. In contrast, areas of high net productivity—more than 150 grams of carbon fixed per square meter per year—are commonly dominated by large phytoplankton that sink rapidly, and as much as 50% of total production is new.

The presence of recycled production can be misleading. As with the watchmaker repeatedly assembling and disassembling the same watch, phytoplankton in the surface layer can appear busy and productive as they recycle their ingredients. But this recycled productivity is tenuous. Because a recycled cell can be produced only when an old cell dies, a recycled population of primary producers cannot increase its number. Furthermore, because recycling is never 100% efficient, nutrients are lost to the deep sea as the biological pump takes its toll, and recycled productivity must inevitably decrease. These detrimental effects can only be reversed through influx of new nutrients.

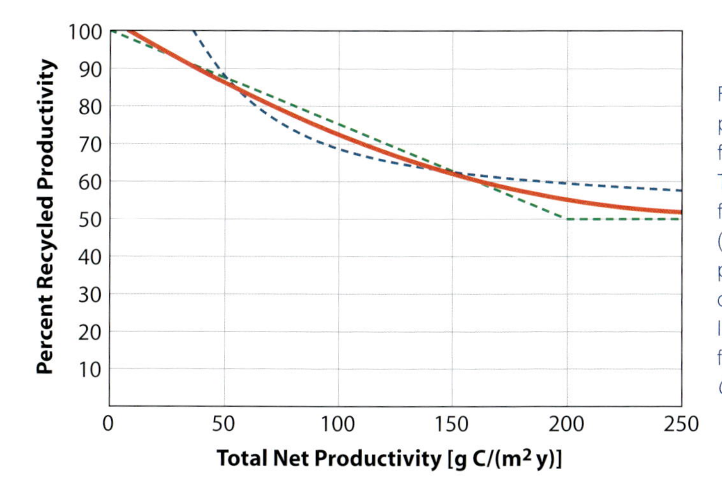

Figure 6.8. The fraction of overall production that is recycled decreases as a function of increasing overall productivity. The red trend line is fit to data (green line) from R. W. Eppley and B. C. Peterson (1979), Particulate organic matter flux and planktonic new production in the deep ocean (*Nature* 282:677–680), and (blue line) R. L. Iverson (1990), Control of marine fish production (*Limnology and Oceanography* 35(7):1593–1604).

Summary

In summary, the growth and reproduction of living cells takes nutrients out of solution in the ocean's surface layer, and as these cells sink, die, and decompose, these nutrients are transported below the thermocline by the biological pump. Recycling in the surface layer can slow the rate of the biological pump, but can never bring it to a complete halt. Because the two-layered arrangement of the ocean is stable, it is difficult for nutrients, once sunk, to be transported back into the surface layer. As a result, most surface waters are nutrient-poor (figure 6.1). We return yet again to the fact that the two-layered ocean is a lousy life-support system.

Meanwhile, Back at the Oxygen Graph . . .

We now have an explanation for nearly all the facts depicted in figure 6.1. The one exception concerns the graph of oxygen concentration (figure 6.1F). We noted previously that the oxygen content of the ocean is high at the surface, where water is in intimate contact with the air, and high in the deep ocean, which is ventilated by the thermohaline circulation. Why, however, should the oxygen concentration be low at middle depths?

Having explored the biological pump, an answer is now readily at hand. We have seen that there can be a steady rain of fecal pellets, marine snow, phytoplankton, and dead animals out of the surface layer, and that this material rots as it sinks. Now, the bacteria responsible for decomposition need oxygen to perform their work: they respire. Therefore, as decaying matter sinks into the ocean interior, oxygen is taken up by bacteria from the water. As a result, the oxygen content of water gets lower and lower as you travel down below the thermocline. At some depth, however, all the organic matter that can rot has rotted, and subsequently no more oxygen is absorbed. Below that depth, the oxygen concentration does not get any lower, and in fact it gradually increases with increasing depth as one moves into the depths of the ocean where oxygen is delivered by thermohaline circulation.

The net result of all this is an oxygen minimum zone, typically centered at a depth of 500–1000 meters. Oxygen concentrations in this zone can be surprisingly low, down to only 2–4% of the concentration in surface waters. This level of oxygen is too low for many animals to survive, and the few organisms that inhabit these depths have evolved strategies for coping with the lack of this important gas. Recall, for instance, that denitrifying bacteria thrive in the absence of oxygen by using nitrate as a substitute. The presence of the oxygen minimum zone thus contributes to the flux of nitrogen gas from the ocean back into the atmosphere.

Conclusion

We have now returned to the point at which we started this chapter, and as a summary we revisit our initial characterization of the ocean (figure 6.9). A typical cross section of the ocean has two layers. The upper layer is light, warm, well oxygenated, and nutrient-poor, and the lack of nutrients makes it a difficult place

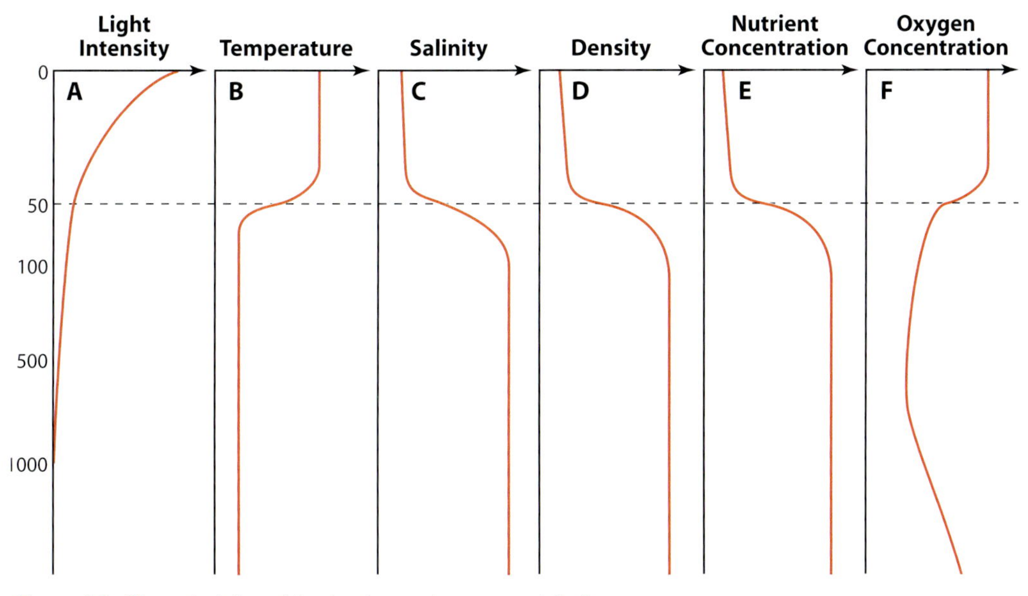

Figure 6.9. Characteristics of the two-layered ocean, revisited.

for phytoplankton to live. The deep layer has plenty of nutrients, but no light, so it too is a difficult place for phytoplankton to live: without phytoplankton to harness solar energy, animals cannot survive. Thus, because of the separation of light and nutrients, the two-layered ocean is an ineffective life-support system.

We also have a mechanistic explanation for each of the characteristics shown in figure 6.9. In other words, we know, at least to a first approximation, why the ocean is this way. What we do not know is how life solves this dilemma. In some special circumstances, as we will see, there are indeed solutions. None of them is ideal, and several of the mechanisms have the tendency to break down from time to time. But they are all we have, and they are what keep life in the ocean going, so we will look at them in detail.

These details rely on some simple but nonintuitive physics, however. So, before we can explore the solutions to the dilemma of the two-layered ocean in chapters 8 and 9, we must digress in chapter 7 to learn about the Coriolis acceleration.

Further Reading

Mann, K., and J. R. Lazier (2006). *Dynamics of Marine Ecosystems* (3rd edition). Blackwell Publishing, Malden, MA.

Sarmiento, J. L., and N. Gruber (2006). *Ocean Biogeochemical Dynamics*. Princeton University Press, Princeton, NJ.

Pond, S., and G. L. Pickard (1993). *Introductory Dynamical Oceanography* (3rd edition). Elsevier Butterworth-Heineman, Oxford.

Appendix

Calculating the Energy of Thermocline Displacement

Our goal here is to calculate the energy required to increase the depth of the thermocline by mixing water up from below. Before we can do that, however, we need to take a few preliminary steps. First, we explore the concept of gravitational potential energy.

Consider the apparatus shown in figure 6.A1, a bucket of water attached to a rope that runs through a pulley. The bucket initially sits on the ground, but if we apply sufficient force with the rope, we can lift it upward. The force required to raise the bucket is its *weight*, the product of its mass m and the acceleration of gravity g (9.81 meters per second squared):

$$\text{weight} = mg \qquad \text{(Eq. 6.1)}$$

If we lift the bucket to a height z above the ground, the energy we expend is equal to the force we exert times the distance over which that force moves the bucket:

$$\text{energy} = \text{weight} \times \text{distance} = mgz \qquad \text{(Eq. 6.2)}$$

The energy we expend in raising the bucket is stored as gravitational potential energy. In other words, because of its position above the ground, the bucket has the potential to do work. For example, we could couple the pulley to an electrical generator. If we then released the rope, the bucket would fall, turning the generator and providing electrical energy.

Now consider the apparatus shown in figure 6.A2, a tank of water resting on the ground. Because the water has weight and is held above ground by the tank,

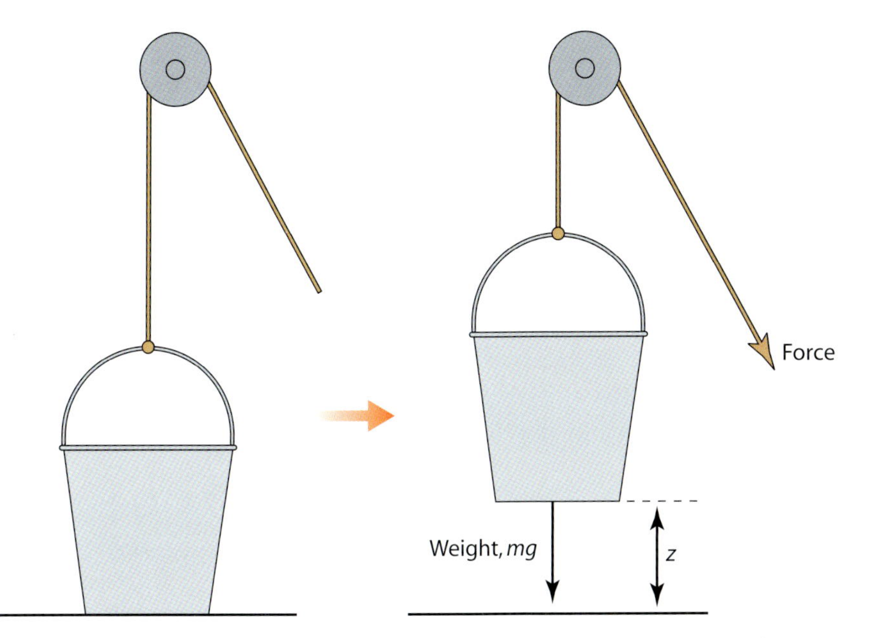

Figure 6.A1. A bucket suspended from a rope provides an example of gravitational potential energy.

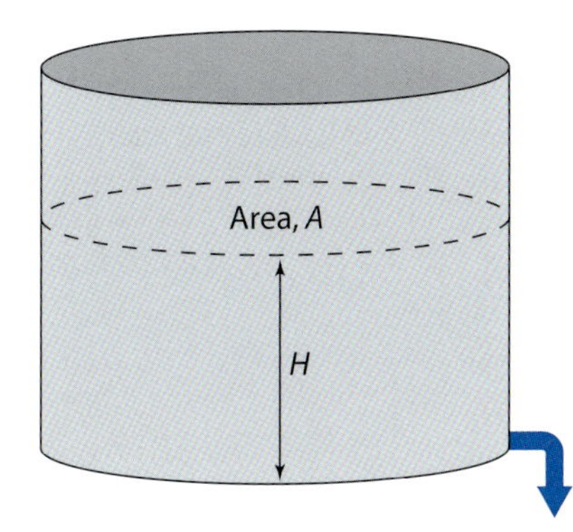

Figure 6.A2 Variables used to calculate the potential energy of water in a cylindrical tank.

it has gravitational potential energy. For instance, if we installed a drain in the bottom of the tank, we could allow the water to flow downhill past a water wheel, and we could use that wheel to turn an electrical generator.

How much potential energy does the water in the tank have? Because water at the top of the tank is far from the ground, it can fall a substantial distance and therefore has substantial potential energy. In contrast, water at the bottom of the tank is already at ground level, and because it cannot fall any further, it has no potential energy at all. By averaging over the whole depth of the water column, we find that the overall potential energy of water in the tank is equal to its total weight times the distance from the bottom of the tank to the midpoint of the water column, that is, to its center of gravity.[3] In other words, if the total

[3] If the tank had a cross section that varied with distance from the bottom, the center of gravity of the water column might be located elsewhere. But here we need concern ourselves only with the simple situation of constant cross section.

depth of the water column is H, the gravitational potential energy of water in the tank can be expressed as

$$\text{potential energy} = \text{weight} \times \text{distance to midpoint} = mg\frac{H}{2} \qquad \text{(Eq. 6.3)}$$

We can take this analysis a step further by calculating m, the mass of water. If the tank has constant cross-sectional area A, the volume of the water column is AH. Multiplying this volume by the density of water ρ, its mass per volume, gives us the water's mass:

$$m = \rho AH \qquad \text{(Eq. 6.4)}$$

Thus, the potential energy of water in the tank can be expressed as

$$\text{potential energy} = mg\frac{H}{2} = (\rho AH)g\frac{H}{2} = \frac{\rho gAH^2}{2} \qquad \text{(Eq. 6.5)}$$

The taller the column of water and the greater its density, the greater its potential energy.

With this background, we now can tackle our original question: how much energy is required to increase the depth of the thermocline by mixing water up from below? Consider the system shown in figure 6.A3, a simplified version of the ocean. In a tank with cross-sectional area A, a layer of low-density fluid, representing the ocean's mixed layer, sits on top of a layer of higher-density fluid, representing the ocean interior. The surface layer has thickness a and density ρ_1, and the deep layer has thickness b and density ρ_2 ($\rho_2 > \rho_1$). What is the gravitational potential energy of this "ocean"?

We approach this question in two parts. First, we can calculate the potential energy of the lower layer in exactly the same fashion we did for our previous example of water in a tank:

$$\text{initial mass of lower layer} = m_2 = \rho_2 Ab \qquad \text{(Eq. 6.6)}$$

$$\text{initial potential energy of lower layer} = m_2 g\frac{b}{2} = \rho_2 Abg\frac{b}{2} = \rho_2 gA\frac{b^2}{2} \qquad \text{(Eq. 6.7)}$$

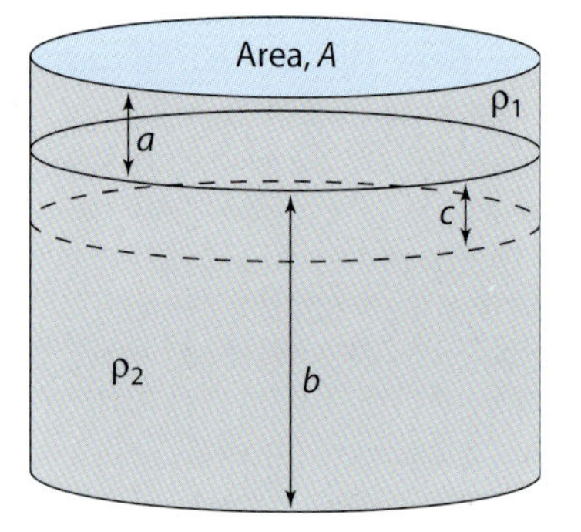

Figure 6.A3. Variables used to calculate the energy required to depress the thermocline by mixing.

similar calculation can be carried out for the top layer; however, in this case we note that the midpoint of the top layer is a distance $a/2$ above the top of the bottom layer, and therefore a distance $b + a/2$ above the ocean floor:

$$\text{initial mass of upper layer} = m_1 = \rho_1 Aa \qquad \text{(Eq. 6.8)}$$

$$\text{initial potential energy of upper layer} =$$

$$m_1 g\left(b + \frac{a}{2}\right) = \rho_1 Aag\left(b + \frac{a}{2}\right) = \rho_1 gAab + \rho_1 gA\frac{a^2}{2} \qquad \text{(Eq. 6.9)}$$

The overall potential energy of the ocean is the sum of that for the upper and lower layers:

$$\text{initial overall energy} = \rho_1 gAab + \rho_1 gA\frac{a^2}{2} + \rho_2 gA\frac{b^2}{2} \qquad \text{(Eq. 6.10)}$$

Now let's suppose that we stir the ocean's surface, and as a result, high-density water from a thin stratum of the lower layer is mixed with the water of the upper layer. The thickness of the stratum mixed upward is c, as shown in figure 6.A3. After this mixing, the surface mixed layer is thicker than before (by c), and the ocean has a new potential energy. In effect, some of the dense water from the lower layer has been lifted upward, and that requires an increase in the potential energy of the system.

The new potential energy is calculated in the same fashion as before. The lower layer, still with density ρ_2, is now $b–c$ thick, and its new midpoint is $(b–c)/2$ above the seafloor. Thus:

$$\text{final mass of lower layer} = m_2 = \rho_2 A(b-c) \qquad \text{(Eq. 6.11)}$$

$$\text{final potential energy of lower layer} =$$

$$\rho_2 gA(b-c)\frac{b-c}{2} = \rho_2 gA\left(\frac{b^2}{2} - bc + \frac{c^2}{2}\right) \qquad \text{(Eq. 6.12)}$$

The upper layer now has thickness $a+c$, and its midpoint is located at $(a+c)/2$ above the new top of the lower layer (which lies at $b–c$). Thus,

$$\text{new location of upper center of gravity} =$$

$$(b-c) + \frac{a+c}{2} = b + \frac{a}{2} - \frac{c}{2} \qquad \text{(Eq. 6.13)}$$

We now need to calculate the new mass of the upper layer, taking into account the addition of mass from stratum c. This new mass is equal to the old mass ($\rho_1 Aa$) plus the mass of the newly mixed stratum ($\rho_2 Ac$):

$$\text{new upper layer mass} = \rho_1 Aa + \rho_2 Ac \qquad \text{(Eq. 6.14)}$$

Using the new midpoint and mass for the upper layer, we can calculate its potential energy:

$$\text{final potential energy of upper layer} =$$
$$m_{new}g\left(b+\frac{a}{2}-\frac{c}{2}\right)=(\rho_1 Aa+\rho_2 Ac)g\left(b+\frac{a}{2}-\frac{c}{2}\right) \qquad \text{(Eq. 6.15)}$$

We now add the post-mixing potential energies of the upper and lower layers to arrive at the new total potential energy:

$$\text{final total potential energy} =$$
$$(\rho_1 Aa+\rho_2 Ac)g\left(b+\frac{a}{2}-\frac{c}{2}\right)+\rho_2 gA\left(\frac{b^2}{2}-bc+\frac{c^2}{2}\right) \qquad \text{(Eq. 6.16)}$$

The difference between this post-mixing total potential energy (equation 6.16) and the pre-mixing potential energy (equation 6.10) is the energy required to mix stratum c into the upper layer. When the dust settles from the algebra, we find that

$$\text{change in total potential energy} = \frac{gacA(\rho_2-\rho_1)}{2} \qquad \text{(Eq. 6.17)}$$

The greater the difference in density between the two layers, the more energy is required to mix them. Similarly, the thicker the stratum c we mix into the upper layer, the more energy is required. Somewhat surprisingly, the mixing energy also depends on the initial thickness of the upper layer, a. This effect is due to our assumption that mass taken from stratum c is mixed throughout the upper layer. The thicker the upper layer, the farther some of this mass must be lifted against gravity, and consequently, the larger the potential energy.

One last step brings us to our goal. Dividing both sides of equation 6.17 by A gives us a measure of mixing energy per surface area of ocean:

$$\frac{\text{change in total potential energy}}{\text{area}} = \frac{gac(\rho_2-\rho_1)}{2} \qquad \text{(Eq. 6.18)}$$

To give this result some tangibility, let's assume that the upper layer is initially 50 meters thick ($a = 50$ meters), and we mix up a stratum 1 meter thick from below the thermocline ($c = 1$ meter). If the initial difference in density between upper and lower layers is 2 kilograms per cubic meter, 491 joules of energy are required to accomplish this mixing for each square meter of ocean.

The Coriolis Effect and Its Consequences

The last chapter ended with a grim picture of a typical ocean: where there is sufficient light to support photosynthesis, there aren't enough nutrients. Ocean physics causes this problem, and as we will see in chapter 9, ocean physics provides at least a partial solution.

To understand the way in which life obtains its necessary nutrients, we must first work our way past a classic hurdle in the study of oceanography: the Coriolis acceleration. This intriguing phenomenon is an important and unavoidable consequence of the earth's rotation. Yet for slow-moving creatures such as ourselves who interact with the world on a small scale, manifestation of the Coriolis acceleration is subtle, and as a result, the concept is almost entirely nonintuitive.

There are two strategies to dealing with the Coriolis acceleration. The first is to treat the effect as three magic rules:

(1) *Any object moving horizontally on earth's surface has its trajectory deflected: to the right in the northern hemisphere, to the left in the southern hemisphere.*
(2) *The faster an object moves, the greater its tendency to deflect.*
(3) *The tendency to deflect is greatest at the poles and decreases to zero at the equator.*

These rules are simple and easy to remember, and when applied in an oceanographic context, they give correct answers. If you find these rules satisfying, feel free to skip ahead to the section on geostrophic flows later in this chapter.

If, however, you find yourself asking, "But *why* is motion deflected right or left? And *why* does it matter how fast you move?," a second strategy is available—to understand the mechanism behind these rules, even if the mechanism is nonintuitive. Explanation of the Coriolis acceleration requires us to deal with physics in a bit more depth than we have in previous chapters, but bear with me—the result is worth the effort. We begin with Sir Isaac Newton.

Newton's Laws of Motion

In 1687, Newton published his *Principia Mathematica*, a momentous volume in which he proposed a theory of how forces interact with matter. In developing this theory, he established the basis for much of modern physics and invented calculus along the way. At the core of Newton's mechanics are three basic laws.

The First Law. *Unless an object feels an unbalanced force, its state of motion does not change.* We need to consider two aspects of this law. First, what matters here is not the total force on an object, but the imbalance of force. Imagine two groups of people pulling against each other in a tug of war. A large tensile force is exerted on the rope connecting the groups, but unless one group pulls harder than the other, thereby throwing the force out of balance, the rope does not move. An unbalanced force is often referred to as a *net* force.

Next, we note that motion involves a combination of speed and direction. For instance, we could describe the state of motion of an automobile as moving east at 100 kilometers per hour. If the car speeds up or slows down, its state of motion changes. If the car maintains the same speed but changes its direction, again its state of motion changes.

Thus, Newton's first law tells us that, any time we see a change in the speed or direction of an object, or both, we know an unbalanced force is acting. For present purposes, we are more concerned about changes in direction than we are about changes in speed. A simple way to think about this is to apply a "curve criterion": if we find an object moving along a curved path, which means it is changing direction, we know that a net force must be acting, regardless of whether its speed is constant or not.

The Second Law. *The acceleration of an object, the rate of change of its state of motion, is directly proportional to the net force applied and is inversely proportional to the object's mass.* In other words, acceleration equals force per mass. For an object of given mass, the more net force we impose, the faster the object's speed, direction, or both will change. Conversely, for a given imposed force, the more massive the object, the slower its state of motion will change.

The Third Law. *If object* A *exerts a force on object* B, *object* B *exerts an equal and opposite force on object* A. For example, as you stand on the floor, your weight imposes a force acting down on the floor boards. At the same time, however, the floor exerts a force pushing up on the bottom of your feet. The two forces are equal in magnitude, but opposite in direction.

Circular Motion

Newton's three laws apply to all motions and all forces. For present purposes, however, we concern ourselves only with the force acting on objects as they move in a circle, a motion we choose because of earth's rotation. As you sit reading this book, you are likely to be more or less stationary relative to the ground below you. But in reality, both you and the ground are moving. Because the earth spins on its axis once per day, every object fixed on earth's surface is

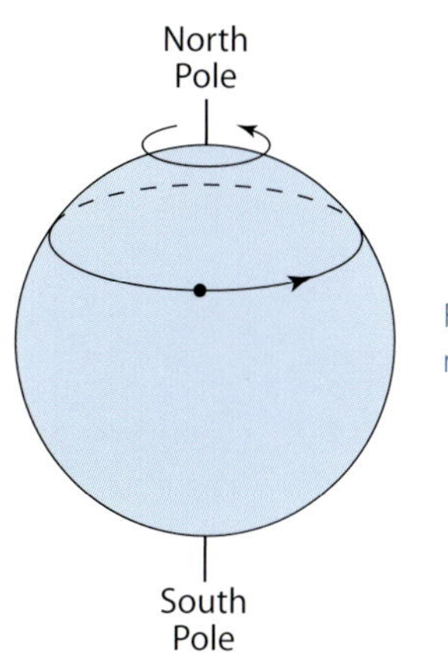

Figure 7.1. Objects attached to the earth move in circles as the world turns.

actually in constant motion relative to the rest of the universe, and that motion traces out a circle (figure 7.1). To avoid confusion, I will use the term "stationary" for objects that are fixed with respect to earth's surface, even though they move with respect to the universe.

Since stationary objects on earth actually move in circles, we can readily apply the curve criterion: because these objects—you, me, the water in the ocean—are constantly changing their direction of motion, they must be acted upon by a net force. The characteristics of this force are most easily understood if, rather than thinking of objects moving as the earth rotates, we first concentrate on a phenomenon of smaller scale: a pail of water attached to a rope.

Centripetal Force

Imagine yourself swinging this pail rapidly around your head. As you twirl the pail (an object with mass), it moves with a constant speed, but travels in a circle. As before, the curve criterion tells us that a net force is continuously required to nudge the pail of water into this circular path. In this case, the source of the force is obvious: it is the tug of the rope that forces the pail to move in a circle.

Now, as the pail swings, one end of the rope travels with it while the other end remains at the circle's center (figure 7.2A), and this allows us to specify the direction of the force imposed on the pail. As noted above, tension in the rope pulls the pail and water into circular motion. The tension in any rope acts along the rope's length, and, in this case, the rope connects the pail to the circle's center. Thus, as the pail travels in a circle, the force imposed by the rope continuously pulls the pail toward the center of that circle.

There is nothing particularly special about a pail swinging on a rope. Indeed, we can generalize from the force experienced by the pail to conclude that the force required for *any* circular motion is always directed at the center about which an object moves. This center-directed force is called a *centripetal* ("center-seeking") force.

Let us now return to the large scale of objects traveling on earth's surface. Exactly the same physics apply. As an object (let's say a small volume of water

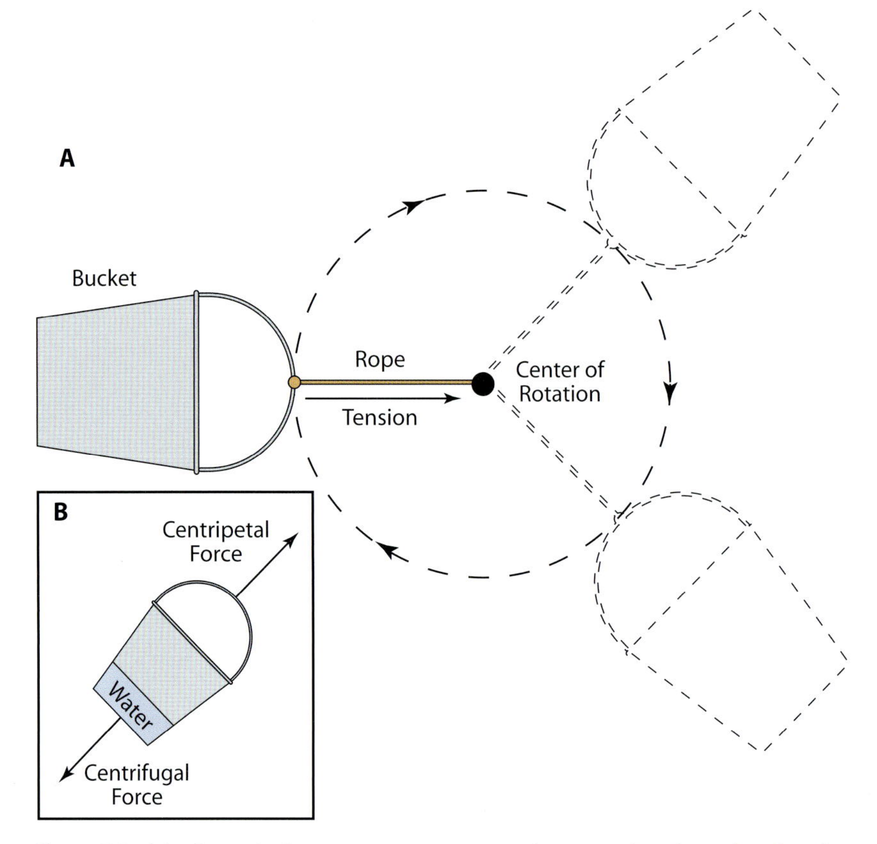

Figure 7.2. A bucket swinging on a rope serves as a demonstration of centripetal and centrifugal forces. (A) The entire system. (B) Details of the water in the bucket.

in the ocean) moves with the earth, it travels in a circle (figure 7.1) and therefore must be acted upon by a centripetal force.

The most obvious difference between water in the ocean and water in a spinning pail is that, for the ocean, there is no rope to provide the centripetal force. Instead, it is the inward tug of gravity that keeps the ocean moving in its circular path.

The Center of the Earth vs. the Center of Rotation

At this point we touch upon a subject that can potentially cause confusion. On earth, the force of gravity pulls down toward the center of the planet. In fact, the pull of gravity is more or less how we define "down." In contrast, the centripetal force acting on a stationary object at the earth's surface pulls inward toward the center of the circle about which the object moves. Often the center of the earth and the center of motion are not the same, and their separation is important.

The discrepancy between the center of the earth and the center of the circle of motion can be seen by looking at the earth from two different perspectives (figure 7.3). First, we look down from space at the North Pole, and follow the

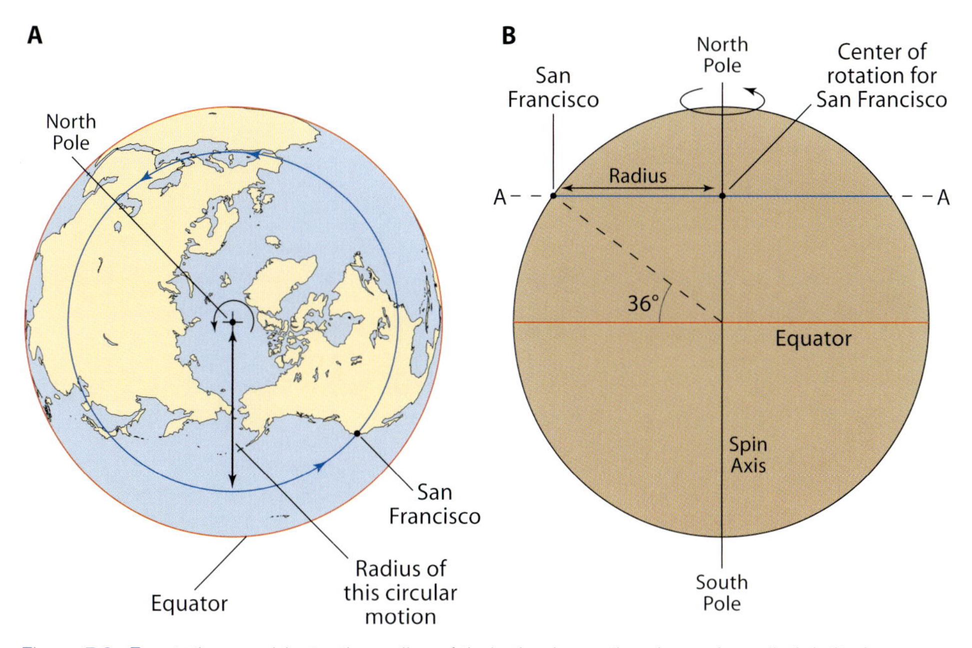

Figure 7.3. For stationary objects, the radius of their circular motion depends on their latitude. (A) A view from above the North Pole. (B) A north-south cross section through the earth, showing that the center of the earth is often not the center of rotation.

path of a bit of water in, say, San Francisco Bay as it moves with the earth for a day (figure 7.3A). As the earth spins, the water clearly travels in a circular path, with its center on the axis about which the earth rotates.

Now look at the earth in cross section (figure 7.3B). Again, we are concerned with water in San Francisco Bay. Imagine the water's motion as the earth rotates. It comes out of the page at you and then travels back in, always at 36° North. This motion defines the plane, shown by the solid line *A–A*, in which the centripetal force must act. But note that the plane intersects earth's axis at a point well north of earth's center. Thus, centripetal force acting on water in San Francisco Bay is not directed toward the center of the earth. Indeed, for any object not at the equator, the centripetal force acting on it is not directed toward the earth's center.

Gravity as Centripetal Force

We now encounter a second point of potential confusion. Gravity provides the centripetal force to keep objects stationary, that is, fixed with respect to the earth's surface. But gravity acts toward the center of the earth, while centripetal force (as we have just seen) is typically directed toward a different point. How can gravity provide centripetal force when its pull is not toward the center of rotation? To answer this question, we need to consider the vector nature of forces.

Imagine yourself walking across a lawn. Because you are thinking about physics rather than paying attention to where you are going, you run face-first into a clothesline stretched between two trees, and the force imposed by the line knocks you onto your backside. Where did that force come from?

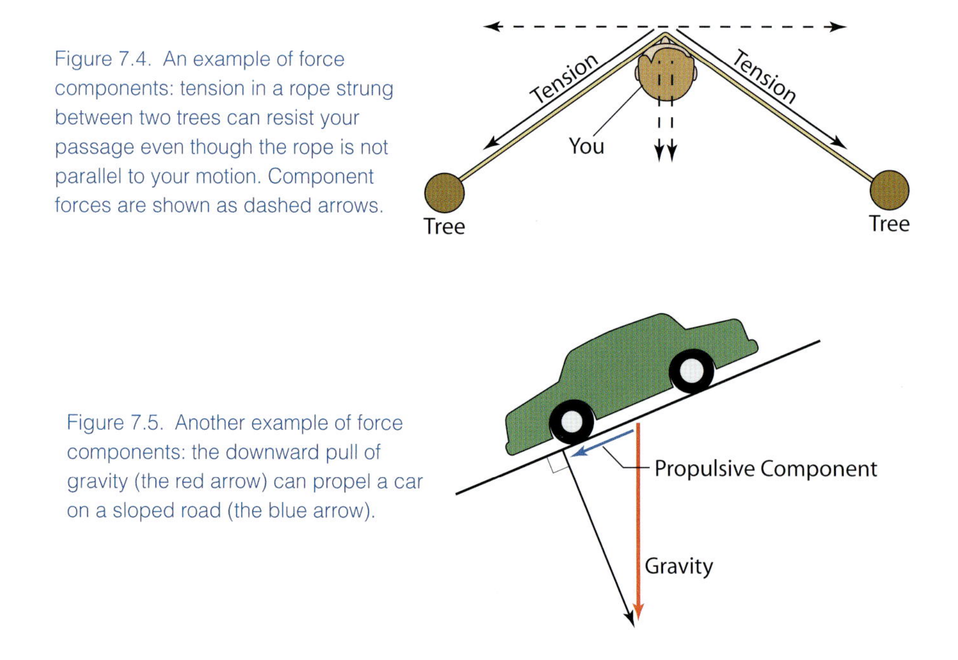

Figure 7.4. An example of force components: tension in a rope strung between two trees can resist your passage even though the rope is not parallel to your motion. Component forces are shown as dashed arrows.

Figure 7.5. Another example of force components: the downward pull of gravity (the red arrow) can propel a car on a sloped road (the blue arrow).

This scenario is shown from above in figure 7.4. As you impact the clothesline, it stretches, and tension forms in each half of the line, centered on your head. Each of these tensile forces can be decomposed into perpendicular components. For example, I have decomposed the overall force of tension in the clothesline into two forces: one directed against your head and one parallel to the line connecting the trees to which the clothesline is attached.

In this case, we don't particularly care about the component of force parallel to the line between the two trees. The reaction to this force pulls on the trees (Newton's third law), but that is the trees' concern, not ours. In contrast, the other component force, the one directed against your head, is important to us: it answers our original question. This component of the overall tensile force is the force that knocks you over. So, even though the overall tensile force acts along the length of the clothesline, a component of that force is available to push you in another direction, in this case back whence you came.

Another example may help clarify this concept. Imagine yourself coasting down a hill in your car (figure 7.5). Gravity pulls vertically downward toward the center of the earth, so why is the car instead propelled forward? We can decompose the gravitational force into the two components shown. One component acts perpendicular to the road's surface; this component keeps the car glued to the tarmac. The second acts parallel to the surface, directed downhill. It is this second component of gravity that pulls the car forward along the road. Thus, even though the tug of gravity is downward, one of its components is available to propel the car forward. The magnitude of the propulsive component of gravity depends on the slope of the road. If the road were horizontal, the propulsive component would disappear.

We can now apply these ideas to objects traveling with the earth (figure 7.6A). Again, let's consider water in San Francisco Bay. As it moves in its circular path, this water is acted upon by a centripetal force supplied as a component of the overall gravitational force.

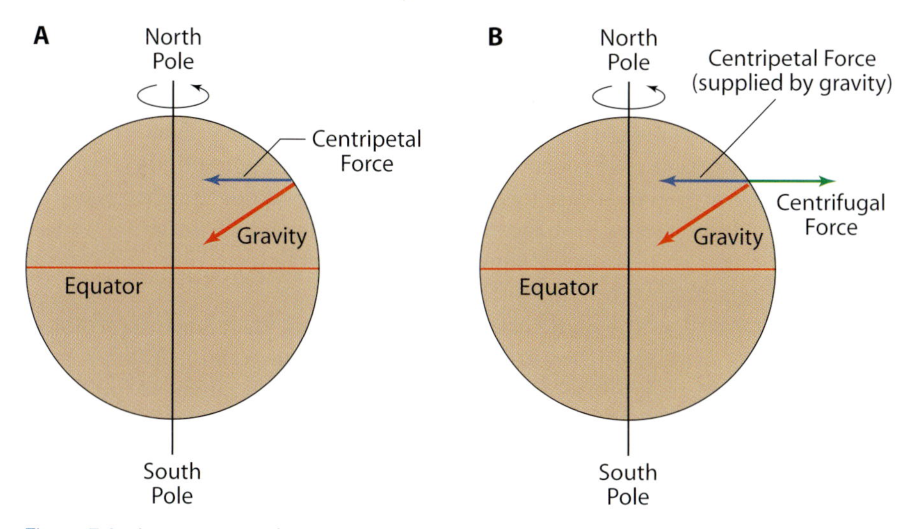

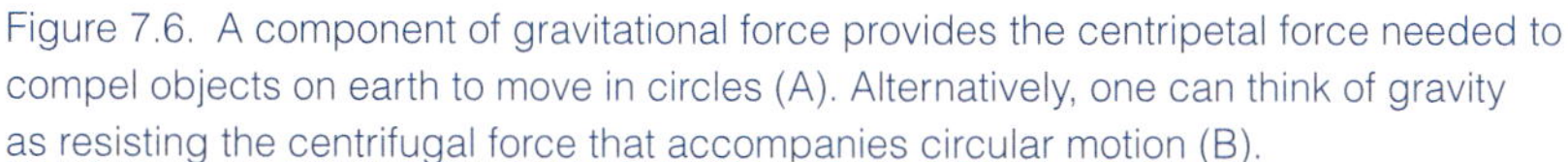
Figure 7.6. A component of gravitational force provides the centripetal force needed to compel objects on earth to move in circles (A). Alternatively, one can think of gravity as resisting the centrifugal force that accompanies circular motion (B).

This statement is true everywhere with the exception of the two poles, where gravity would act perpendicularly to centripetal force, and therefore could supply no resistance. Fortunately, these polar exceptions do not create a problem. Directly at either pole, an object sits right on the earth's axis and rotates in place rather than traveling in a circle. Because it doesn't move in a circle, no centripetal force is required. Thus, at the two points where no component of gravity is available to provide a centripetal force—the poles—no centripetal force is necessary.

Let's take a moment to summarize the facts we have learned: (1) Because—with the exception of the poles—any stationary bit of water in the ocean is actually moving in a circle, it must be acted upon by a centripetal force. (2) This force is provided by gravity.

How Big Is the Centripetal Force?

Having established the concept of centripetal force, our next job is to figure out how large it is. To simplify the process, we will not deal with force itself. The centripetal force exerted on a volume of water depends on the volume's mass. For instance, in the swinging-pail system we considered earlier, the more water there is in the pail, the larger the centripetal force required as the pail swings at a given speed. To avoid always having to specify the mass of each object we deal with, we instead deal with the force *per mass*. If you glance back at Newton's second law, you will see that we already have a word for "force per mass": *acceleration*. In other words, the centripetal force per mass for an object moving in a circle equals the acceleration of that object toward the circle's center.

So, how large an acceleration are we talking about? The magnitude of centripetal acceleration (the force per mass) is equal to the square of angular velocity ω, which we will define in a moment, multiplied by R, the radius of the circle about which the object travels.

$$\text{centrifugal acceleration} = \omega^2 R \qquad \text{(Eq. 7.1)}$$

The radius we are talking about here is shown in figure 7.3A and B. It is the length of the line connecting the object to earth's axis in the plane of the object's motion: the object's "rope." This means that the radius is longest for water at the equator (in which case the radius of the circle equals earth's equatorial radius, 6378 kilometers), and the radius gets shorter and shorter the closer water is to either the North or South Pole. For San Francisco at 36° North, $R = 5150$ kilometers.

Angular velocity, as the name implies, measures how fast something rotates. Any stationary object on the earth travels through one complete circle—2π radians—in a day—86,400 seconds—so it has an angular velocity of 0.000073 radians per second.[1]

Using these values for R and ω, we see that the centripetal acceleration of water in San Francisco Bay is about 0.027 meters per second squared.[2] Now, the acceleration due to gravity is about 9.81 meters per second squared, so the centripetal acceleration of water in San Francisco Bay is relatively small, only about 0.27% that of gravity. Because the centripetal acceleration is so small, we humans tend not to notice it, but as we will see, it is nonetheless very important.

Note that centripetal acceleration varies with latitude. At the equator, where the radius of circular motion is greatest, the centripetal acceleration, 0.034 m/s^2, is about 25% greater than it is in San Francisco. And, at the North or South Pole, where the radius of circular motion is zero, there is no centripetal acceleration at all.

In summary, we have established that because the earth rotates, water in the ocean is acted upon by a centripetal force per mass. This acceleration varies with latitude: it is largest at the equator and vanishes at the poles.

Centrifugal Force

We now make use of Newton's third law to describe the reaction force that inevitably accompanies circular motion. Consider again the rope and pail shown in figure 7.2. If the rope (object A) pulls inward on the pail (object B), we know that the pail must pull outward on the rope. Thus, because the pail feels a center-directed centripetal force, the rope feels an outward-directed *centrifugal* ("center-fleeing") reaction force. This same centrifugal force can be exerted on other objects as well. For example, as you hold the rope and swing the pail, you yourself feel a centrifugal force as the pail pulls on the rope.

The same line of reasoning can be applied to the interaction of the pail and the water it contains (figure 7.2B). As with any mass, water in the pail has a tendency to travel in a straight line, and it follows its circular path only because the bottom of the pail applies a centripetal force pushing it in toward the center

[1] Radians, a measure of angle, are equal to the ratio of arc distance on a circle divided by the circle's radius. The circumference of a circle of radius R is $2\pi R$, thus the angle subtended by a complete circle is 2π. For a graphical depiction of this explanation, see figure 2.A1.

[2] Note that, as a ratio—the arc distance along the circumference of a circle divided by the length of the circle's radius—radians are pure numbers, so it is legitimate to drop "radians" from the units of the final answer.

of the circle. And, like the rope and your hand, the centripetal force of the pail on the water is accompanied by a centrifugal reaction force in the water that pushes back on the pail's bottom. Because of this centrifugal force, the water congregates in the bottom of the pail, and, if you spin the pail fast enough, the water does not spill.

To physicists, this centrifugal force is a "pseudo" force. From a vantage point outside the system of the rope and spinning pail, it is clear that there is no force pulling out on the water. But *within the system*—that is, to the water in the pail—the centrifugal force appears real.

As you and I stand on earth, we are "within the system," and as a result, to us centrifugal force can feel real. Let's explore a concrete example: how your weight would vary if you visited the North Pole and the equator. Usually, we think of weight as the force imposed on a mass by the acceleration of gravity. By this reasoning, your weight, in newtons, is equal to your mass, in kilograms, multiplied by the acceleration of gravity, 9.81 meters per second squared, and this is indeed the weight a scale would register at the North Pole. However, at the equator, earth's rotation creates a centrifugal force acting in the direction opposite that of gravity (figure 7.7), thereby decreasing your weight. Centrifugal acceleration at the equator is 0.034 meters per second squared, and the resulting centrifugal force decreases the effect of gravity by 0.34%. Thus, a person who weighs 700 newtons at the North Pole, where there is no centrifugal acceleration and therefore no centrifugal force, weighs slightly less (697.6 newtons) at the equator.

Water that is stationary on the rotating earth is also within the system, and subject to an apparent centrifugal force directed away from its center of motion. A particularly relevant example is presented in figure 7.6B. Here the green arrow shows the centrifugal reaction to the centripetal force depicted in figure 7.6A. Note that because centripetal acceleration varies with latitude, centrifugal force varies in the same fashion: it is largest at the equator and decreases to zero at the poles.

Let's pause briefly to review our results. Water stationary on earth's surface actually travels in a circle, and is therefore subject to centripetal acceleration. The reaction force that results—centrifugal force—is directed away from the

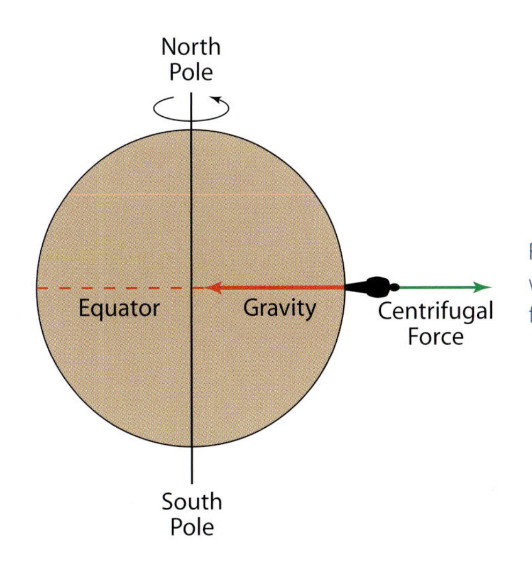

Figure 7.7. Centrifugal force reduces the weight of an object at the equator. The force vectors are not drawn to scale.

water's center of rotation. Centrifugal force is zero at the poles, and maximal at the equator.

Consequences of Circular Motion

One might well question the consequences of these conclusions: if water in the sea feels a centrifugal force, won't it try to move "out" the same way water in a washing machine moves out during the spin cycle? Yes, indeed it will.

This situation is shown in figure 7.8A, where centrifugal force is again shown as a green arrow. We decompose the centrifugal force into two components, one perpendicular to the earth's surface and one parallel to it. The perpendicular component represents a tendency for an object to fly upward, away from the earth; gravity easily resists this component. In contrast, the component of centrifugal force *parallel* to the earth's surface is perpendicular to gravity's line of action, therefore gravity cannot act to resist it. We have already encountered an appropriate analogy: if a car sits on a level road, a road whose surface is perpendicular to gravity, gravity can propel the car neither forward nor backward. The inability of gravity to resist this component of centrifugal force is important because, without gravitational resistance, the component of centrifugal force parallel to the earth's surface can cause water to move toward the equator. We can repeat this analysis for water in the southern hemisphere (figure 7.8B). Again, water feels a component of centrifugal force parallel to the earth's surface, and again this force pulls water toward the equator. Thus, we predict that everywhere on earth water tends to move away from the poles and toward the equator due to the urging of centrifugal force.

Let's follow this line of logic to its end. If centrifugal acceleration were allowed to act unchecked, all the water in the oceans would pile up at the equator. This would drain the ocean basins and form the largest mountain range on

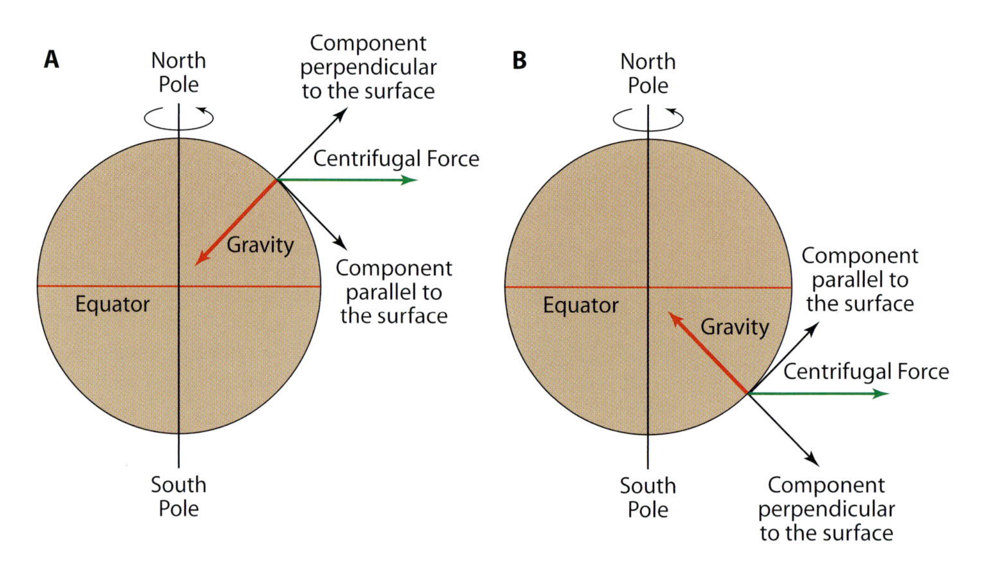

Figure 7.8. A component of centrifugal force compels water to move toward the equator in both the northern (A) and southern (B) hemispheres.

earth, albeit a mountain range made of water. In reality this doesn't happen, so there must be some aspect of earth's physics we have neglected. Something must act to counteract the centrifugal tendency to drive water toward the equator.

Water Flows Downhill

Again gravity comes to the rescue, although in this case somewhat indirectly. Let's begin by looking at the change of pressure with depth in a column of water. There are two points we need to consider.

First, pressure is defined as force divided by the area over which the force is imposed. In the ocean, the force creating pressure in the water is due to the weight of the atmosphere and water pushing down. In essence, if we want to calculate the pressure in the water at a given depth below the surface, we measure the weight of the water and air above a given area (figure 7.9). That weight, divided by the area, yields the pressure. For example, directly at the water's surface, just air pushes down. The column of air above a square meter of ocean surface weighs 1.01×10^5 newtons, so atmospheric pressure at sea level is 1.01×10^5 newtons per square meter. If we move our square meter down into the water, we have to include, in addition to the weight of the air, the weight of the water above our area. As a consequence, the deeper we go, the higher the pressure.

Next, we note that fluids, such as water and air, tend to move from areas of high pressure to areas of low pressure. For example, when you turn on a faucet, water flows out because the pressure in the pipe leading to the faucet is higher than the pressure in the air outside.

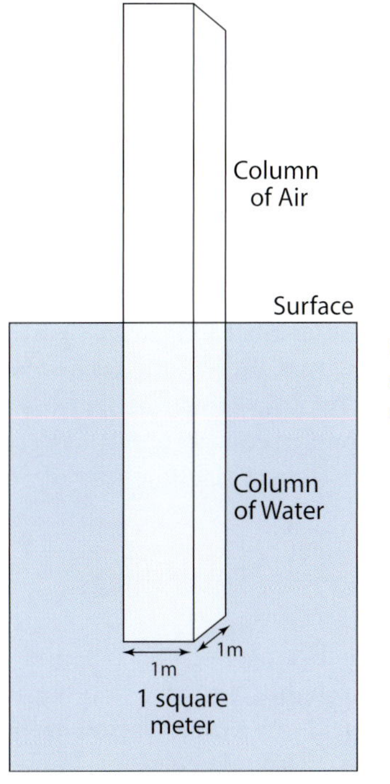

Figure 7.9. The pressure (force per area) at a given depth in the ocean is exerted by the weight of air and water above that depth.

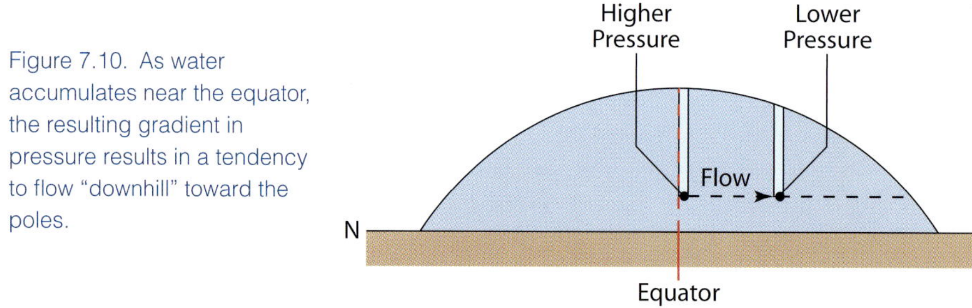

Figure 7.10. As water accumulates near the equator, the resulting gradient in pressure results in a tendency to flow "downhill" toward the poles.

Let us now apply this information to our supposed equatorial mountain range, shown in cross section in figure 7.10. The range has a peak in the middle (at the equator) and slopes down toward each of the poles. Compare the pressure at the two points shown. One is directly under the peak, and one is off to one side, but both points are at the same elevation. That is, they are the same distance away from the earth's center. Because the sides of the mountain slope down, the point directly under the peak has a taller column of water above it, and therefore feels a higher pressure, than the point lying to the side. Because of this difference in pressure, there is a tendency for water to flow away from the peak. And because the peak is directly at the equator, this means that there is a tendency for water to flow away from the equator, toward the poles.

This process should be familiar; it is the mechanical basis for water flowing downhill. Note, however, that the mechanism we have just described does not require the earth's surface to be tilted, as is the case for water flowing down the tilted bed of a mountain stream. As shown in figure 7.10, the surface of the solid, spherical earth is perpendicular to gravity, and therefore has no "tilt." Nonetheless, because the surface of the *water* in our aqueous mountain is tilted, the north-south variation in pressure causes water to move away from the equator.

Thus, just as centrifugal acceleration tends to form a mountain of water by moving water equatorward, gravity, in the disguise of pressure, resists that tendency by forcing water to accelerate downhill, which in the case of an equatorial mountain range means that water tends to flow toward the poles. When both accelerations act at the same time, as in reality they do, the net result is a compromise: enough water moves from the poles to the equator to create a sufficient counter-tendency to flow downhill, and as a result, the ocean around the earth's girth has a slight but distinct bulge.

Newton deduced the mathematics of the compromise, at least to a first approximation, in the same book in which he laid out his three laws of motion. According to his calculations, at equilibrium the ocean at the equator should be about 20 kilometers deeper than at the poles (figure 7.11A).

The Fluid Earth

Once again we find a conflict between theory and reality. As we learned in chapter 2, the oceans are roughly the same depth everywhere (about 4 kilometers), and the deepest point in the ocean (the Mariana Trench) is neither at the equator nor anywhere near 20 kilometers deep. These facts are at odds with the

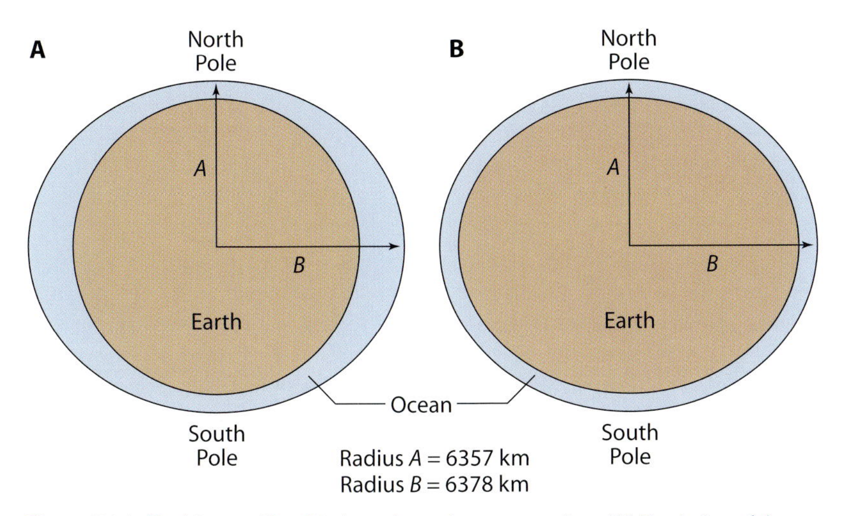

Figure 7.11. Earth's centrifugal bulge, shown in cross section. (A) The bulge of the ocean on a spherical earth. (B) The constant-depth ocean on a bulging earth.

theory we have just explored. Should we doubt Newton? No, he got it right; there really is a 20-kilometer bulge at the equator, 21.5 kilometers to be exact. But the bulge is in the solid part of the earth rather than in the oceans.

Much of the core of the earth is molten, and therefore fluid. Even other parts of the earth—the mantle and the crust—can, over long periods of time, flow as if they were liquid. This ability to flow underlies (literally) the movement of tectonic plates and the continents they support. If we accept the fact that the "solid" parts of the earth can flow, it is a short jump to suppose that they flow in response to centrifugal acceleration in the same fashion that we have just proposed for water in the sea. In other words, over the course of geological time, the shape of the solid earth has come to equilibrium with the centrifugal forces exerted by its spin, with the result that the distance from the center of the earth to any point on the equator is 21.5 kilometers longer than the distance from the center to either of the poles.

The size of earth's equatorial bulge should be kept in perspective. Compared to the equatorial radius of the earth (6378 kilometers), a 21.5-kilometer bulge is quite small (0.3%). If the earth were a billiard ball, you probably wouldn't notice that it was out of round. On the other hand, as we will see, the equatorial bulge plays an important role in the Coriolis effect.

What happens to the oceans on this bulging earth? Nothing, really. The oceans form a layer of more or less constant thickness over the bulging earth (figure 7.11B). In terms of the forces exerted, it does not matter to a small volume of ocean water whether the bulge beneath it is formed of water or basalt. Centrifugal force still tends to move water toward the equator and gravity still tends to move water downhill toward the poles. In this case, "downhill" can be taken in a more literal sense—traveling poleward involves moving down the hill formed by the bulging solid earth—but the physics are the same. Given the current shape of the earth, stationary water everywhere in the ocean is at equilibrium between centrifugal and gravitational accelerations.

The Equilibrium Platform

Because the bulging shape of the earth allows water in the oceans (and any other object, for that matter) to be at equilibrium between gravitational and centrifugal forces, this shape is known as an *equilibrium platform*. Upon this platform, winds blow and ocean currents flow, and as we will see, the properties of the equilibrium platform affect these motions.

It will be easiest, however, if for the moment we again step away from the real ocean and deal with an analogy. Let's imagine that we could put a thin, frictionless, rigid coating over the surface of the earth. This coating would have the same shape as the idealized bulging earth, ignoring local topography, and therefore it would have the shape of an equilibrium platform. But because it is rigid and smooth, we don't have to worry about flexure of the coating or any friction it might exert on an object.

Now let's place a bowling ball on our equilibrium platform. We put the ball down, make sure it is stationary, and then let it go. In what direction will it move? Well, if we really have made our coating into the shape of an equilibrium platform, the ball will not move at all. Yes, gravity pulls the ball toward the center of the earth, but this force is resisted by the rigid surface, which supplies the same kind of reaction force that keeps you from falling through the floor. Yes, the ball has a tendency to roll downhill, away from the equatorial bulge and toward the poles, but this tendency is offset by the component of centrifugal force acting to push the ball toward the equator. With our smooth, rigid surface and bowling ball, we are clearly dealing with the same mechanics just described for water in the oceans.

The reason for constructing this analogy will become clear when we take the next step and create a hypothetical laboratory analogue of the earth's equilibrium platform. Bear with me for a moment as we work our way through this project.

First, we shrink the earth down to a size that will easily fit on a lab table. Then imagine, if you will, unzipping the earth's smooth equilibrium surface along the equator. We now transfer the salient properties of the earth's equilibrium platform to a shape that is easier to deal with. Take the northern half of this unzipped surface and flatten it out as shown in figure 7.12. The North Pole is in the center, and the equator forms the distal rim of this new shape.

Actually, "flatten out" is not exactly what we do. To maintain the essence of the physics we have been dealing with, we bend our transformed surface so that gravity can counteract centrifugal force. Working with this small object in the lab, we ignore the gravitational pull of the surface on itself, and rely on the lab's gravity instead. So, unlike the real equilibrium platform, where gravity always acts toward the center of the earth, in our analogue, gravity pulls downward toward the lab floor on all parts of the surface. We can cope with this shift by making the equatorial rim of the new shape higher than its North Pole. That is, we bend the surface, turning it inside out, so that it looks like a bowl rather than a dome (figure 7.11B). In this way, it is downhill from anywhere on the equator to the North Pole, just as it is on earth, and any object placed on the surface will tend to move poleward under the urging of gravity.

We now need a source of centrifugal force. To this end, we mount our earth analogue on an axle through its North Pole and allow it to spin around this axle (figure 7.12A). As a result of this spin, there is a centrifugal acceleration for objects stationary relative to the analogue, and this acceleration tends to

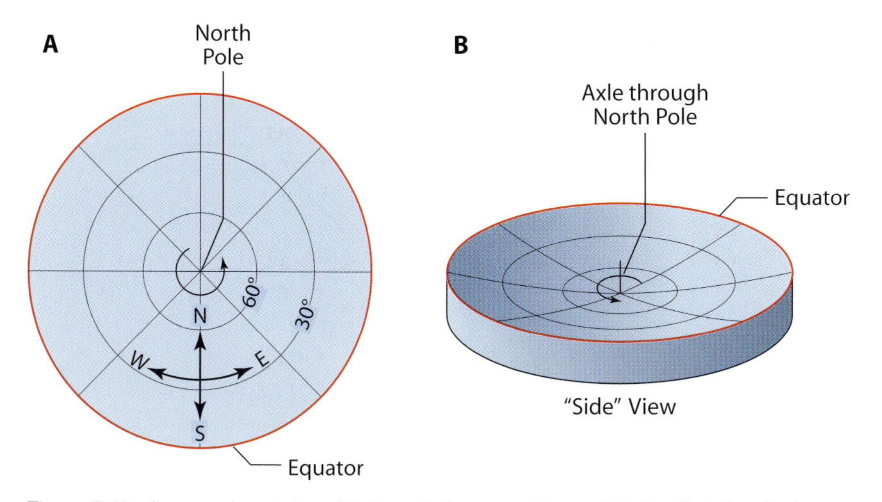

Figure 7.12. An experimental equilibrium platform used to explain the Coriolis effect. (A) A view from above the North Pole. (B) A perspective view showing the curvature of the platform.

force objects outward, away from the pole and toward the equator, just as centrifugal force does on the real earth.

If we are careful, we can adjust our new shape so that, for a given angular velocity, the poleward pull due to a component of gravity, set by the slope of the surface, is just offset by the equatorward pull due to a component of centrifugal force. Because centrifugal force gets larger the farther an object is from the axis of rotation (eq. 7.1), the slope of our surface, and the resulting gravitational tendency to move downhill, must likewise increase.

When all these adjustments are finished, we are left with an object that looks for all the world like a roulette wheel (figure 7.12B). In fact, a well-designed roulette wheel can be an equilibrium platform if spun at the proper rate.

Before playing with our new toy, let's review its analogy to the earth. The North Pole is at the wheel's center, and the equator at its edge (figure 7.12B). That means that lines of longitude run radially from the wheel's center to the periphery. Moving "north" entails moving downhill along one of these meridians toward the center of the wheel; moving "south" entails moving uphill along a meridian toward the rim of the wheel. Lines of constant latitude form circles centered on the North Pole, and "east" and "west" are as shown. The discussion that follows will be more easily digested if you take a moment to orient yourself on this analogue of earth.

In which direction should our analogue spin? Well, the sun rises in the east and sets in the west because the earth spins from west to east. Thus, for our roulette-wheel earth to be analogous to the real earth, it must spin counterclockwise when viewed from above its North Pole, as indicated in the figure.

Explaining the Coriolis Effect

With these preparations complete, we can now begin to experiment. We give the wheel the angular velocity appropriate to make it an equilibrium platform and carefully place a small ball stationary on its surface. In practice this would

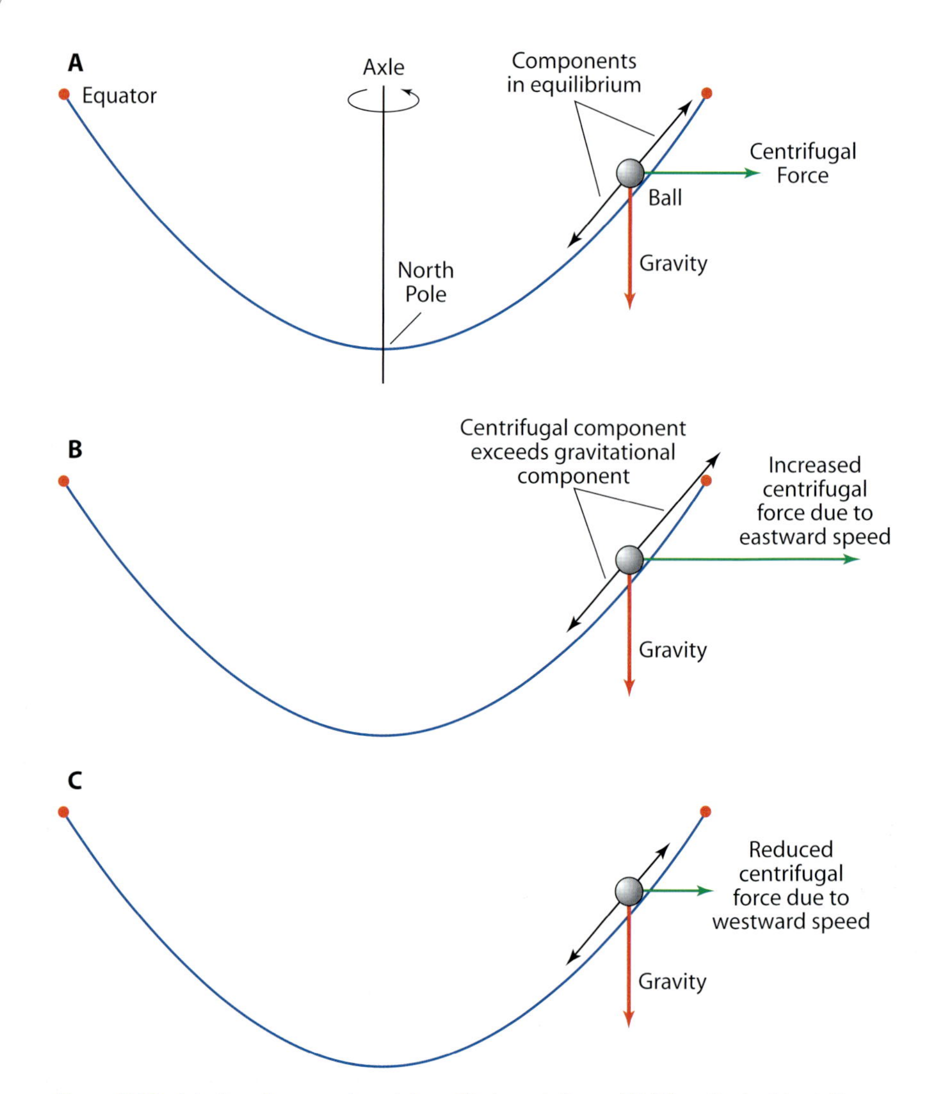

Figure 7.13. A ball on the experimental equilibrium platform. (A) When the ball is stationary relative to the platform, the components of force due to gravity and centrifugal force offset each other. (B) When the ball moves in the direction of spin (into the page), the component of increased centrifugal force overwhelms the constant component of gravity, and the ball moves upslope toward the "equator." (C) When the ball moves opposite the direction of spin (out of the page), the component of reduced centrifugal force is overwhelmed by the component of gravity, and the ball veers toward the "pole."

be difficult—you'd have to move your hand round and round to match the speed of the wheel—but we won't let that stand in our way. The ball is somehow put into position, held stationary relative to the wheel's surface, and then let go. What happens?

If you have followed the argument to this point, you should realize that the ball on our roulette wheel is in a situation exactly analogous to the bowling ball on the real earth. The roulette wheel is an equilibrium platform (figure 7.13A). By moving with the wheel, the ball experiences a component of centrifugal force just sufficient to hold it in place against the component force due to gravity tending to pull it downhill. In fact, this equilibrium exists wherever we place the ball on the wheel, hence the term "equilibrium platform."

East-West Motion

With our next experiment, we finally approach our ultimate goal, an understanding of the Coriolis effect. First, let's place the ball on the wheel as before, stationary on the equilibrium platform. In this case, however, after placing the ball on the wheel, we give it a shove directly to the east, the direction in which the wheel is spinning. In response to our push, the ball attains an eastward velocity relative to the surface of the wheel. What else happens?

Well, because the ball is moving in the same direction that the wheel is spinning, west to east, if it stays at this latitude, it will complete a circle before the wheel itself does. To see this, imagine yourself on a merry-go-round. You are sitting on one of the horses, and are therefore stationary relative to the merry-go-round's platform, and have just waved to some friends who are standing on the ground beside the ride. If you stay on the horse, it will take you a few seconds to get back to where your friends are. If, however, you hop off the horse and walk forward on the merry-go-round, in the direction in which it spins, you complete a circle and arrive back at your friends before your horse does.

The same thing applies to a ball on our equilibrium platform. By giving the ball an eastward velocity, we have decreased the time it takes to make one circuit, and we have thus increased the ball's angular velocity. Furthermore, from equation 7.1, we know that any increase in angular velocity results in an increase in centrifugal acceleration. So, by giving the ball an eastward velocity, we have increased its centrifugal acceleration, and thereby the centrifugal force it feels (figure 7.13B).

This increase in centrifugal force breaks the ball's equilibrium. As long as the ball was fixed with respect to the platform, the centrifugal force acting on it was just sufficient to keep if from rolling downhill. Because the ball is now moving eastward with respect to the platform, the uphill component of centrifugal force exerted on it has increased, and it now exceeds the constant downhill component of force due to gravity. Under the influence of this unbalanced force, the ball accelerates uphill.

Now, uphill on our wheel is equivalent to southward on the earth. Thus, when we shove our ball to the east, the ball actually veers to the right and acquires some velocity to the south. Interesting.

What happens if, rather than pushing the ball eastward, we push it to the west? In this case, the ball's velocity is subtracted from that of the wheel, and it takes it longer to complete a circle than does the wheel itself. In other words, by shoving the ball to the west, we have decreased its angular velocity. The decrease in angular velocity results in a decrease in centrifugal acceleration and a consequent decrease in the centrifugal force pulling the ball uphill. The ball's equilibrium is again broken. In this case, however, because the ball now feels less centrifugal force pulling it uphill than gravitational force pulling it downhill, the ball rolls downhill (figure 7.13C). Downhill (inward) on our wheel is analogous to northward on the earth. So the net result of giving the ball a velocity to the west is for the ball to veer to the right and begin moving to the north. Apparently, on this equilibrium platform, *any motion east or west results in a tendency to veer to the right.*

North-South Motion

What happens if the initial motion of the ball is to the north or south? Before tackling this question, we need to digress briefly to review one more characteristic of circular motion. Consider a solid surface rotating with constant angular velocity (figure 7.14); our roulette-wheel earth is an example. Because the surface is solid, every point on it completes a circle of motion in the same period of time. But points farther from the axis of rotation trace larger circles than points closer in, and, as a consequence, travel a longer distance in completing their circles. Traveling a longer distance in the same period of time means that the farther a point is from the center of rotation, the faster it must travel. With this idea in mind, let's return to our experiments.

We begin the next experiment by placing a ball and a target on our roulette-wheel earth (figure 7.15). Both are on the same meridian, with the target due south of the ball. As explained above, the target has a larger eastward velocity than does the ball because the target is a larger distance from the axis of rotation.

Now, give the ball a shove to the south, directly at the stationary target. By pushing the ball southward, we do not affect its eastward velocity (which is relatively small), but we do impart to it some speed toward the target. In the time it takes for this moving ball to reach the latitude of the target, both have traveled to the east. But, because the ball has a smaller eastward velocity, its eastward motion lags behind that of the target. In other words, even though we initially shoved our northern ball directly south, *it appears to veer to the right* and ends up west of the point at which we aimed.

Let's repeat the experiment, but this time we reverse the initial locations of ball and target, and give the ball a shove to the north (figure 7.16). Because the ball now has the larger eastward velocity, by the time it reaches the latitude of the target, it has pulled ahead, arriving at the northern latitude somewhere to the east of the spot at which it was aimed. In other words, any attempt to

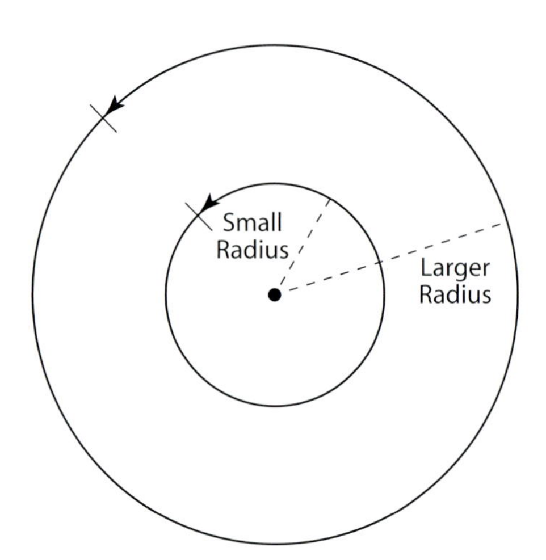

Figure 7.14. The farther an object is from the center of rotation, the larger the radius of the circle it travels. As a result, stationary objects near the poles have a slower eastward velocity than stationary objects near the equator.

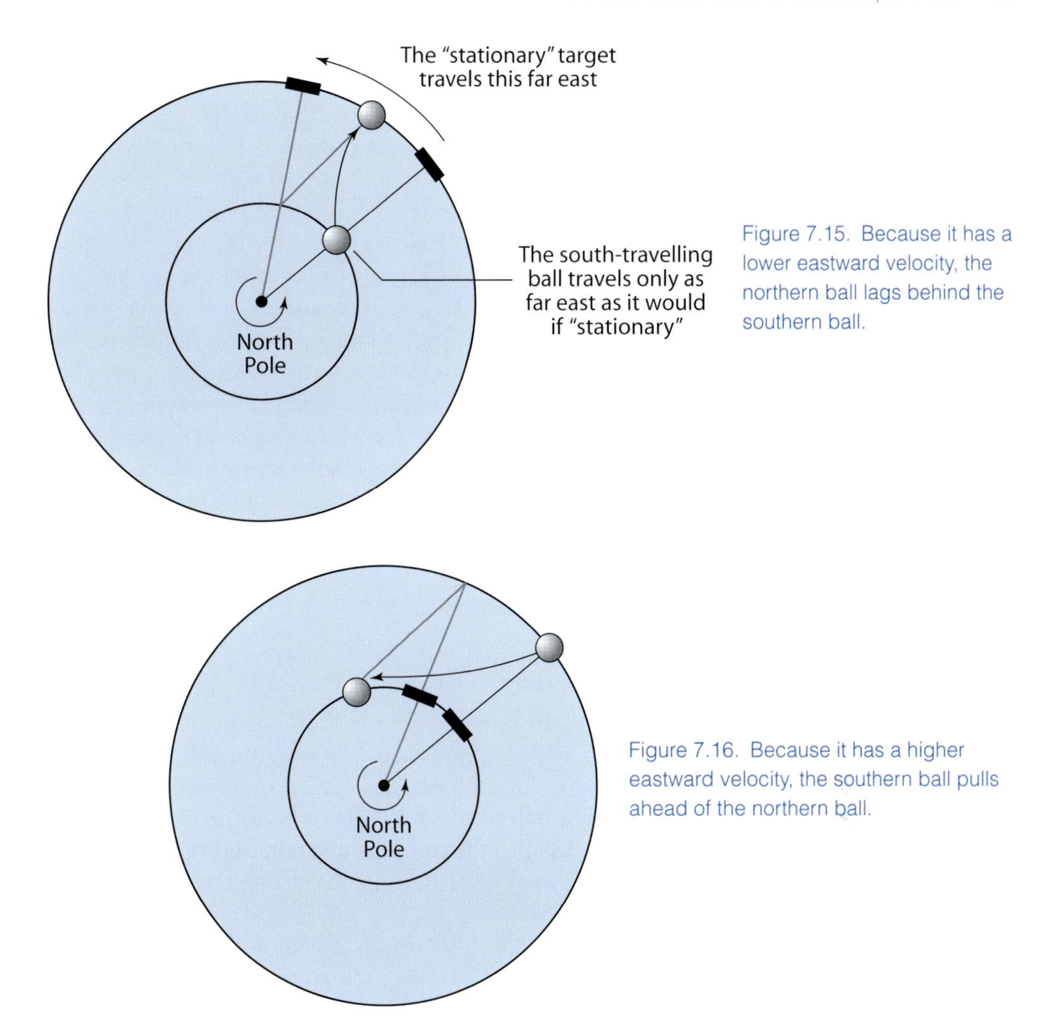

Figure 7.15. Because it has a lower eastward velocity, the northern ball lags behind the southern ball.

Figure 7.16. Because it has a higher eastward velocity, the southern ball pulls ahead of the northern ball.

propel a ball due north ends with the ball veering to the right in an eastward direction.

There is an obvious pattern to these results. On an equilibrium platform rotating in the direction we have described, any induced motion (to the east, west, north, or south) ends up veering to the right. This tendency to veer is the *Coriolis acceleration*, first described in 1835 by Gaspard-Gustave de Coriolis, a French engineer and mathematician.[3] Although, as we will discuss below, the details are slightly different on the real earth than they are on a roulette wheel, the same basic physics apply: *in the northern hemisphere, any horizontal motion of air or water results in a tendency to veer to the right*. This is the first magic rule we mentioned at the beginning of this chapter.

There is one other aspect of the Coriolis acceleration that we can deduce from these experiments. Let's return to the experiment in which we try to propel

[3] In addition to describing the acceleration to which history has attached his name, Coriolis (1792–1843) introduced the modern concepts of "work" and "kinetic energy" into the lexicon of physics. In the same year in which he published his ground-breaking work concerning motion on rotating bodies, he published another treatise on the *Mathematical Theory of the Game of Billiards*.

a ball either west or east. In this case, we deduced that the ball tends to veer north-south because the changed equatorward centrifugal acceleration no longer equals the constant poleward pull of gravity. The centrifugal acceleration changes because movement east or west has altered the angular velocity of the ball. Let us extend this argument.

The faster the ball moves, the greater the change in its angular velocity. As a result, the faster the ball is propelled east or west, the greater the change in the centrifugal acceleration acting on it, and the greater the tendency to veer. In other words, from our experiments, we can conclude that the faster an object moves east or west on an equilibrium platform, the more forceful the Coriolis effect. Although it requires a more mathematical explanation (see the appendix to this chapter), a similar conclusion can be reached for north-south motions of the ball. In summary, *the faster an object moves on an equilibrium platform, the greater the Coriolis acceleration*. This is the second of the magic rules presented at the beginning of this chapter.

The Southern Hemisphere

So far, we have focused our attention on the northern hemisphere. What about the southern hemisphere? We could wander through the whole process again, starting with the southern half of our unzipped earth, but we would find that only one relevant fact is different. When viewed from above the South Pole, the earth rotates clockwise, in the opposite direction of the earth viewed from above the North Pole. Reversing the roulette wheel's direction of rotation, and preserving all other properties of the platform, we find that *horizontal motion, regardless of initial compass heading, veers to the* left *in the southern hemisphere.*

The different directions of the Coriolis acceleration in northern and southern hemispheres has led to the wonderful fiction that the direction water spins when it drains from a bathtub or when a toilet is flushed is opposite in the two hemispheres. Apparently, there are even con artists in equatorial Africa who (for a small fee) will show you that water spins one way as it exits a washtub, and then step ten feet across the equator to show you that the spin is in the other direction. Hogwash. As we will soon see, the forces exerted by the Coriolis acceleration are quite small, and at the scale of water in a toilet or tub, these forces are easily overcome by even the slightest preexisting swirl in the water. If a toilet in Australia flushes with opposite spin from a toilet in England, the difference is due to the bowl's design rather than to the Coriolis acceleration.

A Bit More History

Here we deal with the Coriolis acceleration as it applies to motion of water and air, but I would be remiss if I didn't digress for a moment to note the role of the Coriolis acceleration in military history. As we have seen, the tendency to veer either to the left or to the right depends on the speed of an object. The faster the object moves, the larger its acceleration to the side. Thus, a bullet fired from a gun ought to experience a substantial Coriolis acceleration, and indeed bullets do. However, when a soldier shoots at his enemy, the bullet is in flight for such a short time that, even though it is accelerating rapidly to the side, it does not have time to accrue much sideways velocity before it hits its mark, and its deviation is unnoticeable. As a result, for many centuries, the Coriolis acceleration could be ignored by soldiers and sailors. This changed in World War I, however, with the advent of guns capable of firing shells great distances. The largest of these weapons were the massive siege guns built to launch shells across the trenches in France. The most famous of these siege guns was the German's "Big Bertha," which could propel a shell more than 20 miles. Initially, the Germans found that when they shot at the French to their south, they missed their targets by a substantial margin: the shells kept ending up to the west of where they were aimed. Only when they corrected for the Coriolis acceleration did their accuracy improve. There are possibly apocryphal tales that when, in 1982, Britain fought Argentina over the Falkland/Malvinas Islands in the southern hemisphere, their gunnery was initially off because their tables corrected for a veer in the wrong direction.[4]

A Few Details

So far, we have discussed the Coriolis acceleration in very general terms: there is a tendency for moving objects to veer to the right in the northern hemisphere and to the left in the southern hemisphere. The faster an object moves, the greater the tendency to veer. For future purposes, we need to be more specific. Is the magnitude of this effect the same everywhere within a hemisphere? No, it is not, and we take a moment to understand why.

Let's compare the consequences of motion near the poles to the consequences of motion near the equator. As we have done before, let's compare the size of the circles that stationary objects travel at these two locations. Both objects complete a circle in one day, but the object near the equator travels a much greater distance, so its speed along its circular path must be greater. For example, at a latitude of 20° North or South, a stationary object on the earth's surface is 5987 kilometers from earth's axis (figure 7.17A). When we multiply this radius by 2π, we find that the object travels 37,616 kilometers in a day as it completes its circle, a speed of 1567 kilometers per hour. At a latitude of 70° North or South, a stationary object is 2179 kilometers from earth's axis (figure 7.17A), and it travels 13,691 kilometers in a day, a speed of only 570 kilometers per hour.

These are the speeds of stationary objects. Now, we let our objects at each of these latitudes move east or west, for example at a speed of 20 kilometers per hour. At latitude 20°, this velocity relative to the earth's surface amounts to a 1.3% change in angular velocity, whereas at latitude 70°, a 20-kilometer-per-hour velocity causes a 3.5% change in angular velocity, more than twice that at the lower latitude.

Recall that centrifugal force is proportional to angular velocity (equation 7.1). As a result, the smaller relative change in angular velocity at lower latitude

[4] The British refer to these islands as the Falkland Islands, and consider them British territory. The Argentines refer to the islands as the Malvinas, and consider them part of Argentina. This basic difference of opinion led to the war.

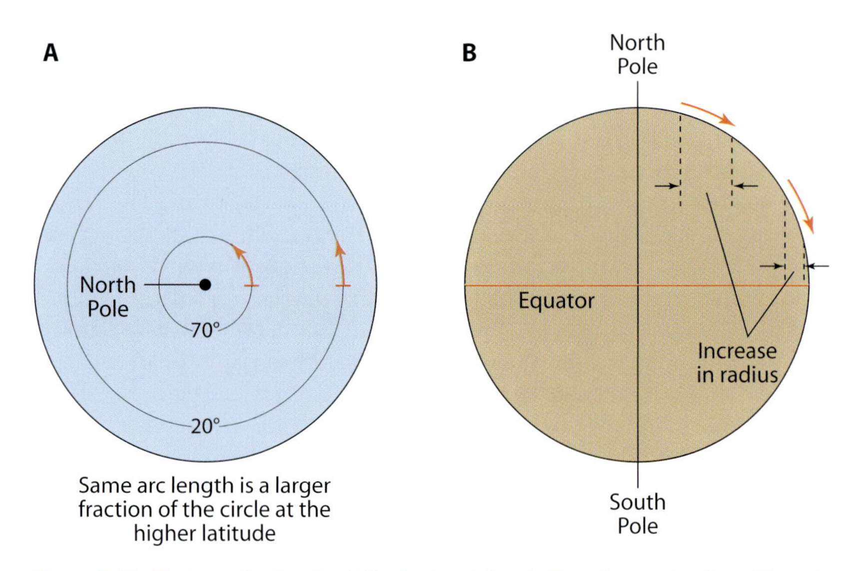

Figure 7.17. Factors affecting the latitudinal variation in Coriolis acceleration: (A) east-west motion, (B) north-south motion. See the text for the details of the explanation.

results in a smaller relative change in the centrifugal force acting on an object at that latitude. This smaller relative change in force in turn causes a smaller tendency to veer. In other words, *the lower the latitude, the smaller the magnitude of the east-west Coriolis acceleration.*

What about north-south motion? In this case, the Coriolis acceleration is due to the difference in east-west speed of objects at different latitudes. For example, we have seen that an object moving south in the northern hemisphere lags behind the stationary objects it travels past, and therefore it veers to the right. The magnitude of this tendency depends on how fast the east-west speed of stationary objects changes as one moves north or south, and this, in turn, depends on how north-south motion affects the distance an object is from earth's axis.

To see this, consider an object at a latitude of 70° (figure 7.17B). If the object travels 10 kilometers to the south along earth's surface, it increases its distance from earth's axis by 9.4 kilometers. In contrast, at a latitude of 20°, because of the earth's curvature, a journey 10 kilometers south increases an object's distance from earth's axis by only 3.4 kilometers (figure 7.17B). As we have seen, the difference in speed between two objects stationary on the earth's surface is proportional to the difference in their distance from earth's axis. Thus, because an object moving south at low latitude gains less distance from the earth's axis than an object moving the same distance south at high latitude, the more southerly object accrues a smaller difference in speed between it and the stationary objects it travels past, and has a lesser tendency to veer. In other words, *the lower the latitude, the smaller the magnitude of the north-south Coriolis acceleration.*

How low can the Coriolis acceleration go? Consider what happens at the equator. Here, as elsewhere, east-west motion relative to earth's surface changes the centrifugal force acting on an object, but because, at the equator, centrifugal force is directly in line with gravity, any change in centrifugal force cannot cause an object to move north or south. Thus, east-west motion at the equator incurs no tendency to veer either left or right. What if we move north-south at the equator? As we have just discussed, the magnitude of the Coriolis acceleration acting on objects moving north-south depends on a change in the distance from earth's axis. An object moving from 1 kilometer south of the equator to 1 kilometer north of the equator (or vice versa) would find itself at the same distance from earth's axis, and as a result, its tendency to veer east or west would be zero. These considerations show that *directly at the equator, there is no Coriolis tendency to veer either left or right.*

In summary, the tendency to veer to the right in the northern hemisphere or to the left in the southern hemisphere is greatest at the poles and decreases to zero at the equator. This is the third magic rule cited at the beginning of the chapter. As we will see, this latitudinal variation in Coriolis acceleration has an important effect on ocean currents.

A Brief Review

You have just ingested a substantial bite of physics. Perhaps your digestion will benefit from a review of the salient themes. Because the earth spins, objects stationary on its surface experience interacting gravitational and centrifugal forces and latitudinally varying east-west speeds. As a result, any object moving horizontally relative to the earth's surface veers from its original path: to the right in the northern hemisphere, to the left in the southern hemisphere. This is the Coriolis acceleration, and the faster an object moves, the greater its tendency to veer. There is one important exception to this general rule. Directly on the equator, an object can move without any tendency to veer sideways. In the next chapter, we will see that the Coriolis acceleration and its induced motions can locally help to overcome the dilemma of the two-layered ocean.

Geostrophic Flows

Before leaving the subject of the Coriolis acceleration, we need to introduce one of its consequences.

A Hill in the Ocean

Let's consider the flow that would occur if we were magically to mound up a huge hill of water in the middle of the North Pacific Ocean (figure 7.18A). When we dispel the magic by which the hill was formed and return to the physics of reality, what happens? Well, as we have noted, under the pull of gravity, water flows downhill. For a hill as symmetrical as the one we have formed, water can flow downhill by moving in any direction radially away from the hill's peak. Water to the north of the peak flows downhill to the north, water to the east of the peak flows downhill to the east, etc. (figure 7.18B). As the water flows away from the peak, the hill subsides, and eventually disappears.

Or it would disappear if it weren't for the Coriolis acceleration. As water begins to flow away from the peak, it picks up speed, begins to feel a Coriolis

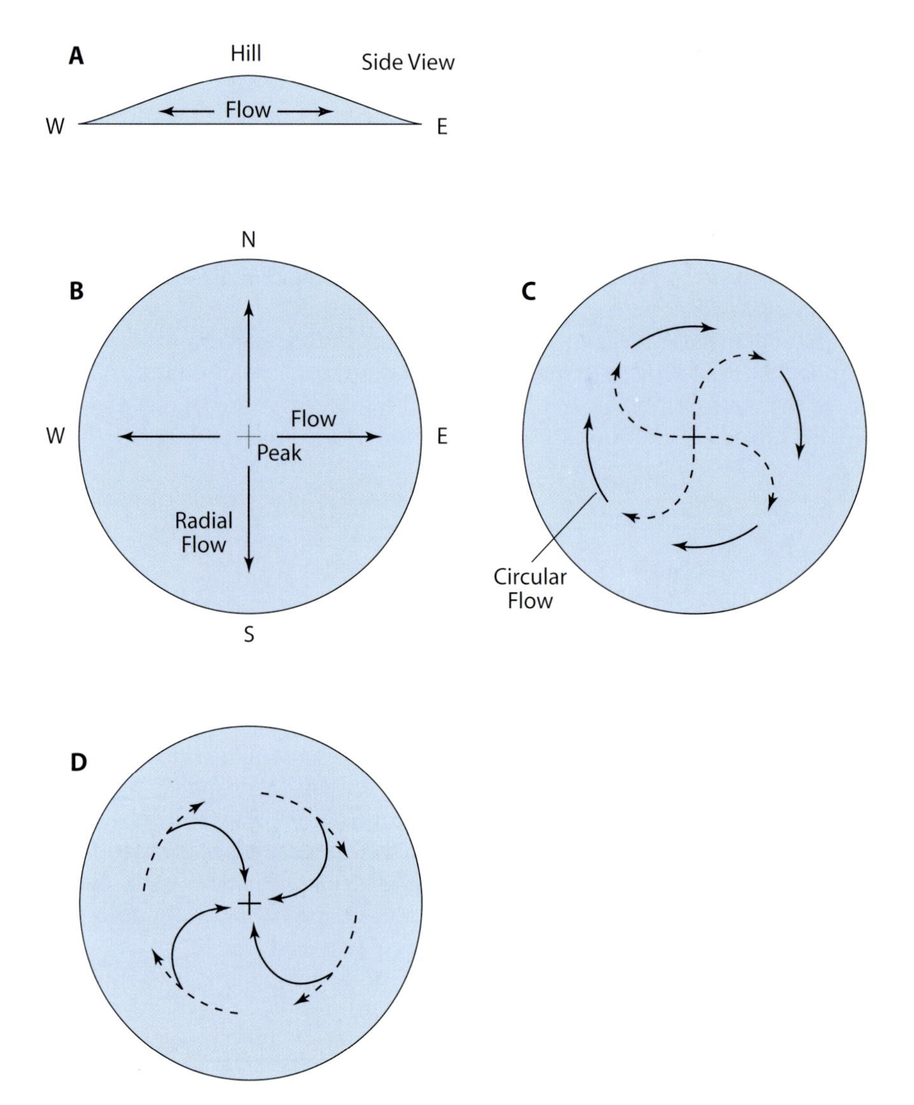

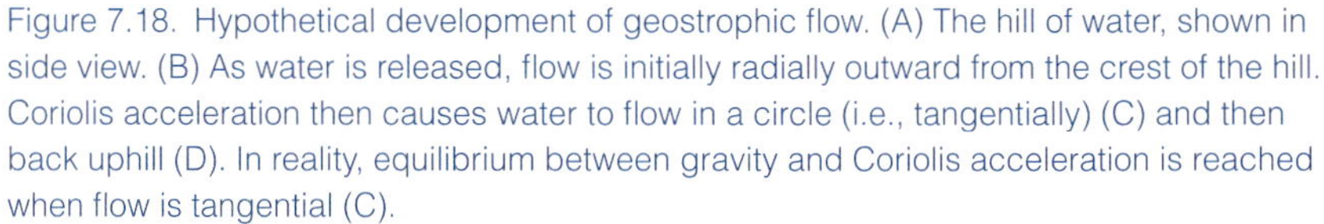
Figure 7.18. Hypothetical development of geostrophic flow. (A) The hill of water, shown in side view. (B) As water is released, flow is initially radially outward from the crest of the hill. Coriolis acceleration then causes water to flow in a circle (i.e., tangentially) (C) and then back uphill (D). In reality, equilibrium between gravity and Coriolis acceleration is reached when flow is tangential (C).

acceleration, and (because the hill is in the northern hemisphere) deflects to the right. As a result, shortly after flow begins, water that was traveling northward now flows east, water that was flowing eastward now flows south, and so forth. Because of the Coriolis acceleration, radial flow has been transformed into circular flow, with water moving in a clockwise direction around the hill (figure 7.18C).

But why stop there? The flowing water is still acted upon by the Coriolis acceleration, with the result that flow again deviates to the right. The flow to the east, which was originally flow to the north, now veers to the right and flows to the south (figure 7.18D). Flow to the south, which was originally flow to the east, now flows to the west. In every case, under the continued action of the Coriolis acceleration, the original flow eventually reverses direction. But by reversing direction, flow is now uphill. If this reversed flow were allowed to continue, it would re-form the hill that we started with.

In reality, the process never gets that far. In attempting to divert water uphill, the Coriolis acceleration opposes the acceleration of gravity, an acceleration that provides continual impetus for water to move downhill. In accord with what is becoming a general theme of this chapter, these two accelerations come to an equilibrium. It turns out that if water flows in a circuit around a hill, as depicted in figure 7.18C, the gravitational tendency for water to flow downhill is exactly offset by the its Coriolis-induced tendency to flow uphill. This pattern of motion is known as *geostrophic flow*, and the equilibrium of forces that leads to its formation is known as *geostrophic equilibrium*.

I should note that perfect geostrophic flow as described here is an ideal that is seldom seen in nature. As water flows downhill, it loses some energy to friction, and as a result, it never quite gets up enough speed so that the Coriolis acceleration can completely counteract gravity. Thus, unless energy were supplied to overcome that lost to friction, flow in a real hill of water would slowly spiral outward rather than moving in the closed circuit shown in figure 7.18C. We will return to this spiral effect when we deal with global patterns of wind.

Pressure Revisited

We can also consider the hill of water from a different perspective by harking back to our earlier discussion of pressure. Before we created the hill, water at the ocean's surface was at atmospheric pressure. By piling water on top of the ocean to make the hill, we have increased the pressure of the water under the hill. The pressure is highest under the peak of the hill, decreasing as we move radially outward, until at the hill's foot water is again at atmospheric pressure. We can thus think of the flow of water away from the peak not as flow downhill, but instead as flow from an area of high pressure—under the peak of the hill—to an area of lower pressure anywhere along the hill's base.

This perspective is especially useful when we realize we can apply the same ideas to air. If one were to build a mound of air in the atmosphere, one would create under the mound an area of high pressure at the earth's surface (figure 7.19). As a result, surface winds would initially blow radially out from the center of high pressure toward the lower pressure at the hill's periphery, but (as with flow in water, and ignoring the effects of friction) these winds eventually

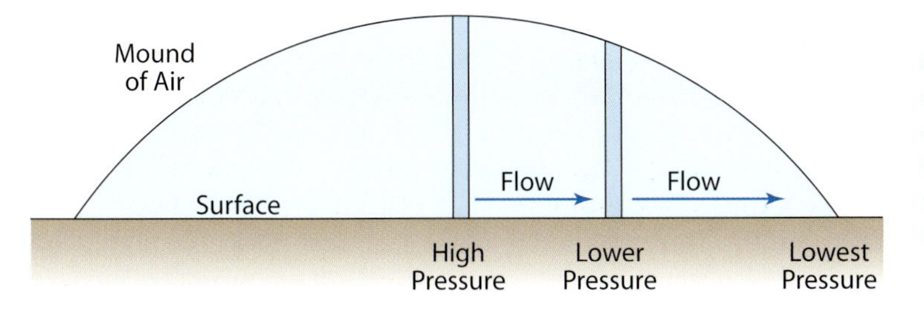

Figure 7.19. In the absence of Coriolis acceleration, wind flows down the pressure gradient, away from the crest of a hill of air.

Figure 7.20. Flow of air in a northern hemisphere hurricane. Air, which initially flows inward toward the storm's eye, veers to the right to form the winds depicted here. The photograph shows Hurricane Isabel north of Puerto Rico. Photo from the Visible Earth project at NASA (http://veimages.gsfc.nasa.gov/5862/Isabel.A2003257.1445.1km.jpg).

end up flowing in circles as they reach a geostrophic equilibrium. In other words, in the northern hemisphere, winds flow in a clockwise fashion around a center of high pressure.

A more familiar manifestation of this process becomes apparent when we reverse the locations of high and low pressure. If, instead of piling air up to create a center of high pressure, we scoop air out to make a center of low pressure, the directions of flow reverse. Initially air moves radially in toward the central low, but in the northern hemisphere, it is deflected to the right by the Coriolis acceleration. In this case, the direction of geostrophic flow is counterclockwise, as found in a hurricane or typhoon. The eye of a hurricane is an area of

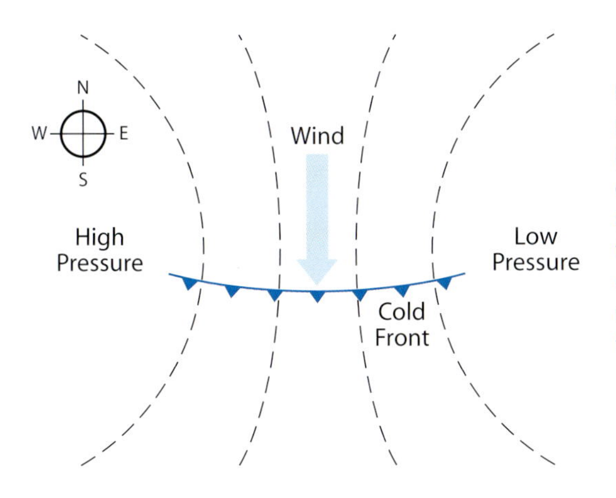

Figure 7.21. Geostrophic wind flows parallel to lines of constant pressure (isobars, the dashed lines), although one would intuitively expect it to blow across isobars from high pressure to low pressure. In this hypothetical example, cold air from the north moves with the wind, creating a cold front that straddles isobars.

intensely low pressure. As air flows inward toward the low pressure center, the eye, it swirls counterclockwise in the northern hemisphere (figure 7.20). So, as a hurricane moves westward to approach the east coast of the United States, the first winds felt at the coast come from the north. To atmospheric scientists, this flow swirling around a central low pressure is known as *cyclonic flow*.

As always, the directions of flow reverse in the southern hemisphere, so there cyclonic flows swirl clockwise. This reversal of direction is one reason why hurricanes never cross the equator. To do so, the storm would have to die out and then reform with winds in the opposite direction. In fact, the swirl created by the Coriolis acceleration is intrinsic to hurricanes' formation. At the equator, where there is no Coriolis acceleration, hurricanes can't get started.

If you are an aficionado of the weather maps published in daily newspapers, you may well have noticed this type of geostrophic effect without realizing its cause. Figure 7.21 shows a typical weather pattern in the northern hemisphere, with an area of high pressure to the west and an area of low pressure to the east. Without knowledge of the Coriolis acceleration and geostrophic equilibrium, one might easily expect the winds in this situation to blow west to east, from high pressure to low. In fact, wind blows from north to south in what (at a larger scale) is the counterclockwise direction. The southerly flow of cold northern air forms a cold front, depicted by the saw-toothed line.

If the Coriolis acceleration can have these powerful consequences at the scale of individual storms, one can only imagine the effects at the scale of the entire ocean. We will venture to this larger scale in the next chapter.

Summary

Rotation of the earth affects the motion of air and water: in the northern hemisphere, as fluids move they veer to the right; in the southern hemisphere they veer to the left. This tendency to veer (the Coriolis acceleration) depends on latitude. Directly at the equator, the Coriolis acceleration is zero, and it increases to a maximum at the poles. The Coriolis acceleration can lead to strange, nonintuitive effects. For example, in geostrophic flow, the movement of air or water downhill from an area of high pressure is counteracted by an uphill motion driven by the Coriolis acceleration. The result is an equilibrium in which fluid flows at a right angle to the pressure gradient. As we will soon see, the pattern of winds and currents on earth, and consequently the pattern of life in the oceans, would be vastly different if the Coriolis acceleration did not exist.

Further Reading

Pond, S., and G. L. Pickard (1993). *Introductory Dynamical Oceanography* (3rd edition). Elsevier Butterworth-Heineman, Oxford.

Stommel, H. M. (1989). *An Introduction to the Coriolis Force*. Columbia University Press, New York.

Appendix

Calculating East-West Coriolis Acceleration

We desire to calculate the east-west Coriolis acceleration of an object moving south on a rotating platform (figure 7.A1). For simplicity, we use a flat platform, so that any distance traveled to the south corresponds to an equal distance moved away from the center of rotation. On the earth or a roulette wheel, movement south along the surface of the equilibrium platform results in a somewhat smaller movement away from the center of rotation, due to the curvature of the platform (figure 7.17B).

Consider the motion of a ball traveling south from latitude 1 to latitude 2. Latitude 1 lies at a distance R_1 from the center of rotation and latitude 2 lies at a distance R_2 from this center. The north-south distance between the two latitudes is $R_2 - R_1 = \Delta R$. If the ball travels south at velocity V, it takes

$$t = \frac{\Delta R}{V} \qquad \text{(Eq. 7.2)}$$

seconds to travel the distance between latitudes 1 and 2.

Now, the east-west velocity of the ball at latitude 1 is

$$U_1 = R_1 \omega \qquad \text{(Eq. 7.3)}$$

where ω is the angular velocity of the surface, in radians per second. Similarly, the east-west velocity of the target at latitude 2 is:

$$U_2 = R_2 \omega \qquad \text{(Eq. 7.4)}$$

and the difference in these two velocities is

$$\Delta U = \Delta R \omega \qquad \text{(Eq. 7.5)}$$

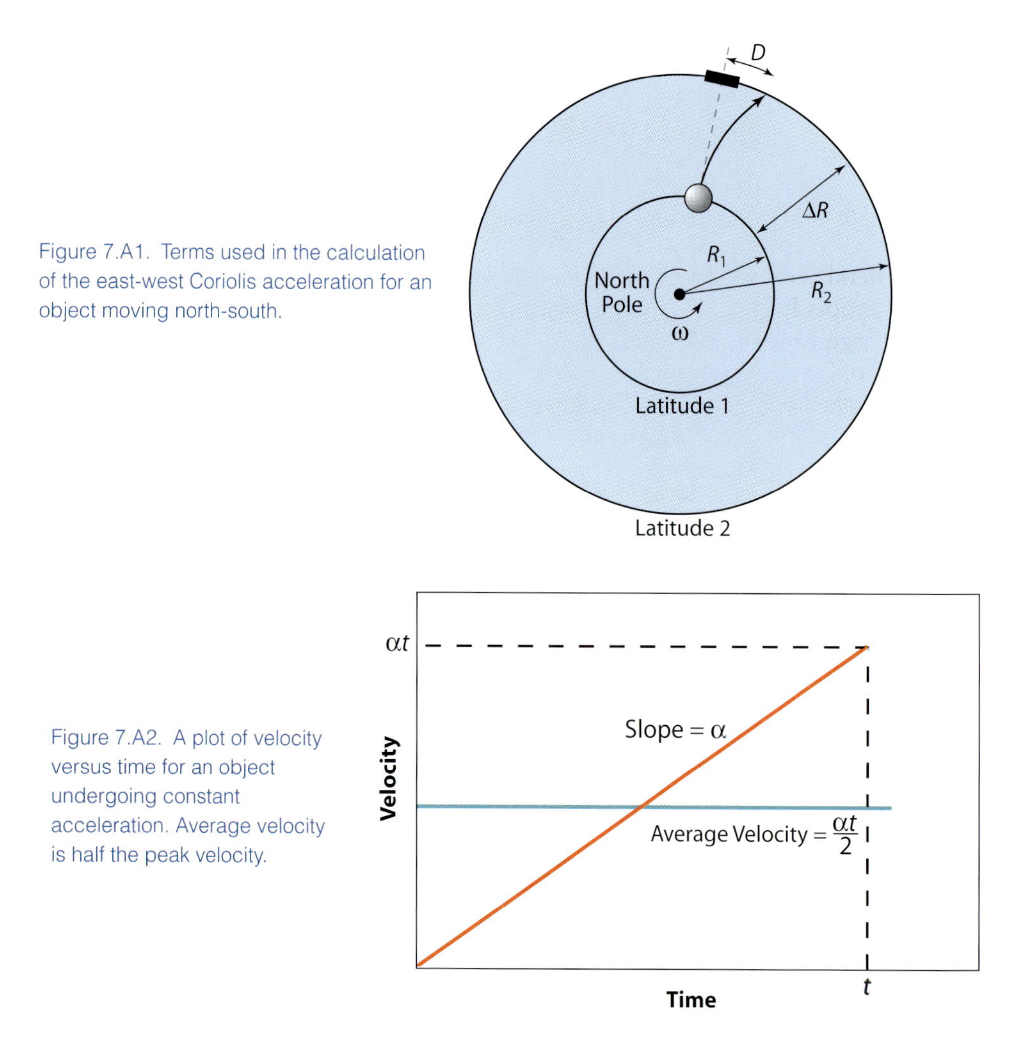

Figure 7.A1. Terms used in the calculation of the east-west Coriolis acceleration for an object moving north-south.

Figure 7.A2. A plot of velocity versus time for an object undergoing constant acceleration. Average velocity is half the peak velocity.

Due to this difference in velocity, the ball traveling south from latitude 1 will lag behind the target at latitude 2 by an amount D equal to

$$D = \Delta U t = \frac{(\Delta R)^2 \omega}{V} \qquad \text{(Eq. 7.6)}$$

We now digress for a moment to derive an important relationship. Consider an object that starts at rest and then accelerates at a constant rate α (figure 7.A2). t seconds after starting out, the object has velocity αt, and its average velocity during its journey is $\frac{1}{2}\alpha t$. Because distance traveled is the product of time and average velocity, the total distance traveled by our accelerating object is

$$D = \frac{1}{2}\alpha t \times t = \frac{1}{2}\alpha t^2 \qquad \text{(Eq. 7.7)}$$

We now apply this formula to the lateral deflection of the south-moving ball. Inserting $\Delta R/V$ for t in equation 7.7, and setting it equal to the lag distance we calculated above (equation 7.6), we see that

$$\frac{(\Delta R)^2 \omega}{V} = \frac{1}{2}\frac{\alpha(\Delta R)^2}{V^2} \qquad \text{(Eq. 7.8)}$$

and solving for α;

$$\alpha = 2\omega V \qquad \text{(Eq. 7.9)}$$

In other words, the lateral acceleration of a ball moving south—α, the Coriolis acceleration—is proportional to V, the north-south speed of the ball traveling southward. A similar argument can be made for a ball traveling northward.

Note that the factor of 2 in equation 7.9 applies only to motion on a flat platform. There is a different coefficient of proportionality for motion on platforms of other shapes, but the direct relationship of lateral acceleration with the velocity of an object is a general feature of motion on a rotating surface.

On earth, the coefficient of proportionality is $2\sin(\phi)$, where ϕ is latitude. At the North or South Pole motion on earth is essentially the same as that on the flat surface used in the calculation above. At the poles, $\sin(\phi) = 1$, and the coefficient of proportionality is 2, in agreement with our calculation. At the equator, $\sin(\phi) = 0$, indicating that there is indeed no Coriolis acceleration there.

Winds and Currents

In chapter 6, we noted that each year the sun delivers a surplus of heat to the tropics and a deficit of heat to the poles. Yet the year-to-year temperature in each region stays roughly the same. Thus, there must be mechanisms for transporting heat from the tropics to the poles and "cold" from the poles to the tropics. Indeed, we described in detail the thermohaline circulation, one mechanism that transports heat in this fashion. There are, however, two additional major mechanisms of heat transport—winds and ocean currents—and they form the subject matter for the next two chapters. Before we drift into this subject, however, we need to digress for a moment to review a few aspects of the physics of gases.

Physics of Gases

Atmospheric Pressure

In chapter 7, we examined the increase in pressure exerted by water as one travels down below the ocean's surface. Here, we extend this logic in the opposite direction, thinking about what happens as we travel up into the atmosphere.

As with pressure in water, the air pressure we experience at earth's surface results from the weight of the column of fluid above. In the atmosphere, this column consists of air alone. So, when we note that air pressure at sea level is 1.0×10^5 newtons per square meter, we mean that a column of air extending above one square meter of ground tips the scales at 1.0×10^5 newtons. We can intuitively conclude that the higher up in the atmosphere we go, the shorter the column of air above us, and the less it weighs. Therefore, air pressure decreases with altitude. For example, the top of Mt. Everest (8848 m above sea level) extends above 2/3 of the earth's atmosphere. The weight of the remaining atmosphere exerts only 1/3 of the pressure at sea level, about 3.3×10^4 newtons per square meter.

This effect is familiar to anyone who has ever flown in an airplane. As the plane ascends, the cabin pressure decreases while the pressure in your middle ear initially stays the same as it was on the ground. The difference in pressure across your eardrum causes it to distend, and that hurts.

The Laws of Gases

The air we have just weighed is a mixture of gases, primarily nitrogen and oxygen. Now, a gas consists of molecules moving around at random, bouncing off each other like billiard balls. The higher the temperature, the faster the molecules move, and the more forcefully they rebound from their collisions. From this behavior we can arrive at several conclusions about the relationship between the temperature and volume of a gas.

First, let's imagine what happens as temperature rises and gas molecules in a group move faster. Because molecules now push each other away more forcefully, the group tends to take up more space. In other words, as temperature increases in the atmosphere, a gas expands. Cooling has the opposite effect. As a gas cools, its molecules move more slowly, they rebound with less force, and the volume of a given group of molecules decreases.

This relationship can be turned around: a change in the volume of a gas can affect its temperature. This effect may be familiar to you. As you expel gas from a can of spray paint, or air from a scuba tank, the can or tank gets cold. That is, as you allow the volume of a gas to increase, the gas's temperature goes down. Conversely, when the volume of a gas is forcefully decreased, its temperature goes up. You might have noticed this effect when using a hand pump to inflate a bicycle tire. As you compress air in the pump, the barrel of the pump can get quite warm.

Next, we note a relationship between the volume and density of a gas. By definition, the density of a material is its mass per given volume. A fixed number of gas molecules in the atmosphere has a fixed mass, but as we have just seen, the volume associated with this mass depends on its temperature. Heating the gas makes it expand. With the same number of gas molecules now spread over a larger volume, there are fewer gas molecules in each cubic meter of gas. Fewer molecules means less mass, and as a result, the gas has a lower density. Conversely, cooling the gas increases its density.

These changes in density can in turn affect the buoyancy of a volume of gas. Buoyancy is the net force acting on an object of one density when it is surrounded by fluid of another density. As we discussed when exploring the lives of phytoplankton, if an object is more dense than the surrounding fluid, it sinks; if it is less dense, it floats. In the atmosphere, buoyancy is most familiar in the context of hot-air balloons. The hot air in the balloon has a lower density than the cool air outside, and as a result, the balloon is buoyed upward. Although it is more difficult to observe, the same effect occurs for air that is not confined in a balloon. As unconfined air is heated, it expands, and its density decreases. The decrease in density leads to a buoyant force, and the air rises. Hot, rising gas is responsible for the shimmer one sees as sunlight passes through the air over a hot road.

These relationships may be familiar, but it is worth making sure that they sit well with your intuition. They will be used at several points in the arguments that follow.

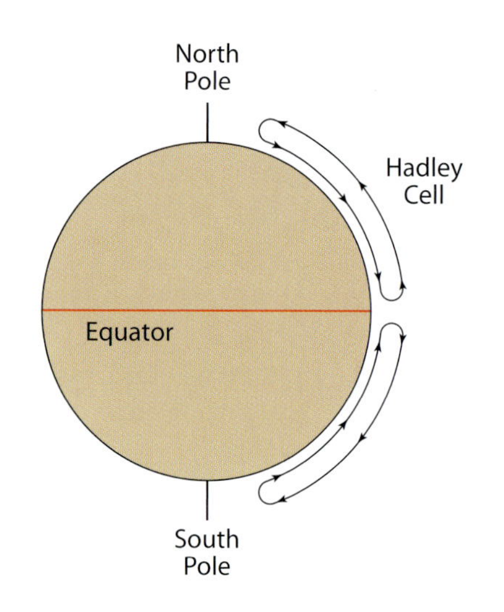

Figure 8.1. Idealized Hadley cells. Warm, moist air rises near the equator; cold, dry air sinks at high latitude.

Part 1. Winds

Ideal Hadley Cells

In 1735, George Hadley (a London lawyer and philosopher) proposed an explanation for the global pattern of air circulation on earth. With a foresightful understanding of latitudinal variation in solar heat delivery, he supposed that air near the equator is heated more than air near the poles. As we have just seen (and as Hadley knew), heated air expands and becomes less dense. As a result, the hot air at the equator, less dense than the air around it, rises like a hot-air balloon. By the same token, air high in the atmosphere near the poles loses heat, becomes more dense, and sinks. Hadley suggested that rising air in the tropics and falling air near the poles are linked together in a grand convection cell, as shown in figure 8.1. This model requires that winds at altitude travel poleward from the equator and that, more importantly for the oceans, winds at the earth's surface flow equatorward from the poles. In honor of its discoverer, this pattern of atmospheric convection is known as a *Hadley cell.*

Although Hadley glimpsed the basic physics of atmospheric circulation, a few details must be added before his ideas can be put to practical use. Taking the Hadley cell as a starting point, let's examine the actual pattern of air flow on earth.

The Density of Humid Air

Near the equator, the large influx of solar radiation heats both the ocean surface water and the atmosphere above it, just as Hadley supposed. As the ocean's surface is heated, water evaporates, and as a result, the warm air above the ocean in the tropics is quite humid. Anyone who has spent time around a tropical sea can attest to the humidity of tropical air.

We already know that, at a given altitude, warm air is less dense than cold air. But in fact, humid warm air is less dense still. The low density of humid air may seem counterintuitive. In North Carolina, where I grew up, we used to describe hot, humid summer air as "heavy." I was surprised, then, to learn as a physics student in college that we had it backwards; our summer air was actually "light."

At a fixed temperature and pressure, a given volume of air contains a constant number of gas molecules.[1] It doesn't matter exactly what these molecules are; it is their *number* that is important. In terms of air's density, however, the composition of the gas molecules does matter. If each molecule in the volume weighs a lot, the overall mass of molecules in the volume is large, and by definition, the gas is dense. Conversely, if we were to replace some of these heavy molecules with lighter ones, keeping the overall number of molecules the same, the gas would be less dense. This replacement of heavy molecules with lighter ones occurs when air becomes humid.

Dry air consists of approximately 79% nitrogen and 21% oxygen. Both molecules are relatively heavy: a nitrogen molecule (N_2) weighs 28 daltons, and an oxygen molecule (O_2), 32 daltons. In contrast, as we have seen, a molecule of water vapor (H_2O) weighs only 18 daltons. Thus, if we replace a nitrogen or oxygen molecule with a molecule of water vapor, the overall density of the air decreases. In fact, fully saturated air at room temperature is about 1% less dense than dry air.

Air in the Tropics

So, being both hot and humid, air over the tropical ocean is less dense than air elsewhere. This low density sets off a chain of four events:

1. Because the air is less dense, it rises, just as a hot-air balloon would rise.
2. As the air rises in the atmosphere, the pressure exerted on it goes down.
3. As the pressure on the rising air decreases, the air expands.
4. And, as air expands, it cools.

Now, cool air cannot hold as much water vapor as hot air can. This is why a can of cold beer set out into hot, humid North Carolina air will quickly be coated in water droplets. Air next to the beer can is cooled, it becomes less able to hold water vapor, and the excess vapor condenses on the can. By the same process, as humid tropical air rises above the ocean, it cools to a temperature at which it is saturated with water, and the water vapor begins to condense. The condensed vapor forms clouds and rain. Again, anyone who has spent time around tropical seas can confirm the frequency with which clouds form and rain falls. In the process, the water that evaporated out of the ocean returns to it.

As we noted in chapter 3, condensation of water vapor is a curious process. We are all familiar with the cooling associated with evaporation of water: your body cools itself on a hot day primarily through the evaporation of sweat. In effect, heat energy is pumped into sweat produced on your skin, and this energy is carried away as the sweat evaporates. The net loss of energy makes you

[1] A relationship first proposed by Amedeo Avogadro (1776–1856), an Italian mathematical physicist. Avogadro's number, the number of atoms in a mole, is named in his honor.

cooler. Conversely, water vapor condensed back out of the air releases the heat that evaporated it. Thus, when clouds form, water vapor condenses, and its latent heat is released. The latent heat released by condensation in tropical clouds increases the temperature of the air. This dry air then expands even more, becomes less dense, and continues to rise.

So, to summarize, solar heating of the tropical sea pumps hot, dry air high into the atmosphere.

Air at Altitude

Eventually, tropical air reaches a height where its density equals that of the air around it, and, as a consequence, it stops rising. However, new air still arrives from below, and continuously displaces the warm dry air at altitude.

Where does the displaced air go? It can't go up because it is denser than the air above it. It can't go down because it is less dense than the air below it. Its only route of escape is sideways, and the path of least resistance is toward the nearest pole. In other words, warm tropical air that rises north of the equator eventually flows northward, and warm air that rises south of the equator eventually flows south.

As this high-altitude air moves poleward, it exchanges heat with its surroundings, gaining heat from the sun and radiating heat back into space. However, by moving away from the equator, the air moves into an area where solar heating is less intense. The air still radiates heat out to space at a rate proportional to its temperature, but it gets less heat back from the sun. As a result, the air cools as it travels to the north or south. As the air cools, it becomes more dense. At a latitude of approximately 30° North or South, the dense, high-altitude air sinks, forming a downdraft.

The Hadley Cell Revisited

We need to consider two relevant consequences of these facts. First, by moving poleward from the equator, air transports heat away from the tropics—exactly the kind of transport that must take place to maintain a more or less constant temperature everywhere on earth. In fact, wind carries about half the required heat flux.

Second, these flows affect atmospheric pressure at the earth's surface. Because air is heated in the tropics, the earth's surface in this region sits under a column of low-density air, and the sea-level air pressure is correspondingly low. Thus, a band of the earth roughly 10° of latitude wide surrounding the area of maximum solar intensity is characterized by low atmospheric pressure, the *equatorial low*. The location of this band varies with the seasons. In contrast, the region between 25° and 35° North or South sits under a column of cooled, high-density air and therefore has a persistent high pressure at sea level, the *subtropical high*.

The difference in sea-level pressure between the subtropical high and the equatorial low provides a mechanism to complete the cycle of flow we have sketched out. At the earth's surface, air moves down the pressure gradient from the subtropical high toward the lower pressure near the equator. For a more quantitative explanation of this type of convective circulation, consult the appendix to this chapter.

Let's follow the cycle around. Air heated in the tropics rises. It then flows poleward at high altitude and loses heat as it goes, finally sinking back to earth as a dense downdraft at a latitude of about 30°. The resulting high pressure at the earth's surface forces air equatorward to start the whole process over again. This motion resembles the original Hadley cell, except that Hadley supposed that air would not fall back toward the earth's surface until it reached a latitude of at least 60° North or South. If the real Hadley cell reaches only to the subtropics, what happens more poleward?

The Ferrel Cell

Here the picture gets a bit messy, but many aspects of global atmospheric circulation can be explained if we suppose that, at latitudes poleward of the Hadley cells, a second pair of cells exists (figure 8.2). Called *Ferrel cells* after William Ferrel (1817–91), a self-taught geophysicist from Tennessee, these hypothetical cells extend from roughly 35° to 60° North and South and rotate in the opposite direction to the Hadley cells.

Let's suppose that the types of pressure difference that drive the Hadley cells also cause the motion of the Ferrel cells. However, in order for the Ferrel cells to rotate in the opposite direction, the positions of the high and low pressure areas must be reversed. In other words, given the area of high pressure in the subtropics, if we want the Ferrel cells to "run in reverse," there should be a center of low pressure at about 60° North and South. The difference in pressure between these subpolar lows and the subtropical highs would cause surface winds to blow poleward. In the northern hemisphere, for instance, surface winds in the Ferrel cell would flow northward from 35° North to 60° North. In the southern hemisphere, surface winds in this second cell would flow southward from 35° South to 60° South. To complete the Ferrel cells in a manner analogous to the Hadley cells, winds at altitude in the Ferrel cells would blow equatorward, joining the downdraft of the Hadley cell at 30° North and South.

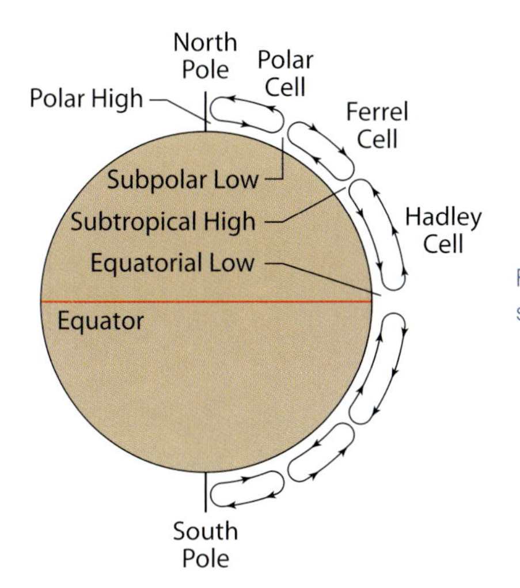

Figure 8.2. A hypothetical three-cell system of winds on earth.

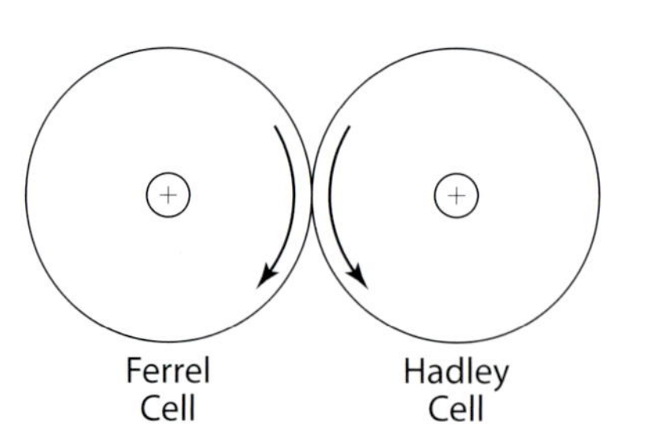

Figure 8.3. The Ferrel cells rotate as if driven by the Hadley cells.

If the Ferrel cells exist, something must cause air to rise in the subpolar low. Because this low occurs where solar heat influx is small, solar heating cannot power the updraft. Instead, flow in the Ferrel cells is driven by large mid-latitude storms[2] and associated phenomena, which complete the heat-transport work of the Hadley cells by carrying warm, moist, subtropical air poleward and upward and drawing cold, dry, subpolar air equatorward and downward. Details of the physics are complex, so for present purposes, let's simply pretend that "friction" with the low-latitude Hadley cells drives these second cells, just as a rotating wheel in contact with another wheel drives rotation of the second wheel in an opposite direction (figure 8.3). This supposition will lead to problems later on—indeed, the whole idea of a defined Ferrel "cell" has problems—but it correctly predicts surface pressures and winds and has intuitive appeal. We will proceed as if the Ferrel cells were real and deal with the problems as they arise.

If the Ferrel cells act as if they are driven by the Hadley cells, could the Ferrel cells contribute to yet another pair of cells farther toward the poles? If so, we can complete our picture of the global wind pattern by hypothesizing a polar high pressure that would drive surface winds equatorward toward the subpolar low.

The overall pattern of circulation would then look like figure 8.4A: three sets of counter-rotating convection cells in each hemisphere, with alternating bands of northward and southward winds flowing between alternating areas of high and low pressure. It is a compellingly simple picture, and some parts of it are even true. However, there are two aspects of this scenario that we need to consider in more detail.

Short-Term Complexity, Long-Term Simplicity

The next time you have a chance, tune in to the Weather Channel and watch their satellite movies of the globe. Does the complex pattern of surface winds evident in these pictures match the simple pattern we have drawn here? Well, no, and there are two reasons why the patterns differ. First, the explanation I have presented so far has not taken the Coriolis acceleration into account. We

[2] Technically, the storms referred to here are extratropical cyclones, the mid-latitude analogue of tropical cyclones (hurricanes and typhoons). Extratropical cyclones differ from their tropical cousins in several respects, most notably that wind shear at these latitudes, the rapid variation in wind speed with altitude, prevents the formation of an eye. Nonetheless, extratropical cyclones are forceful entities, capable of powering circulation in the Ferrel cells.

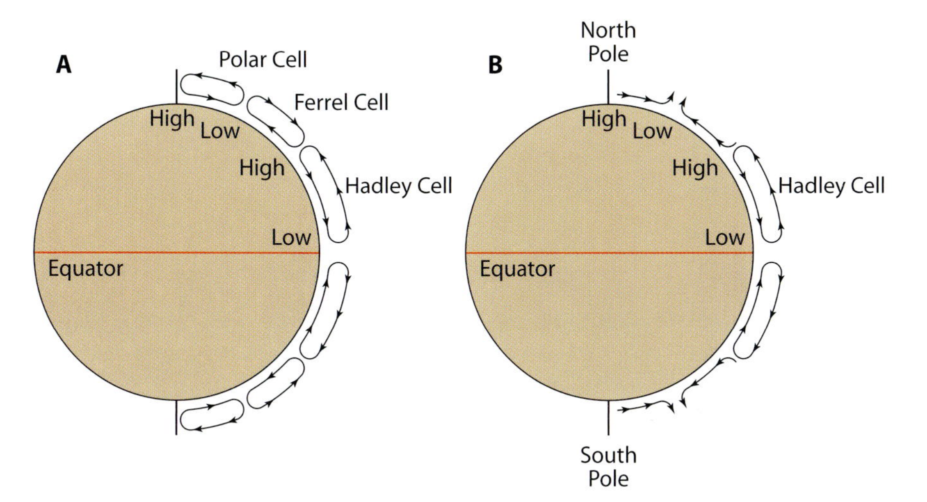

Figure 8.4. The three-cell model of winds on earth (A) is simple and elegant in concept, but only parts of it (B) match reality.

will deal with the Coriolis acceleration presently, but before we do, we need to acknowledge a second reason why our explanation does not match the Weather Channel's movies. The movies show the atmosphere as it varies from hour to hour on one particular day, whereas our explanation describes how the atmosphere moves on average.

At any one time and at any single location on earth, the speed and direction of the wind are difficult to predict. The fickle nature of local winds befuddles sailors and contributes to the skepticism with which we view weather predictions. In contrast, if you step back from the local scale and examine winds over larger areas and longer times, clear patterns emerge. For instance, the fickle hour-to-hour variation in winds at your house makes more sense when viewed in the context of the large-scale flow evident in a day-long satellite movie. This is why television meteorologists use these movies to explain the local weather.

But, while local winds make sense in the context of a satellite movie, the larger-scale flows viewed from space are themselves complex. Storm fronts come and go, often with no readily discernible pattern. When viewed at their own large scale, global flows seen from space seem just as unpredictable as the small-scale flows at your house.

However, if you were to average over several years the complicated swirls and eddies you see in satellite images, much of the instantaneous complexity would again average out, leaving a clearer, simpler pattern. It is this long-term global-scale average pattern that we explore here, whereas we will be able to account for neither the short-term nor the small-scale complexity.

With that said, how much of our idealized scenario (figure 8.4A) matches this global-scale, long-term reality? The low-latitude Hadley cells certainly exist, and our explanation for their existence is correct in its essentials. For the Ferrel cells, long-term records reflect the areas of subpolar low pressure along with the surface winds that blow poleward from the subtropical highs. But the high-altitude winds that would complete the Ferrel cells by flowing equatorward don't seem to exist. In fact, the high-altitude poleward flow typical of the Hadley cells continues right on up through the latitudes where the Ferrel cells should be found. Thus, while the surface aspects of the Ferrel cells exist, the high-altitude aspects don't. If this were a book on meteorology, we would have some explaining to do. For present purposes, we are most concerned with surface winds (because these are the winds that interact with the ocean), and we conveniently ignore the fact that there are loose ends at high altitude.

Evidence of a third, distinct pair of convection cells near the poles is somewhat sketchy. However, as with the Ferrel cells, some aspects of polar cells are present. As we would predict, the surface atmospheric pressure is typically high near the frigid poles, and as a result, there are time-averaged equatorward surface winds that flow southward. These winds meet the surface winds from the Ferrel cells in *polar fronts* around latitude 60° North and South, adding to the updraft associated with the subpolar lows. At altitude, winds near the poles are complex. For example, jet streams—high-altitude, high-speed winds blowing from west to east—generally follow the line of the polar front. It would be difficult to explain the existence of these jet streams, much less their pattern of flow, using Ferrel and polar cells as a model. Again, we ignore these high-altitude loose ends.

When we pare down our hypothetical world to match long-term-averaged reality, we are left with the components shown in figure 8.4B. We will use this model as the basis for subsequent exploration of surface winds.

Enter Coriolis

It is now time to consider the effect of Coriolis acceleration on surface winds. For ease of description, we initially confine ourselves to discussion of the winds in the northern hemisphere. Once we understand these winds, we can extrapolate our findings to the southern hemisphere.

Trade Winds. Let's begin with the winds that blow equatorward from the high pressure area at 30° N. If the earth were not rotating, these winds would blow directly to the south (figure 8.5A). These winds are "northerlies," named (in the tradition of such things) for the direction from which they come rather than the direction in which they blow. However, because the earth does rotate and the Coriolis effect applies, these winds are deflected to the right. So, instead of flowing to the south, they flow to the southwest, at a steady speed of 5–7 meters per second (figure 8.5B).

As noted above, the Hadley cells that account for these winds are reliably present in the long-term average. At the scale of a ship, the result is a source of sail power, and these "northeasterlies" are more familiarly known as the *trade winds*. It is the trade winds that sailors of past times used to travel from Europe and Africa to North and South America.[3]

As the trade winds approach the equator, two things happen. First, the tendency to veer to the right decreases, because, as we have seen, the Coriolis acceleration disappears at the equator. The disappearance of the Coriolis acceleration, along with friction between wind and ocean, explains why the winds

[3] It is commonly thought that the trade winds owe their name to their role in ship-borne commerce. Apparently not. Humphreys, in his classic treatise on the *Physics of the Air*, notes that a now-archaic usage of the word "trade" is as an adverb meaning "consistent, proceeding in a constant direction." Thus, winds that blow trade blow consistently, an apt description of the subtropical northeasterlies.

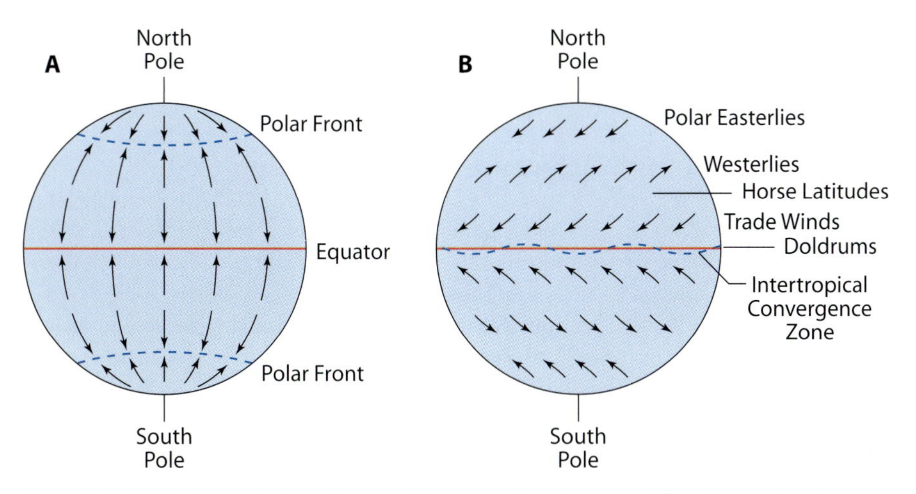

Figure 8.5. (A) Surface winds as they would occur in the absence of Coriolis acceleration. (B) Surface winds in the presence of Coriolis acceleration.

shift only to the southwest, rather than all the way to the west. Second, as the winds enter the area of equatorial low pressure, the impetus for them to flow decreases. Remember that the equatorial low is an area of rising air, in which the winds switch from flow parallel to the ocean's surface (to the southwest) to flow perpendicular to the surface (an updraft). In general, for latitudes within 5° of the center of the equatorial low, the winds are light, 3–5 meters per second, and variable in their direction.

The abatement of the trade winds poses a problem for sailors traveling from one hemisphere to the other. While in the trade winds, a sailing ship can trim its sails appropriately and make good progress north or south. Near the equator, however, the winds slacken, and progress slows. Both for the effect it has on the ship's forward progress and the effect it has on the mental state of the crew, this area is known as the *doldrums*.

The Westerlies. We now shift our attention to the flows just north of 30°. If the earth were not rotating, these winds would flow directly northward (figure 8.5A). Instead, under the urging of the Coriolis effect, they veer to the right, and flow to the northeast (figure 8.5B). These winds, technically southwesterlies, are usually referred to simply as the *westerlies*.

The westerlies also have a role in nautical history. Having sailed before the trade winds to arrive in the New World, sailors of Columbus's day returned to Europe by first sailing north along the east coast of North America, eventually catching the westerlies for a ride home. On the way to the westerlies, however, they had to cross through the subtropical high, roughly 25° to 35° N. As with the doldrums, this area of reduced winds impedes sailors' progress. In this case, the area is known as the *horse latitudes*. One story has it that the name originated as becalmed ships returning to Spain ran low on water and tossed their horses overboard rather than watching them die and rot amidst the crew.

Polar Easterlies. And lastly, we consider flow north of 60°. In the absence of Coriolis acceleration, these winds would flow to the south, away from the polar high pressure and toward the subpolar low. Earth's rotation causes these winds to veer to the right, causing them to flow from northeast to southwest. These are the *polar easterlies.*

The Southern Hemisphere

Polar easterlies, westerlies, horse latitudes, trade winds, doldrums: that is the north-to-south pattern of winds in the northern hemisphere. In the southern hemisphere, two things switch simultaneously. First, the winds that flow from the subtropical high to the equatorial low (the trade winds) flow to the north, not to the south. Second, the Coriolis acceleration makes objects veer in the opposite direction—to the left—in the southern hemisphere. As a result of this double shift, the trade winds in the southern hemisphere blow with a westward component, as do the trade winds in the northern hemisphere. Similarly, the winds from 30° S to 60° S blow to the east, and are called westerlies, as with their counterparts in the north. And winds poleward of 60° S blow to the west: the southern polar easterlies. The southern hemisphere also has a set of doldrums and horse latitudes mirroring those in the north.

The Intertropical Convergence Zone

Now, if the trade winds in the northern hemisphere blow to the southwest and the trade winds in the southern hemisphere blow to the northwest (figure 8.5B), you may be wondering, what happens when they collide? Actually, we already know the answer. As the two sets of winds converge in the doldrums, air is forced upward, away from the ocean's surface, as part of the updraft that powers the Hadley cell. This line, along which the two sets of trade winds meet, is in a sense the meteorological equator. The technical term is the *intertropical convergence zone* (ITCZ), affectionately referred to as "the itch." The ITCZ is readily visible in satellite pictures of the earth, marked by a band of clouds that result from the condensation we discussed earlier.

The ITCZ and the Seasons

Recall that the pattern of heat influx and efflux on earth drives the winds we have just described, and the location of this pattern on the globe varies with the seasons. For example, as a result of the tilt of earth's axis, June 20, 21, or 22 is the summer solstice in the northern hemisphere, the longest day of the year. At noon on the solstice, the sun is directly overhead at a latitude of 23.5° North (the Tropic of Cancer) rather than at the equator (figure 2.2). Thus, we would suspect that, in June, the ocean is heated the most to the north of the equator and that the pattern of winds will have moved northward as well. Conversely, on the winter solstice in the northern hemisphere (December 21 or 22), the sun is directly overhead at 23.5° South (the Tropic of Capricorn), and we would expect the pattern of winds to have shifted to the south.

This north-south shift in the pattern of winds, often documented as a shift in the location of the intertropical convergence zone, does occur, but not to the extent this explanation might suggest. Shifting heat around in both the water of the ocean and the land of the continents requires considerable time. As a consequence, by the time the ITCZ has just begun to shift substantially northward, the northern summer is over, and it is time to shift southward again. So, yes, the intertropical convergence zone shifts with the seasons, but only by about 15°, rather than by the 47° (23.5 × 2) one might expect.

The thermal properties and distribution of land on earth also affect the location of the ITCZ. First, the thermal properties: because soils and rocks have a lower specific heat than water, and because they are not subjected to turbulent mixing, terrestrial surfaces reach higher temperatures than the ocean surface when exposed to the same solar energy influx. As a consequence, air over land can be heated more than air over water. This augmented heating is compounded by the distribution of land. As we learned in chapter 2, well more than half of earth's land area is concentrated in the northern hemisphere. Thus, over the course of a year, air is heated more in the northern hemisphere than in the southern, and as a result, the intertropical convergence zone over the oceans is located, on average, 5° to 10° North rather than around the equator. This slight northward shift will have practical consequences when we deal with the details of air flow at the equator.

Monsoons and the Indian Ocean

This is a good time to introduce a major exception to the rules. The pattern of winds we have so carefully defined can be applied with assurance to the Pacific and Atlantic Oceans, but the Indian Ocean is different. The reasons stem from geography.

Unlike the Pacific and Atlantic, which are spread more or less equally across the northern and southern hemispheres, the Indian Ocean lies mostly south of the equator (figure 8.6). Furthermore, the small fraction of Indian Ocean north of the equator is split in two by the Indian subcontinent. Thus, we cannot expect the winds in the northern Indian Ocean to resemble those in the Pacific or Atlantic, and they don't. The North Indian Ocean is bounded to the north by Asia, the largest continent on earth. As a consequence, its winds are affected more by continental effects than are those over the Pacific or Atlantic.

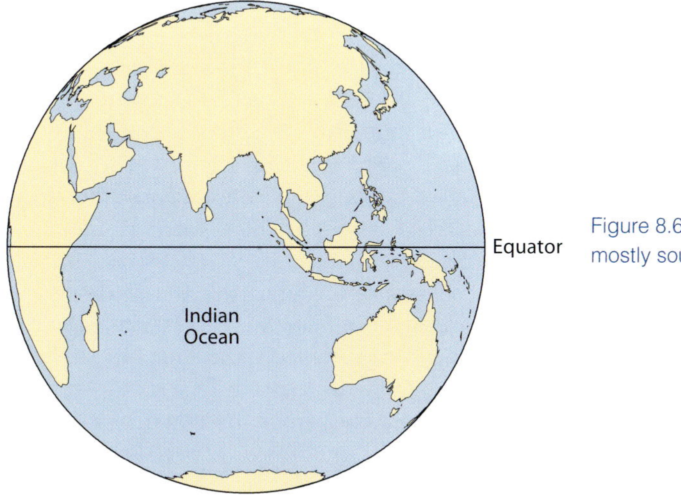

Figure 8.6. The Indian Ocean lies mostly south of the equator.

When it is winter in the northern hemisphere, the land mass of Asia (and especially the Tibetan plateau) is cold, and the cold air above it forms an area of high pressure. In contrast, the equatorial Indian Ocean is warm, forming an area of low pressure. Driven by this pressure gradient, and diverted by Coriolis acceleration, winds flow from east to west across the Indian Ocean, resembling the trade winds in other oceans.

But only in the northern winter. In northern summer, the pattern reverses: the Tibetan plateau heats up, leading to the formation of an area with a pressure lower than that of the tropical ocean. This reversal in the pressure gradient between Asia and the Indian Ocean leads to a reversal in the winds. Thus, during the northern summer, the trade winds of the North Indian Ocean blow from west to east.

These alternating winds are the basis of the monsoons that are such an integral part of life in India. The wintertime winds from Asia are dry, and rain is scarce. The reversal of winds in the summer brings humid oceanic air and the consequent rains. We will revisit the monsoon winds when we explore ocean currents.

Winds over the southern Indian Ocean are similar to those in the South Atlantic or South Pacific. However, the speed of the trade winds in the South Indian Ocean varies seasonally: fast in the southern winter, slow in the southern summer.

Winds and the Transport of Heat

Despite the twists introduced by the Coriolis effect, the high-altitude flow of air in the Hadley cells transports heat out of the tropics, as does the low-altitude poleward flow from 30° to 60°. The equatorward surface flow in the trade winds transports "cold" from temperate to tropical seas, and the polar easterlies carry "cold" away from the poles. Thus, the flow of air in the global pattern of circulation augments thermohaline circulation, and helps maintain constant temperature everywhere on earth. As noted earlier, about half the heat flux on earth is accounted for by the wind.

Deserts

Before leaving the subject of global patterns of wind, we note one effect this pattern has for life on land. Recall that the downdraft of air at a latitude of approximately 30° North and 30° South is fed by air coming from the tropics. In the process of rising, most of the water has already condensed out of that air, so that when the air arrives at the downdraft it is dry. This downdraft of dry air suppresses the formation of clouds and consequently of rain. Therefore, one might expect that land areas at roughly 30° North and South would be drier than other areas on earth.

This is indeed the case. Many of the earth's major deserts (the Sahara and Kalahari in Africa, the Arabian, Iranian, and Takla Makan in Asia, the Mojave, Sonora, and Chihuahua in North America, the Atacama in South America, and the Great Sandy, Gibson, and Great Victoria Deserts in Australia) occur at approximately the latitudes predicted from the downdraft of air in the Hadley cells (figure 8.7).

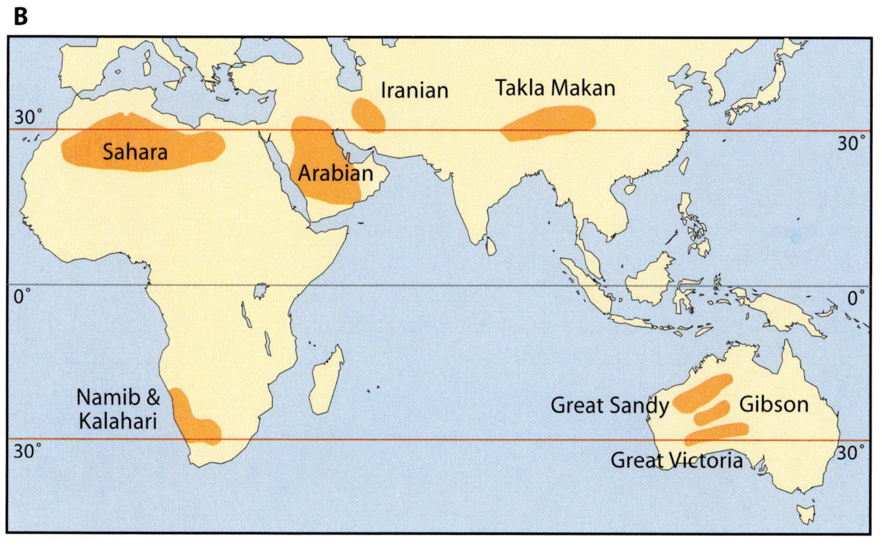

Figure 8.7. Earth's major deserts lies near 30° North and South. (A) Deserts in North and South America. (B) Deserts in Africa, Asia, and Australia.

Summary

The latitudinal variation in solar heating drives a grand heat engine in the atmosphere. Hot air rises in the tropics, forming a low-pressure area, and flows poleward at high altitude, losing heat along the way. Eventually this air becomes cold and dense enough to fall back to earth, forming a high-pressure area at a latitude of about 30° North or South. Surface winds then flow equatorward to complete the Hadley convection cell. Components of a second pair of cells (the Ferrel cells) are found at higher latitudes, driving surface winds that flow north toward the subpolar low, and winds blow equatorward from the polar high. All these surface winds are diverted by the Coriolis effect, forming the trade winds, the westerlies, and the polar easterlies, respectively. As a part of this atmospheric motion, heat is transported away from the tropics toward the poles, augmenting the transport due to thermohaline circulation.

For the oceans, the trade winds, westerlies, and easterlies have an equally important effect: they drive ocean currents.

And a Warning

Before we turn to the subject of ocean currents, we should put our knowledge of winds into perspective. I have painted a picture of the winds with very broad strokes, and have glossed over many details. For example, the equatorial and subpolar lows and the subtropical highs have been described as continuous bands that girdle the earth. In reality, these high- and low-pressure areas tend to occur as alternating, isolated centers rather than as bands. In January, there are persistent centers of low pressure over northern Australia, equatorial South America, and equatorial Africa, interspersed with subtropical centers of high pressure in the Pacific and Atlantic Oceans. A similar alternation of highs and lows occurs at subpolar latitudes, and the jet streams wander north and south among these centers like a skier slaloming through moguls. Similarly, rising and sinking flows in the Hadley cells, which are described here as occurring in continuous bands in the tropics and subtropics, respectively, are in fact more confined. For example, one upward branch of the Hadley cell is centered over the Amazon rain forest, feeding circulation that sinks in either the persistent high-pressure center over Bermuda or the corresponding center in the South Atlantic. The pattern of highs and lows, updrafts and downdrafts changes with the seasons, shifting both north-south and east-west, and at any moment it can be very complex. The basic ideas we have presented here do not explain this complexity, and we should be humble in the knowledge of our ignorance.

Fortunately, we do not need this level of detail to proceed. The simple, time-averaged picture we have explored is true enough for our purposes.

Part 2. Ocean Currents and the Basinwide Tilt

The solar-powered motion of air drives currents in the ocean's surface waters, and we now venture to predict and explain the pattern of these currents. Before we can make these large-scale predictions, however, we need to focus on the small-scale interaction between wind and water.

Friction and Ekman Transport

When wind blows over a stationary surface, either water or land, friction arises between the air and the surface. A familiar manifestation of this friction can be observed as you walk barefoot along a sandy beach on a windy day. The friction between air and sand drags sand particles downwind. You can see the sand move and feel the sting of particles hitting your ankles. The same process occurs as wind blows over water, but with results less visually obvious: the friction between air and water drags the surface water downstream. The faster the wind blows, the larger the frictional force imposed on the water.

This concept of friction between fluids—air and water—can be applied within a single fluid as well. As water at the ocean's surface begins to move under the urging of the wind, it imposes a frictional force on the water just below it. This water then tugs on the water farther down, and in this fashion, the effect of the wind extends down into the water column.

As with any flow on the face of the rotating earth, the water flows induced by wind friction are subject to the Coriolis acceleration. The effect in this case is quite curious. Surface water is initially pulled in the direction of the wind, but the Coriolis effect diverts it a bit to the right in the northern hemisphere or to the left in the southern hemisphere. The actual diversion is about an eighth of a full circle, 45°. This diverted water flow drags with it the water below. Again, due to the Coriolis effect, the water below veers a bit relative to the already diverted flow above. In this fashion, the direction of flow veers more and more from the direction of the wind as one looks deeper and deeper into the ocean (figure 8.8), an effect known as the *Ekman spiral*, after Vagn Walfrid Ekman (1874–1954), a Swedish oceanographer who explained the concept to account for observations made by Fridtjof Nansen on the movement of Arctic ice.

Flow in an Ekman spiral not only changes direction but also decreases in speed as one moves deeper into the ocean. You might well guess that the surface speed of the water must be less than that of the wind. If the water moved at the same speed as the air, there would be no friction between the two fluids and no drag on the water. Similarly, surface water can impose a frictional tug on the water just below only if it moves faster than that water. The net result is that, as

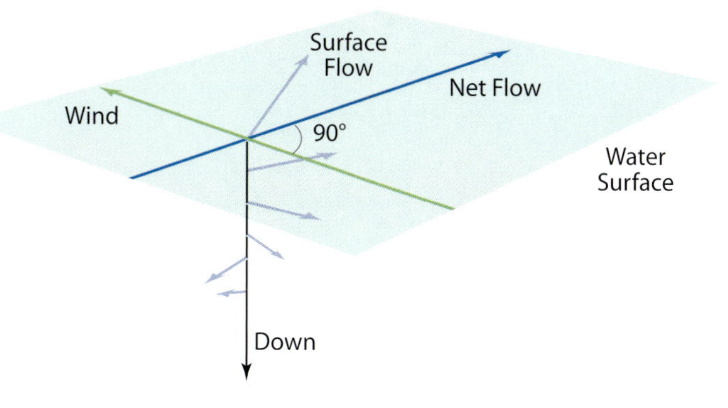

Figure 8.8. Friction with the wind causes water to flow, but Coriolis acceleration causes the direction of flow to spiral with depth. Net flow is perpendicular to the wind. The situation shown here pertains to the northern hemisphere, where Coriolis deflection is to the right.

we move down the Ekman spiral, the flow of water gets slower and slower and eventually vanishes. The total depth of the Ekman spiral—what is often called the depth of the *Ekman layer*—depends on how fast and for how long the wind has blown, but it is never more than a few tens of meters.

If we average the flow over the whole depth of the Ekman spiral, we arrive at an important fact. As wind blows over a level sea, the net direction of water movement *is perpendicular to the wind*: to the right in the northern hemisphere, to the left in the southern. You should be getting used to this sort of nonintuitive conclusion by now. The practical consequences in this case are quite substantial, as we will see.

Ocean Surface Currents

We are now in a position to predict how the global pattern of winds moves water in the oceans. For simplicity, let's focus on a single ocean (the Pacific) in a single hemisphere (the northern); with some care, we'll be able to apply our results to all oceans.

We begin with a review of the winds as we left them earlier in this chapter (figure 8.5). The doldrums, an area of reduced wind activity, are nearest the equator, with the southwestward-flowing trade winds to the north. Still further to the north is a second band of low winds, the horse latitudes, and, in turn, the northeastward-flowing westerlies.

For each of these winds, let's predict the motions induced in the ocean over which they blow. The trade winds blow to the southwest, and their Ekman flow is 90° to the right of that direction. Thus, we expect water between the latitudes of 5° and 25° North to flow to the northwest (figure 8.9). Similarly, because the westerlies blow to the northeast, they induce water motion to the

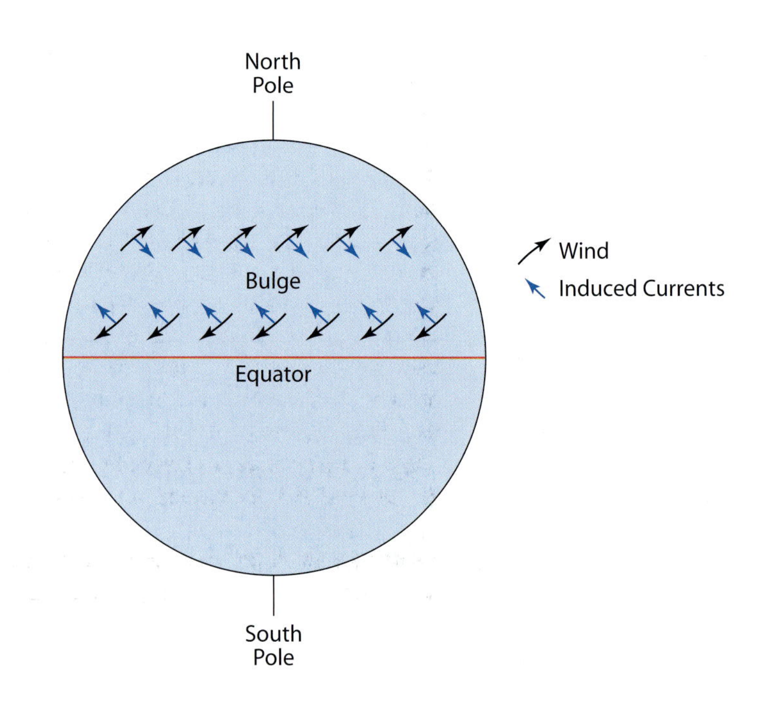

Figure 8.9. Trade winds and westerlies drive surface currents that produce a bulge in the ocean.

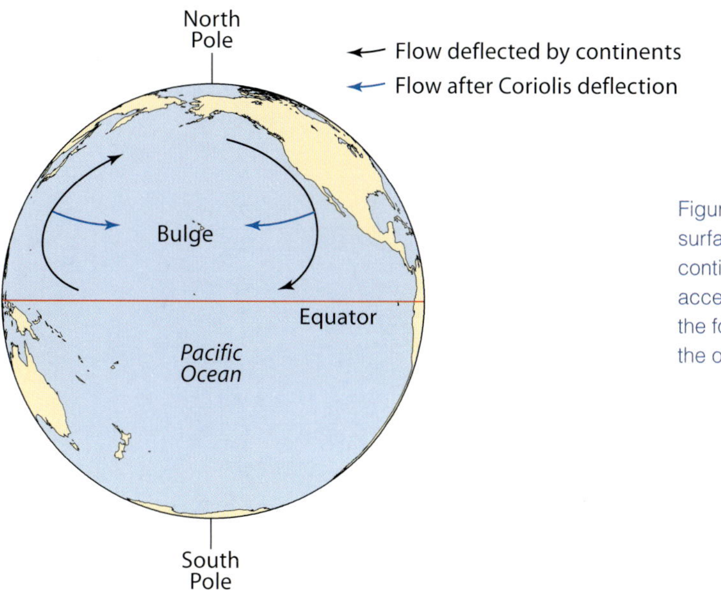

Figure 8.10. Deflection of surface currents by continents and Coriolis acceleration contributes to the formation of a bulge in the ocean.

southeast. In the middle of the horse latitudes, these two flows should collide to form a bulge in the middle of the ocean, roughly at latitude 30° North.

Focus now on what happens as the flow driven by the westerlies arrives at the eastern edge of the ocean. Here, where the North Pacific abuts North America, any eastward flow is blocked by the solid bulk of the continent, and the flow is deflected to the south. The same process occurs where the North Pacific abuts Asia in the west. Water driven northwestward by the trade winds is deflected by the continent northward along the coast.

There are two important aspects of these deflected flows. First, we can connect the north-south flows along the edges of the continents with the east-west flows driven by the trade winds and westerlies to form a large *subtropical gyre*.[4] Water flows northward along the coast of Asia, bending eastward at a latitude of about 45°, southward along the coast of North America, and then westward to arrive back where it started.

Second, as water flows northward along the coast of Asia and southward along the coast of the Americas, the Coriolis effect causes a tendency in each case for these flows to move toward the center of the ocean (figure 8.10). As the Coriolis-driven eastward flow from Asia meets the Coriolis-driven westward flow from North America, a bulge in the ocean results, and this bulge combines with the bulge produced by the trade winds and westerlies to form a hill of water in the ocean.

In summary, winds pulling on the ocean's surface produce a circulating flow that, when acted on by the Coriolis acceleration, forms a hill of water in the center of the North Pacific (figure 8.11).

But this scenario should sound familiar. We dealt with this kind of flow-induced hill in our discussion of geostrophic equilibrium (chapter 7), and the same principles apply here. As the wind-driven currents move in a grand clock-

[4] Gyre: a circular course or motion.

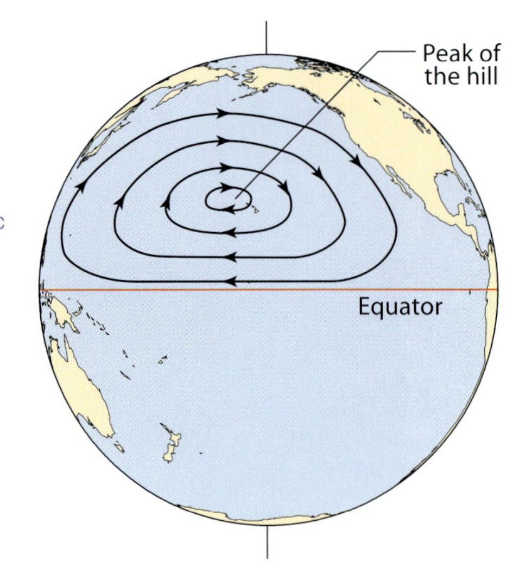

Figure 8.11. An ocean gyre: geostrophic flow that results from the equilibrium of Coriolis acceleration and hydrostatic pressure.

wise gyre around the North Pacific, the Coriolis acceleration pushes water uphill. This tendency is counteracted by gravity, which pulls water downhill. When the hill is about two meters high, the accelerations from gravity and the Coriolis effect just offset each other, and the subtropical gyre comes to a geostrophic equilibrium. Energy supplied by wind is sufficient to overcome losses due to friction, thereby keeping the hill from dissipating.

This circulating flow is indeed the general pattern for the ocean currents in the North Pacific, and the steady and persistent flows that we have outlined have acquired names (figure 8.12). The northward flow along the coast of Asia and Japan is known as the *Kuroshio*. As that current bends to the east, it becomes the *North Pacific Current*, which, as it travels south along the coast of North America, becomes the *California Current*. The cycle is completed by the *North Equatorial Current*, which flows east to west in the tropics north of the equator. Although the different segments of the overall gyre have different names, the gyre is really just one, very big current.

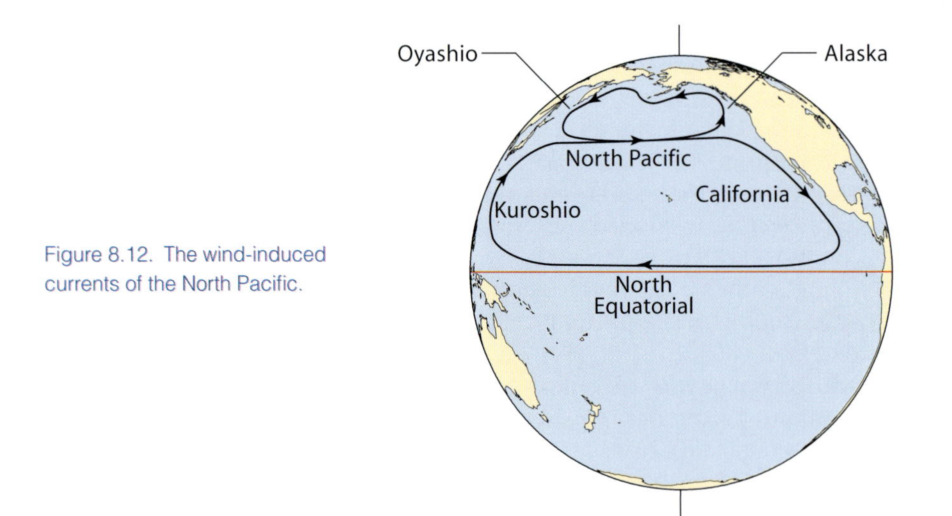

Figure 8.12. The wind-induced currents of the North Pacific.

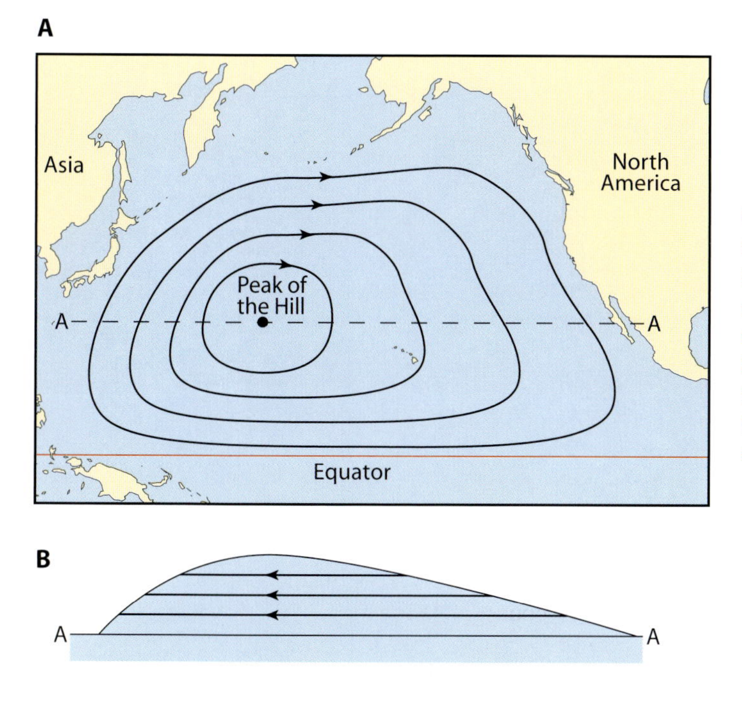

Figure 8.13. (A) The peak of the hill of water in the North Pacific lies west of the ocean's center. (B) The hill of water in east-west cross section. Geostrophic flow is swift on the steep western slope of the hill, slow on the gentle eastern slope.

We have not yet covered one important detail about this gyre. For reasons that are somewhat complex, the peak of the hill in the ocean is not located halfway between Asia and North America. Instead, latitudinal variation in the Coriolis acceleration shifts the peak to the west. As a result, the hill of water in the North Pacific has a steep slope facing Asia and a gentle slope facing North America (figure 8.13). Recall that geostrophic equilibrium occurs when the tendency of water to flow downhill in response to gravity offsets the tendency of water to flow uphill in response to the Coriolis acceleration. The steep hillside facing Asia has a stronger-than-average tendency for water to flow downhill, and as a result, the current must flow faster than average in this region for the Coriolis acceleration to maintain equilibrium. The opposite is true for the hillside facing North America. Given the gentle slope of the hillside there, the current can flow slowly and still have sufficient Coriolis force to maintain geostrophic equilibrium. The westward shift in the bulge thus makes the northward-flowing Kuroshio a swifter current than the southward-flowing California Current. Typical current speeds in the Kuroshio are 1–3 meters per second (2–6 miles per hour); the California Current has an average velocity of only about 0.1 meters per second (about 0.2 miles per hour).

These conclusions can be readily transferred to other parts of the ocean (figure 8.14). For example, in the South Pacific the subtropical gyre flows in the opposite direction. The *East Australia Current* becomes the *South Pacific Current*, which becomes the *Peru Current* (also known as the *Humboldt Current*), which becomes the *South Equatorial Current*. Note that both the North and South Equatorial Currents flow in the same direction, an important point to which we will return later.

The pattern of currents in the Atlantic (figure 8.14B) is similar to that in the Pacific. In the northern hemisphere, the *Gulf Stream* runs swiftly north along

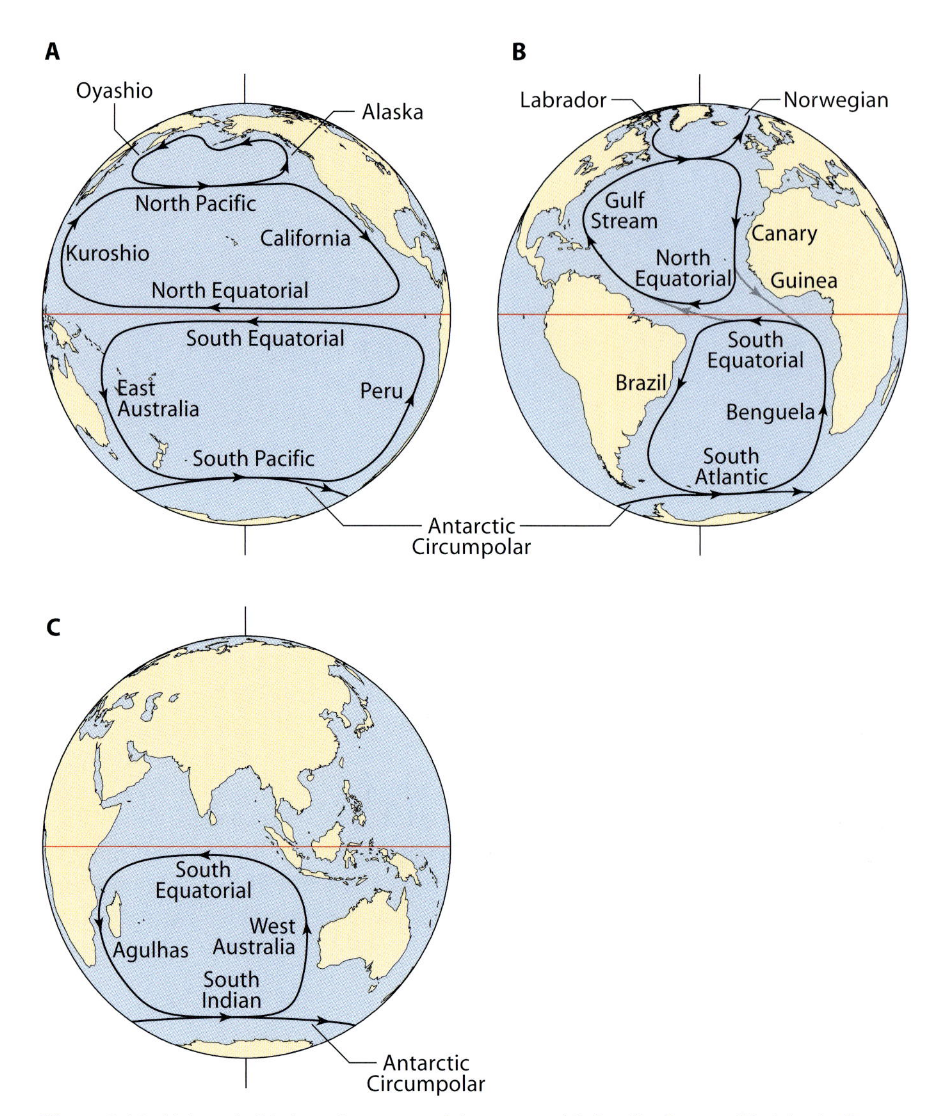

Figure 8.14. Major wind-induced currents of the ocean: (A) Pacific Ocean, (B) Atlantic Ocean, (C) Indian Ocean.

the east coast of North America until it bends eastward at about 45° N and heads for Europe. The *Canary Current* runs south along the European coast and feeds into an Atlantic version of the *North Equatorial Current*. The South Atlantic is home to the *Brazil Current* (off Brazil, surprisingly enough), the *Benguela Current* (off South Africa and Namibia), and another *South Equatorial Current*.

As you might expect, the pattern of currents in the Indian Ocean is different (figure 8.14C). North of the equator, the equatorial current changes direction seasonally, flowing from east to west during the winter and west to east during the summer, when it is known as the *Monsoon Current*. South of the equator, the *West Australia Current* flows north past Australia's western shore and joins the *South Equatorial Current*. The South Equatorial Current always flows from east to west, but its speed varies with the monsoon; it is fastest in the northern winter and slowest in the northern summer. As the South Equatorial Current approaches Africa, it is diverted to the south, forming the *Agulhas Current* that flows along the coast of Mozambique and the eastern shore of South Africa.

The mechanisms that form subtropical gyres are mirrored at high latitudes, where the polar easterlies drive gyres with circulation in the opposite direction. For example, in the North Pacific there is a subpolar gyre rotating counterclockwise (figures 8.12, 8.14A). There are two named legs of this gyre. In the Northeast Pacific is the *Alaska Current,* which feeds into the *Oyashio* in the Northwest Pacific. Flow in the North Atlantic is more complex (figure 8.14B). As the Gulf Stream flows toward Europe it splits, the southern arm forming the Canary Current described above. The northern arm forms the *North Atlantic Current*, which flows past Great Britain and Ireland, becoming the *Norwegian Current*. Beyond this, flow is complicated by the presence of Greenland and surface outflow of water from the Artic Sea, but the south-flowing *Labrador Current* occupies a position roughly similar to that of the Oyashio in the Pacific.

Unlike in the Arctic, where winds over the oceans periodically encounter dry land, winds in the Southern Ocean circle the earth unchecked (figure 8.15). In effect, the westerlies from the Pacific Ocean feed into the westerlies of the Atlantic, which meet those of the Indian Ocean, and then pass back to the Pacific to start all over again. In response, there is a circumpolar flow of water at a latitude of roughly 60° South. The romantic term for this current, the *Westwind Drift*, has recently been supplanted by a more prosaic title, the *Antarctic Circumpolar Current*. The southernmost legs of the subtropical gyres in the Pacific, Atlantic, and Indian Oceans all contribute to the Antarctic Circumpolar Current.

What happens at still more southerly latitudes? There, under the urging of polar easterlies, a second current—the *Eastwind Drift* or *Polar Current*—runs in a westward direction along much of the Antarctic coast.

The Discovery of Ocean Currents

Sailors have long been aware of ocean currents. For example, on one of his trips from Europe to the Caribbean, Columbus noted his ships drifting westward

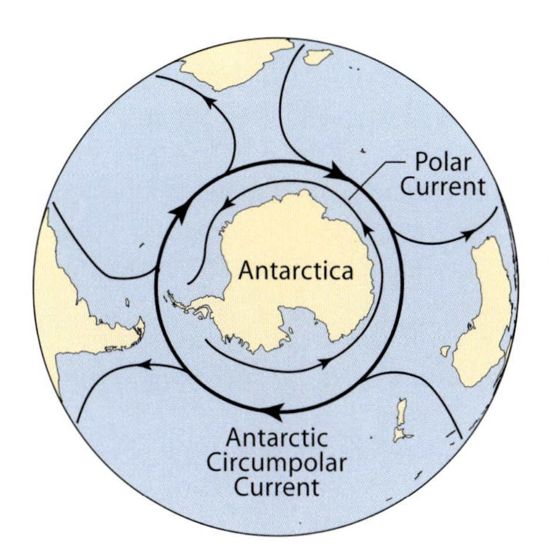

Figure 8.15. Ocean surface currents around Antarctica.

even when the wind was not blowing, an unwitting observation of the Atlantic North Equatorial Current. The Gulf Stream was well known to colonial American sailors, who used the current to minimize the time it took them to sail to Europe. Ships traveling from Philadelphia to London, for instance, arrived a week sooner if, instead of heading on a direct route across the Atlantic, they diverted north, keeping to the warm water of the Gulf Stream and North Atlantic Current. Ships sailing the other direction arrived sooner if they kept to the cold water south of these currents. Benjamin Franklin learned of these tricks from his nephew Timothy Folger, a Nantucket whaling captain, and, in his capacity as assistant colonial postmaster, published a map of the Gulf Stream as a guide to mariners.

The study of drifting objects provided more detailed information about major ocean currents. For instance, Prince Albert the First of Monaco (1848–1922) spent much of his life (and a substantial fraction of his fortune) sailing to various points in the Atlantic to release sealed bottles and beer barrels, each with a note inside. When these drifters came ashore, the notes, written in ten languages, asked that the finder contact the Prince. Knowing when and where the bottle or barrel was cast adrift and when and where it came ashore, Prince Albert could plot the direction and average speed of the current that had transported the drifter.

Experiments of this sort can even happen by chance. During World War I, more than 200,000 explosive mines were laid in the waters off Europe. Some of these mines inevitably broke their moorings and went adrift, and their subsequent arrival onshore was a matter of note: each mine contains several hundred pounds of high explosive. When and where the mine entered the water could be deduced from its type.

Perhaps the grandest of all inadvertent drifter experiments have been carried out in the past two decades. Much transoceanic transport is currently conducted by container ships, whose decks are stacked high with 40-foot-long metal boxes. In storms, these shipping containers have a nasty habit of parting company with the ship. For instance, on May 27, 1990, the *Hansa Carrier* was sailing from Korea to the United States when it ran into a severe storm. Twenty-one shipping containers were lost overboard, five of which contained Nike-brand running shoes. As the containers broke open, 61,280 shoes were released, and the great experiment was on. After more than 220 days at sea, 1600 of the shoes were recovered along the coast of Vancouver Island, Washington, and Oregon, allowing scientists to track the path of the Kuroshio, North Pacific, and California Currents. Two years later, a few shoes eventually arrived at Hawai'i, evidence that the California Current does indeed flow into the North Equatorial Current.

To explain the precise pattern in which the shoes drifted, researchers made careful measurements of how the shoes floated. For instance, they noticed that slight differences in left and right shoes exposed them differently to the wind, and caused them to take different paths. This helped to explain the fact that some beaches received mostly right shoes and others mostly left shoes. Resourceful beachcombers organized swap meets to match up pairs of shoes, which were quite wearable despite their long submergence.

More recently, major spills of floating bathtub toys, hockey gear, and plastic Lego™ blocks have added to our understanding of ocean currents in both the Atlantic and Pacific. From these and other observations, we now know that it takes about 14 months for water to make a full circuit of the North Atlantic gyre, 54 months for the much larger circuit of the North Pacific gyre.

Ocean Currents and Heat Transport

Let's return one final time to the transport of heat on earth. As ocean gyres turn, they carry heat with them. For instance, the Gulf Stream and Kuroshio carry warm surface water from the tropics northward toward the pole. The California and Canary Currents bring cold water from high latitudes southward to the tropics. Thus, ocean surface currents transport heat in the same direction as thermohaline circulation; indeed, the poleward return flow of the thermohaline circulation is carried by the wind-driven surface currents. The combined transport of ocean water accounts for about half the latitudinal heat flux on earth. Taken together, winds, thermohaline circulation, and wind-driven surface currents provide the overall mechanism for transport of excess heat delivered to the tropics toward the poles, and these three flows maintain the latitudinal equilibrium of temperature on earth.

I should note again that one can think of the entire motion of the winds and currents as one giant, heat-driven engine. The engine in your car provides mechanical energy, what thermodynamicists call "work," by taking the heat of exploding gas and transferring that heat to the colder air outside the engine. Similarly, a steam engine takes hot steam and condenses it to water, providing mechanical energy. In the process of transporting heat from the tropics to the poles, the heat engine of the earth provides the mechanical energy to circulate the air and the oceans. We will see in the next chapter that the ability of life to survive in the oceans is a side effect of this heat engine.

The Basinwide Tilt

Imagine the following experiment. You fill your bathtub half full with water and then add a couple of inches of cooking oil on top. The oil floats on the denser water, and as long as you don't disturb the tub, both layers remain horizontal in response to the downward tug of gravity (figure 8.16A). In other words, in this undisturbed state, the oil layer has the same thickness everywhere in the tub.

Now take a fan, and install it at one end of the tub so it blows along the surface of the oil toward the opposite end. How does the wind affect the configuration of oil and water?

As with wind moving over the ocean, air moving over the surface of the oil drags the oil in the direction of the wind. (Your bathtub is too small for the Coriolis acceleration to be important.) Because oil is transported by the wind, the surface of oil at the upwind end of the tub is lower than it was before the wind was applied, and the oil layer there is thinner. At the other end of the tub, oil piles up, making the oil surface higher and the oil layer thicker. As a result of the downwind shift of oil, the interface between oil and water, which was horizontal in the absence of wind, now has a tilt (figure 8.16B). After the fan has blown for a while, the system reaches an equilibrium with a stable tilt at the interface. If you were to turn off the fan, gravity would level things out again, and the oil would flow back whence it came.

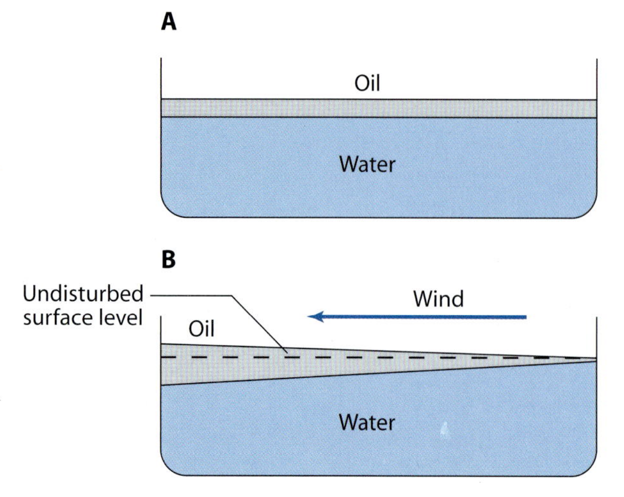

Figure 8.16. A bathtub experiment exploring the effects of wind stress. (A) In the absence of wind, the low-density surface layer is horizontal. (B) Wind pulls the low-density surface layer to one end of the tub. As a consequence, the surface layer is now tilted: it is thick at the downwind end of the tub and thin at the upwind end.

As you probably suspect, this experiment is analogous to what happens in the ocean. The layer of oil represents the warm water above the thermocline, and the water in the tub represents the cold water of the ocean interior. The fan is an analogue for the trade winds and any other winds that flow east to west near the equator.

As we discussed above, the trade winds drive the warm surface water near the equator in a westward direction. Some of this westward flow is diverted north in the northern hemisphere to produce the Kuroshio and the Gulf Stream and south in the southern hemisphere to produce the East Australia, Brazil, and Agulhas Currents. Where warm water runs into a continent, however, some water builds up in place, as oil did in the bathtub. In fact, water in the western end of the equatorial Pacific is about 40 centimeters higher than water in the eastern end.

In addition, as water piles up, it pushes the thermocline down to a depth of about 200 meters, forming what is descriptively known as a *warm pool.* And as with the oil-water interface in the tub, the thermocline has an east-west tilt across the ocean basin. This *basinwide tilt* is characteristic of the tropical waters of the Pacific and Atlantic Oceans (figure 8.17). The seasonal shift in winds

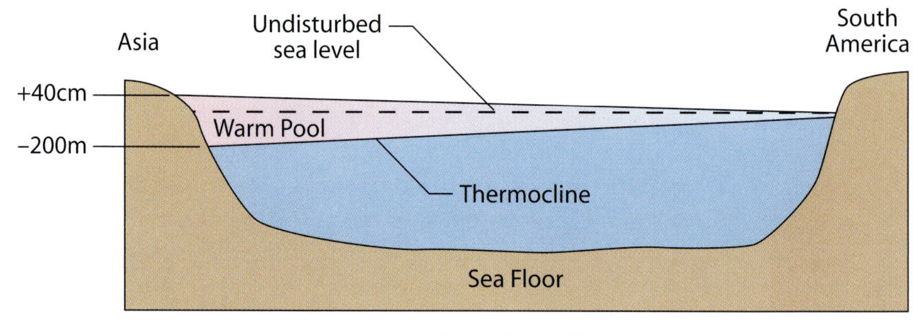

Figure 8.17. The basinwide tilt of the thermocline in the Pacific. Sea level is higher, and the thermocline deeper, in the warm pool at the western end of the ocean.

in the Indian Ocean prohibits the effective formation of a basinwide tilt and warm pool there.

Note that the warm pool and the associated basinwide tilt in the Pacific and Atlantic are a result of the trade winds, and we might expect to find them only in those areas where trade winds blow: roughly from 10° to 25° North and 0° to 25° South. We would not expect to find the warm pool and a tilted thermocline in the area of the doldrums. But in fact, the warm pool and tilt are found all the way from 25° North to 25° South, with no gap in the middle. Why isn't there a gap?

First, recall that the location of the trade winds varies with the seasons. For instance, in the northern summer, when the intertropical convergence zone is well north of the equator, the "southern" trade winds may straddle the equator, and during this time, a warm pool can be established there. The second, and perhaps equally important, reason is that once an equatorial warm pool is established, it is self-perpetuating. To see why, let's return to the graph of the basinwide tilt and think of the consequences it might have for the atmosphere.

In the western part of the equatorial ocean, surface water is warm. After all, the warm pool is located there. In contrast, in the eastern part of the ocean, surface water is relatively cool, warm water having been blown west to make the warm pool. Warm water to the west and cold water to the east set the stage for the same sort of convection process that Hadley envisioned. The warm pool heats the air above it, which rises. The cold water of the eastern ocean cools the air above it, and the air sinks. Connecting these two flows are winds that blow from west to east at high altitude, and more importantly, winds that blow from east to west at the ocean's surface (figure 8.18). These winds are called *zonal winds*: because they have no north-south component, they stay within the same zone of latitude.

Unlike the Hadley cells, which transport heat across latitudes, the transport of heat by equatorial zonal winds does not tend to even out the ocean temperature—in fact, it does just the opposite. Westward-flowing zonal winds drag warm water with them, augmenting the warm pool, and thereby reinforcing the zonal winds. In other words, after the wandering trade winds have initiated a warm pool, the surface winds of the resulting equatorial convection cell maintain it.

The equatorial convection cell we have just described is called a *Walker cell*, after Gilbert Thomas Walker (1868–1958), a British mathematician and head of the Indian Meteorology Service, who first proposed the concept. Walker cells are characteristic of the equatorial Pacific and Atlantic Oceans. As we will see in the next chapter, the winds of the Walker cells have important biological consequences.

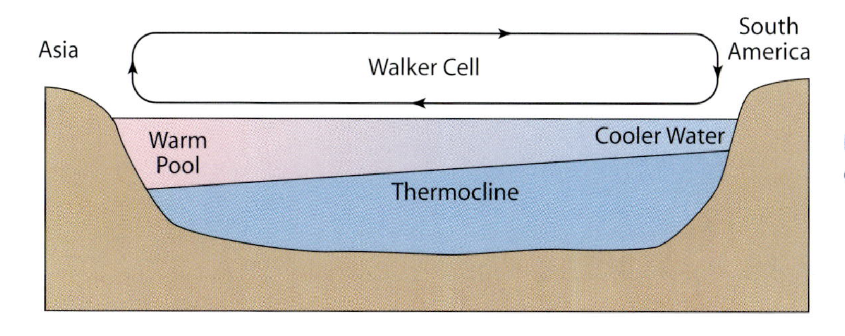

Figure 8.18. The Walker cell of zonal winds.

Summary

We end this chapter by summarizing what we know about the two-layered ocean. The warm surface layer is well lit but nutrient-poor, whereas the deep layer is dark and nutrient-rich. The two-layered system is stable, but as we have just learned, the thickness of the warm upper layer varies from place to place. In the eastern parts of tropical seas, the trade winds pull the thermocline nearer to the surface. The opposite occurs at the western end of the ocean, where the warm pool forces the thermocline down. In the next chapter, we will explore how this pattern can be exploited to allow the oceans to support life.

Yet Another Warning

The description we have developed for ocean currents is accurate as far as it goes, but it is far from complete. For instance, we have only scratched the surface of equatorial flow. There is a deep current along the equator, the *Equatorial Undercurrent*, that flows counter to the surface currents we have described, and there is also a shallow current, the *North Equatorial Counter Current*, that flows from west to east approximately 10° north of the equator. Neither of these equatorial countercurrents has a major effect on oceanic biology, so we ignore them.

The list of details such as these is intriguing and endless, but it won't be examined here. We have the big picture, and it is time to see what it can tell us.

Further Reading

Humphreys, W. J. (1940). *Physics of the Air.* McGraw Hill, New York.

Open University (2001). *Ocean Circulation* (2nd edition) Butterworth-Heinemann, Oxford.

Appendix

Convective Circulation

Convective circulation in the atmosphere is a concept that initially seems intuitive. If you heat air near the equator it rises. If you cool air at latitude 30° North, it descends. And it seems to make sense that you can couple these vertical motions into a grand cycle. Air moves toward the equator along earth's surface to replace the rising air, and air moves toward the poles at altitude to replace the falling air. But the more you think about it, the less intuitive this motion becomes. Why should air lofted in the northern hemisphere tropics necessarily flow northward? What establishes the pressure gradient that causes air at the surface to move toward the equator?

A simple experiment serves to illustrate the physics. Consider two cylinders side by side, each containing the same mass of water (figure 8.A1A). A pipe (with ends labeled 1 and 2) connects the two cylinders just below the surface of the water, and another pipe (with ends 3 and 4) connects them at their bases. As long as water temperature is the same everywhere in the system, conditions are identical in both cylinders, and there is no tendency for water to flow.

Now, let's close a valve in each pipe, severing the connection between cylinders, and heat the water in the cylinder on the left. As the temperature rises, the water expands, and the surface of the water rises (figure 8.A1B). Because pipe end 1 is now farther below the water's surface than it was before, the pressure at that point is higher than it was before, and higher than the pressure at end 2, which has not changed. Although the volume of water in the left-hand cylinder has increased, the *mass* of water has not changed, so the pressure at pipe end 3 is not altered by the rise in temperature.

Given this situation, we now open the valve in the top pipe (figure 8.A1B). What happens? Pressure at pipe end 1 is higher than that at end 2, so water flows from left to right in the top pipe. But as water flows from the cylinder on

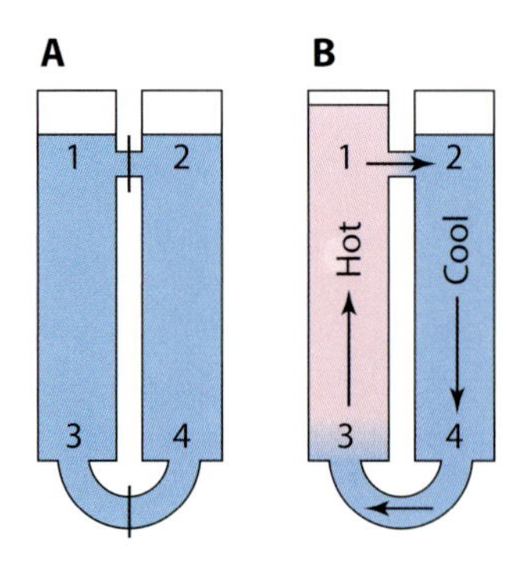

Figure 8.A1. Two cylinders of water are connected by pipes near their tops and at their bases. At the start of the experiment (A), both cylinders are at the same temperature and valves close the pipes. During the experiment (B), the cylinder on the left is maintained at a higher temperature than the cylinder on the right, and convective currents flow.

the left into the cylinder on the right, the mass of water in the right-hand cylinder increases, thereby increasing the pressure at pipe end 4. Because pipe end 4 is at higher pressure than pipe end 3, if we now open the valve in the lower pipe, water flows in the pipe from right to left. In other words, by heating water in the cylinder on the left, we have created a convective circulation between the two cylinders: an updraft in the left-hand column, flow from left to right at the top of the water columns, a downdraft in the right-hand column, and flow from right to left at the bottom. Note that we could achieve the same results by cooling the water in the right-hand cylinder. In either case, circulation continues as long as the temperature is different between the cylinders.

I have chosen water as the fluid for this experiment because it is easy to locate its surface. But the same physics apply if the cylinders contain columns of air. Heating the air in the left-hand cylinder would cause it to expand, and the resulting change in pressure would drive circulation through the pipes.

The analogy to convective flow in a Hadley cell is straightforward. The hot cylinder here is analogous to the tropics, the cold cylinder to a latitude of 30°. Note that the system as outlined here provides additional explanation for the presence of low surface pressure under an area of updraft (e.g., the pressure at pipe end 3), and high surface pressure under an area of downdraft (e.g., pipe end 4).

Solutions to the Dilemma

For three chapters now, we have developed the theme of a two-layered ocean: a warm, well-lit, and nutrient-poor layer at the surface with a cold, dark, and nutrient-rich layer below. The separation of light and nutrients impedes phytoplankton growth and reproduction, and the resulting lack of net productivity at the bottom of the food chain adversely affects marine consumers. If the oceans are such an awful life-support system, why wasn't life in the sea snuffed out long ago?

Fortunately, there are several mechanisms that circumvent the two-layered ocean, ranging from processes that apply locally or for short periods, to large-scale phenomena that continuously provide nutrients to surface waters. We begin with the local and intermittent, and work our way up to the large and continual.

Local Effects: The Shallows

The seafloor forms the ultimate barrier to sinking in the ocean. Upon arrival at the seafloor, any organic particles that have not completely decomposed are held in place, where they may continue to rot, eventually releasing carbon and nutrients into the surrounding water. In much of the ocean, the seafloor is so deep that these remineralized nutrients and carbon cannot readily be mixed back to the surface. In shallow parts of the ocean, however, the seafloor prevents organic particles from sinking beyond the reach of surface mixing.

As we have seen, wind interacts with the ocean's surface, creating turbulent eddies in the surface waters, and these eddies distribute heat throughout the water above the thermocline. In shallow parts of the ocean, turbulent mixing can stir water all the way to the ocean floor. In addition, in many shallow areas, wind-induced mixing is augmented by the interaction of currents with the bottom. Just as wind blowing over the water's surface creates turbulence in the water, the flow of a current over the solid seafloor creates turbulence that stirs the water column. As a combined result of winds and currents, turbulent mixing in ocean shallows re-suspends nutrients released at the seabed, and these areas can be continuously productive. For example, the Grand Banks, a group of underwater plateaus east of Newfoundland, commonly have depths of only 25 to 100 meters. Exposed to the turbulence from the North Atlantic's frequent storms and mixing from both the Labrador Current and Gulf Stream, the shallow water column is well supplied with nutrients. The resulting primary productivity formed the basis for one of earth's richest fishing grounds. For centuries, cod and haddock from the Grand Banks fed multitudes of Europeans and Canadians, and the Grand Banks would still support a great fishery today were it not for egregious overfishing by humans. We will return to the tragedy of overfishing in chapter 11.

Other productive ocean shallows include the Yellow Sea and the East China Sea, the Arafura Sea and Gulf of Carpenteria between New Guinea and Australia, and the extensive shelves adjacent to Brazil and southern Argentina.

Short-Term Effects: Storms

But what about the rest of the ocean, where the depth of the surface mixed layer is not limited by the seafloor? Here, temporal variation in mixing comes into play.

Consider a typical patch of ocean where, just below the thin, warm upper layer, cold, nutrient-rich water is available. If the wind over this patch of ocean increases in strength (during a storm, for instance), the resulting increase in turbulent mixing forces the thermocline down. In the process, cold, nutrient-rich water previously below the thermocline is entrained into the warm surface water and subsequently mixed throughout the surface layer. Thus, the strong winds of intermittent storms can locally introduce nutrients into the upper layer of the ocean.

A short-lived biological bonanza results. The upper layer is typically warm, well lit, and well oxygenated; as a life-support system, it lacks only nutrients. So, when nutrients are supplied by turbulent mixing, biological activity increases in the mixed layer. Typically, it is the phytoplankton just above the thermocline, those closest to the source, that benefit most from turbulent injection of new nutrients. As they multiply, these phytoplankton form a *subsurface chlorophyll maximum*, which is often made more evident by the photoinhibition of phytoplankton in the bright light near the water's surface. Small animals in turn eat the abundant phytoplankton, and themselves grow, reproduce, and are eaten.

Note that when nitrogenous nutrients are mixed up into the surface layer, they are typically in the form of nitrate (see chapter 4). *Prochlorococcus*, the dominant form of phytoplankton in many nutrient-poor waters, cannot use nitrate as a nitrogen source, so it cannot participate in the initial feast. In contrast, large phytoplankton (diatoms and dinoflagellates, for instance) readily take up nitrate, and it is their growth that results in increased production. Only after the large phytoplankton have been eaten, and some of their nitrogen excreted as ammonia, can *Prochlorococcus* partake of the newly arrived nutrients.

However, nutrients stirred up by a typical storm can support only a temporary increase in biological activity. As the wind dies, the thermocline becomes shallower, the biological pump exports nutrients downward, and the two-layered ocean—with its inherent biological limitations—reestablishes itself. Thus, storms and their associated turbulent mixing can provide only local, intermittent relief from the dilemma of the two-layered ocean.

Wind-driven turbulent mixing can occur anywhere in the ocean, and in the major ocean gyres this mixing is the primary means by which nutrients are moved up into the surface layer. The nutrients thus provided are sufficient for these areas of the ocean to export from the surface layer about 5–10 grams of newly fixed carbon per square meter per year. For the sake of comparison, we will use the midpoint of this range, 7.5 grams of new production per square meter per year, as the "basal" rate of new ocean production, the rate allowed by the basic two-layered structure of the sea.[1]

As I have hinted, this basal rate of production is really quite low. There is 1 gram of carbon in every 2.5 grams of carbohydrate, such as sugar, so the basal oceanic annual productivity of 7.5 grams of carbon per square meter is equivalent to the production of roughly 19 grams of sugar per square meter per year. The average human being eats the caloric equivalent of 500 grams of sugar every day, 182,500 grams per year. So at this rate of productivity, it would take nearly 10,000 square meters of ocean surface (2.4 acres) to produce enough food to keep one person fed in sustainable fashion.

In contrast, productive areas on land (a field of beans, for instance) can fix 500 to 700 grams of carbon per square meter per year, 67 to 93 times the production of ocean gyres, and it takes only about 300 square meters (0.07 acres) to produce enough food for a single person. As advertised, the two-layered ocean makes a lousy farm.

Seasonal Variation: Temperate Seas

In our quest for relief from the dilemma of the two-layered ocean, we have so far considered only short-term, small-scale effects of turbulent mixing. Longer-term and larger-scale effects derive from seasonal variation in light and weather.

Consider temperate seas at 45° to 50° North or South, the latitude of Canada, Japan, or Chile. In summer, this part of the ocean is characterized by a typical two-layered structure: a shallow mixed layer, depleted of nutrients, floating on a cold, nutrient-rich interior. Let's follow this temperate sea through the rest of the year.

In autumn, day length shortens and the air cools, and as a result, the upper layer of the ocean loses heat to the atmosphere. Eventually, the temperature of the upper layer approaches that of the lower layer, and the thermocline disappears. We noted in chapter 6 that we are often interested not in the thermocline itself, but in the other properties of seawater for which the thermocline is an index. In this case, our primary concern is water density. As surface water cools,

[1] Productivity values cited here are from P. G. Falkowski et al. (1998), Biogeochemical controls and feedbacks on ocean primary productivity, *Science* 281: 200–206. As new information becomes available, values may change slightly.

its density approaches that of water below the thermocline. Recall that it is the density difference between the warm upper layer and the cold lower layer that makes the thermocline a barrier to mixing. As a consequence, when the upper layer cools sufficiently so that the two layers have nearly the same density, this mixing barrier is removed, and turbulence created by the wind can stir water to a much greater depth than previously possible. As a result, when the thermocline disappears in autumn, nutrient-rich water that had been trapped below the thermocline throughout the summer mixes up to the surface. Winter storms and their strong winds increase the intensity of turbulent mixing, facilitating the seasonal introduction of nutrient-rich waters into the surface layer.

One might suppose that production should then increase. However, two problems typically stifle this potential productivity. First, turbulent mixing is a two-edged sword: the increased mixing that brings nutrients to the surface also mixes phytoplankton down into the depths. This deep mixing reduces the time any individual phytoplankton cell spends in the light, thereby decreasing its ability to take up nutrients and photosynthesize. Second, as autumn progresses, days shorten, and less light impinges on the ocean surface. Together, these two effects prevent phytoplankton from taking advantage of the autumnal import of nutrients. Life is grim, and it stays that way until the next spring.

But spring, ah, spring! As days get longer in springtime, abundant light becomes available for photosynthesis, and the thermocline is reestablished, reducing the depth of mixing. Bathing in the nutrients delivered up in the fall and basking in the vernal sun, phytoplankton in the surface layer are suddenly able to grow. The result is a *spring bloom* of phytoplankton, a biological explosion that travels up through the food chain as herbivores feast on the abundant phytoplankton and are themselves eaten by predators. The spring bloom in temperate seas is one of the largest events in global productivity.

However, as with all explosions, the spring bloom is short lived. It persists only until early summer, when the biological pump once again depletes the surface waters of nutrients. Shortly, we are back to the two-layered ocean, where we started. This seasonal cycle is depicted in figure 9.1.

In the course of this cycle, each square meter of temperate ocean typically fixes about 50 grams of new carbon, a rate about seven times the basal rate found in major ocean gyres. The relatively high rate of productivity in temperate seas is testimony to what can happen when the thermocline disappears, even if

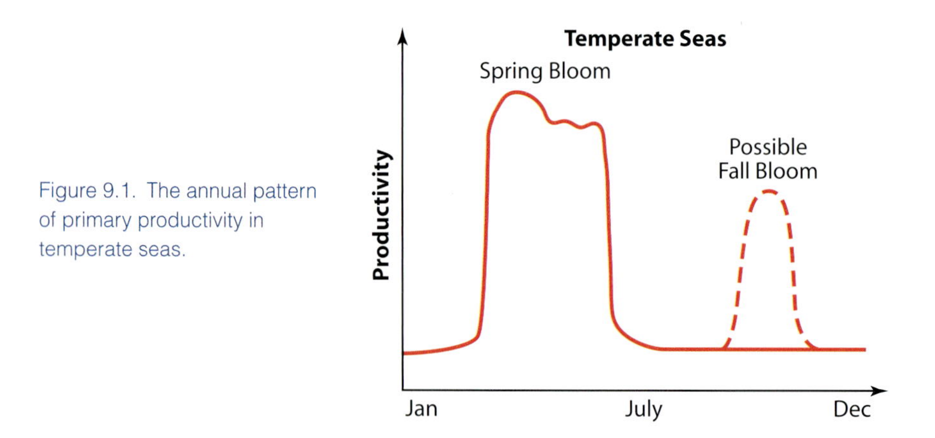

Figure 9.1. The annual pattern of primary productivity in temperate seas.

only on a seasonal basis. In exceptional cases, the annual new production in temperate seas can be as high as 120 grams of carbon per square meter per year.

Occasionally, the spring bloom in temperate waters is echoed by a bloom in the fall. If, due to early storms or cold weather, the thermocline weakens before light levels get too low, nutrients and light may temporarily co-occur in the surface waters in autumn. But this bloom is less reliable, and usually smaller in magnitude, than the typical spring bloom.

Seasonal Variation: Arctic Seas

As we have just seen, in temperate waters the occurrence of a plankton bloom depends on both seasonal availability of nutrients and seasonal variation in light. In the Arctic, effective solar heating occurs for such a short time each year that the thermocline never has a chance to really get established, and nutrients are continuously available in surface waters. In this case, productivity depends primarily on two factors: light and mixing. In winter, storms are common, mixing is intense, and light levels are low. As a consequence, productivity is low. In summer, however, mixing is benign, light levels are high, and phytoplankton are productive. Thus, Arctic waters do not undergo the boom-and-bust cycle of a spring bloom. Instead, productivity closely tracks the seasonal variation in light intensity. This pattern is shown in figure 9.2.

Arctic waters can be among the most productive in the ocean. Even though light availability is seasonal, the continuous abundance of available nutrients allows each square meter of ice-free Arctic Ocean to export about 100 grams of carbon each year, roughly 13 times the basal rate of the major ocean gyres. In a few locations, annual new productivity reaches 150 grams of carbon per square meter.

Note that this scenario applies only to the Arctic. Productivity in the Antarctic is governed by a different set of rules, and will be discussed below.

Upwelling

As we have just seen, the dilemma of the two-layered ocean can be temporarily circumvented in temperate and arctic seas, but most of the time most of the ocean is characterized by a persistent and stable two-layered structure. Life in

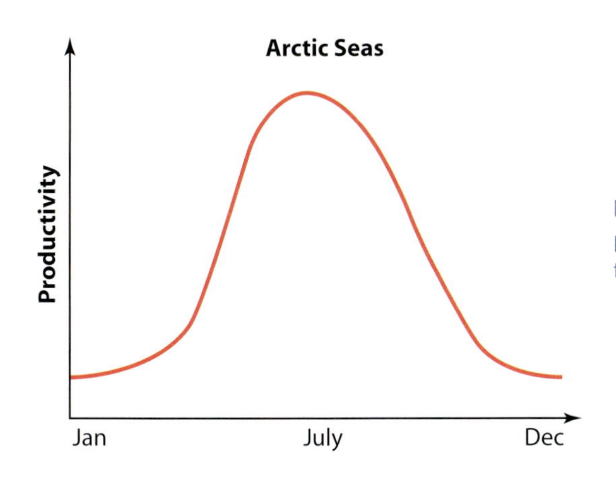

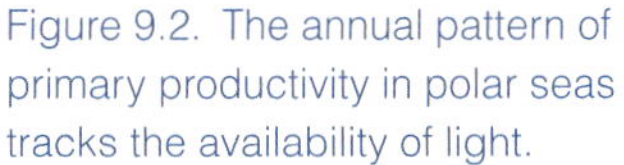
Figure 9.2. The annual pattern of primary productivity in polar seas tracks the availability of light.

the seas as a whole would be severely limited if it could rely only on the short-term and relatively small-scale effects we have described so far. Is there a larger-scale and more persistent mechanism that delivers nutrients to the surface layer?

Equatorial Upwelling

Consider the effects of the trade winds shown in figure 9.3, a situation typical of the equatorial Atlantic or Pacific Ocean. As noted in chapter 8, due to the tilt of the earth's axis and thermal effects of the continents, the intertropical convergence zone does not lie directly on the equator. Instead, on average, it is shifted 5° to 10° northward. As a consequence, the net wind, a combination of the southern trade winds and the zonal winds of the Walker cell, straddles the equator, blowing generally westward.

This westward flow of air at the equator imposes frictional drag on the surface water, an effect we have dealt with before. However, because we are dealing with water at the equator, this drag has unusual effects.

Directly along the equator, there is no Coriolis acceleration (chapter 7), and water is tugged westward in the direction of the wind, with no tendency to veer. In contrast, even slightly to the north or south of the equator, the Coriolis acceleration operates as we have come to expect, and water responds to the wind by forming a typical Ekman spiral. As always, net Ekman flow is perpendicular to the breeze, but in this case, there is a twist. Water north of the equator veers to the right, whereas water south of the equator veers to the left. In other words, under the urging of the trade and zonal winds, surface water at the equator diverges, an effect shown by the dashed arrows in figure 9.3.

As water at the surface diverges, water from below the surface moves up to take its place (figure 9.4). This is the process of *upwelling* we referred to in earlier chapters. Here, for obvious reasons, the process is called *equatorial upwelling*.

The biological consequences of equatorial upwelling depend on the depth of the thermocline. If the thermocline is sufficiently shallow, water drawn up to the surface comes from below the thermocline. In this case, upwelling delivers nutrient-rich water to the surface layer of the sea, thereby circumventing the dilemma of the two-layered ocean (figure 9.5A). In contrast, if the thermocline is deep, water still wells up, but the upwelled water comes from above

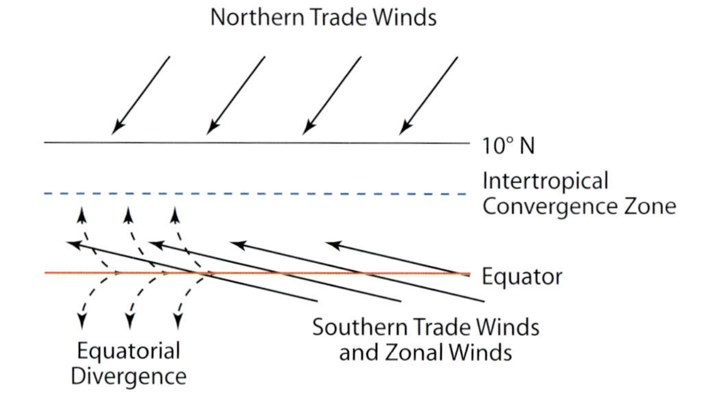

Figure 9.3. The Coriolis acceleration interacts with westward-flowing winds at the equator to cause the equatorial divergence.

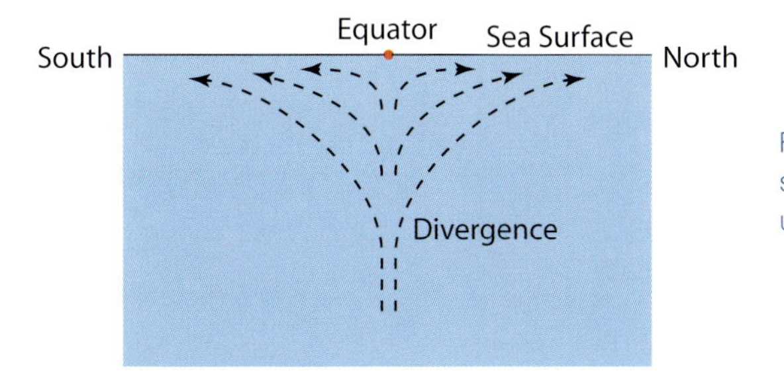

Figure 9.4. A north-south cross section through the equatorial upwelling.

the thermocline, and no new nutrients are delivered (figure 9.5B). Thus, if the thermocline is deep, equatorial upwelling has little biological effect.

We can now make an important prediction by combining this scenario with our understanding of the basinwide tilt in the thermocline (chapter 8). We have just seen that ocean water wells up along the entire equator in both the Pacific and Atlantic Oceans. However, this upwelling effectively delivers nutrients to surface waters only in areas where the thermocline is near the surface. We have also learned that, due to the basinwide tilt, the thermocline is near the surface only in eastern portions of tropical seas. Thus, we would expect to find a band of high phytoplankton productivity along the equator in the eastern part of the ocean, but not in the western part. And in fact this distribution of phytoplankton productivity is exactly what we find in the real ocean: in the eastern equatorial Pacific and Atlantic, there are huge areas of continuously productive water that, without the Coriolis effect, would not exist. Productivity in these areas approximates that in temperate seas: about 50 grams of carbon exported per square meter each year—a rate 7 times that found in the adjacent waters of the major ocean gyres.

The pattern is less pronounced in the Indian Ocean, but as we have repeatedly noted, the Indian Ocean is somewhat atypical.

Coastal Upwelling

Coriolis acceleration leads to upwelling in other areas of the ocean as well. Consider, for instance, the California Current as it runs southward along the

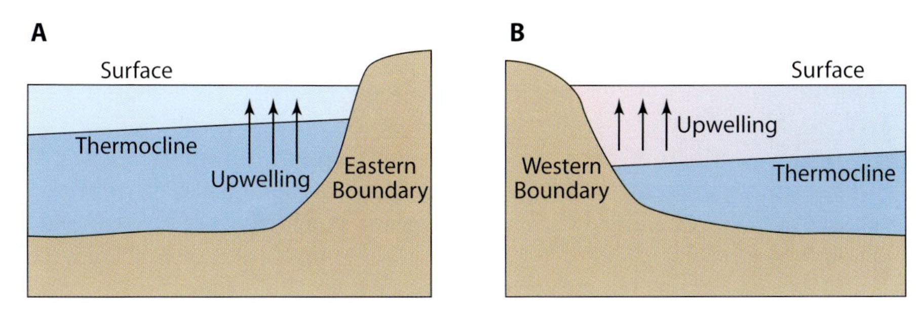

Figure 9.5. The equatorial upwelling in east-west cross section. (A) In the eastern Pacific, where the thermocline is shallow, upwelled water is drawn from below the thermocline. (B) In the western Pacific, where the thermocline is deep, upwelled water is drawn from above the thermocline.

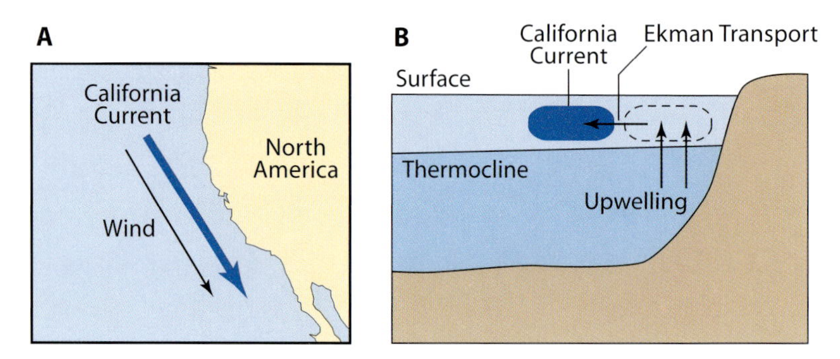

Figure 9.6. Upwelling in the California Current. (A) A map view of the current. In spring and summer winds blow parallel to the current. (B) A cross section through the current perpendicular to shore. The Coriolis acceleration interacts with the resulting increase in current speed to draw the current offshore. As a consequence, water wells up near shore.

west coast of North America (figure 9.6A). Beginning in February and continuing until August, winds blow forcefully from the northwest, paralleling both the coastline and the California Current, and the friction of wind with water increases the speed of surface water in the current. Recall that the Coriolis acceleration is proportional to the speed of an object (chapter 7). So, by accelerating the current, the tug of the wind also increases the Coriolis force on the surface water, and because the California Current is in the northern hemisphere, surface water veers to the right.

We thus arrive at the heart of the matter. For this southward-flowing current, a veer to the right entails movement *away from the coast*. Now, water cannot be drawn out of the land, so as water is pulled offshore by Ekman transport, other water must be drawn up from below to take its place. *Coastal upwelling* results (figure 9.6B).

What are the consequences of coastal upwelling? As we have seen, upwelling is biologically pertinent only if it brings nutrient-rich water to the surface, an effect that depends on the depth of the thermocline. Water in the California Current arrives from the north, having lost heat in its journey across the northern Pacific. This influx of cold water, along with the less-then-tropical influx of heat from the sun, ensures that the thermocline is shallow along the west coast of North America. As a consequence, water brought to the surface by upwelling in the California Current comes from the cold, nutrient-rich water of the ocean's interior.

The introduction of nutrients into the surface layer has the usual effects: phytoplankton bloom and grow, herbivores eat the phytoplankton, and predators eat the herbivores. For the California Current, upwelling season—February through August—coincides with spring and summer in the northern hemisphere. As a result, sunlight is abundant and productivity is exceptionally high. Over a year, each square meter in the California Current upwelling zone fixes 125 to 220 grams of new carbon, equivalent to 2 to 4 times the new productivity of temperate seas and areas of equatorial upwelling, and 15–30 times the new productivity of ocean gyres.

Coastal upwelling occurs commonly along the eastern boundaries of all the world's oceans. For example, there is strong seasonal upwelling associated with both the Peru Current off South America and the Benguela Current off South Africa and Namibia. Because of the upwelling and shallow thermocline with which they are associated, these *eastern boundary currents* are among the most productive areas on earth, supporting many of the world's major fisheries.

One might think that the same process would occur on the western edges of the oceans as well. Indeed, coastal upwelling does occur in association with western boundary currents, but there is a catch. Derived from the equatorial currents of the major ocean gyres, the warm water in western boundary currents pushes the thermocline down along the western boundary of the ocean. As a consequence, water welled up in association with a western boundary current typically comes from above the deep thermocline. Thus, upwelling associated with western boundary currents does little to transport new nutrients to the surface waters, and the coastal waters along the western boundaries of oceans support much less productivity than their counterparts in the east.

Upwelling in the Antarctic

Waters around Antarctica are home to two important currents: the Antarctic Circumpolar Current, which flows from west to east, and the Polar Current adjacent to the continents, which flows from east to west (figure 9.7). This unusual proximity between currents running in opposite directions has important consequences. Because they are in the southern hemisphere, both the Antarctic Circumpolar and Polar Currents tend to veer to the left. Water in the Circumpolar Current thus veers northward, and water in the Polar current moves southward. As a result, where the two currents meet, at approximately 70° South, they diverge (figure 9.8).

At this latitude, surface water is cold, and the thermocline is essentially nonexistent. As a result, this *Antarctic divergence* continually supplies the surface waters of the southern sea with nutrient-rich water from below. During the austral summer, the sun shines both day and night on these rich polar waters, and one would expect the Antarctic divergence to be highly productive, at least as productive as the ocean at equivalent latitudes in the Arctic.

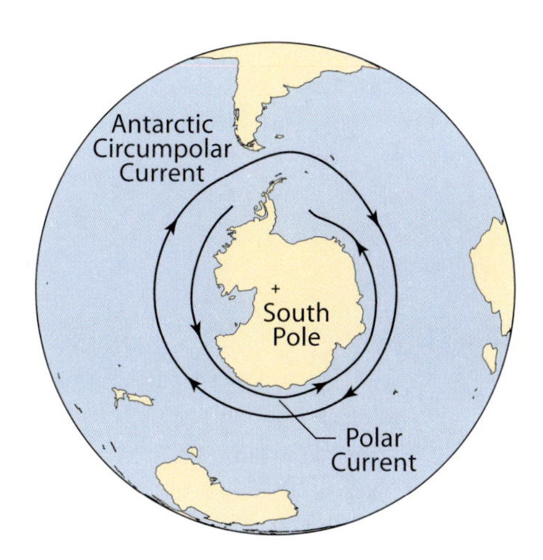

Figure 9.7. Surface currents in the Southern Ocean. The Antarctic Circumpolar and Polar Currents flow in opposite directions.

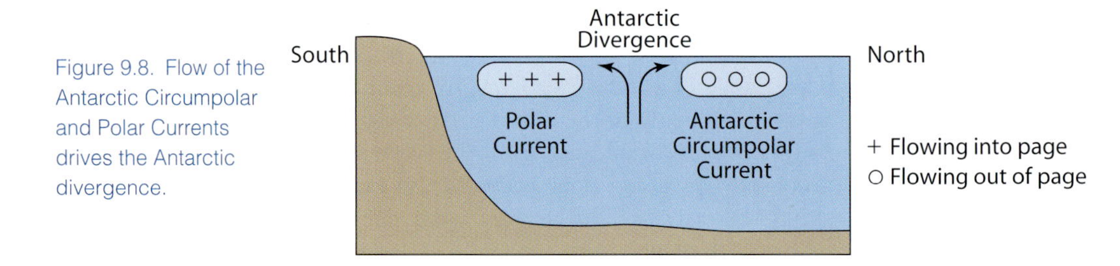

Figure 9.8. Flow of the Antarctic Circumpolar and Polar Currents drives the Antarctic divergence.

However, in reality, annual new production in the southern ocean is low, only 2 to 7 grams of carbon per square meter of ocean surface, a rate comparable to the basal rate of ocean gyres. Why, when supplied with both nutrients and light, would an area of the ocean not be productive? For many years, this question confounded oceanographers. The answer is intriguing, important, and the subject of the next chapter. For the moment, I will remind you that the nutrients supplied by upwelling—primarily nitrate, but phosphate and silicate as well—may not be the only nutrients required by phytoplankton.

Before leaving the Antarctic divergence, we briefly return to our discussion of the thermohaline circulation to add one last detail. Where does the water come from that is upwelled in the Antarctic divergence? Surprisingly, much of it comes from the North Atlantic Deep Water. In other words, cold, dense water that was sunk in the Arctic travels through the deep sea only to be dragged back to the surface in the Southern Ocean. Once at the surface, this water can again be cooled, and it subsequently sinks.

There are two aspects of this process that deserve note. First, we can now draw a more complete diagram of the path taken by water as part of the thermohaline circulation (figure 9.9). North Atlantic Deep Water travels south the entire length of the Atlantic, and wells up in the Southern Ocean, where it mixes with similarly upwelled Antarctic Bottom Water and then subsides and flows north. Water in this return flow is called Antarctic Intermediate Water. Second, upwelling of North Atlantic Deep Water in the Antarctic divergence is relevant to a thought we touched on in chapter 6. The Antarctic divergence is powered by the wind, and this divergence is responsible for bringing North Atlantic Deep Water and Antarctic Bottom Water to the surface. Thus, it is clear that wind provides a major fraction of the energy that drives the thermohaline circulation. Some researchers, concerned that the term "thermohaline" implies that the circulation is driven solely by the density of water as it sinks near the poles, have suggested that this circulation be called *meridional overturning circulation* instead.

An Interim Review

Let's review the territory through which we have traveled. The ocean typically consists of two layers: a warm, well-lit, nutrient-poor layer at the surface and a cold, dark, nutrient-rich layer below. The stable separation of light and nutrients limits the rate at which phytoplankton can grow and reproduce. And because phytoplankton provide the basic source of food for organisms downstream in the trophic river, the separation of light and nutrients restricts life in the sea. This grim

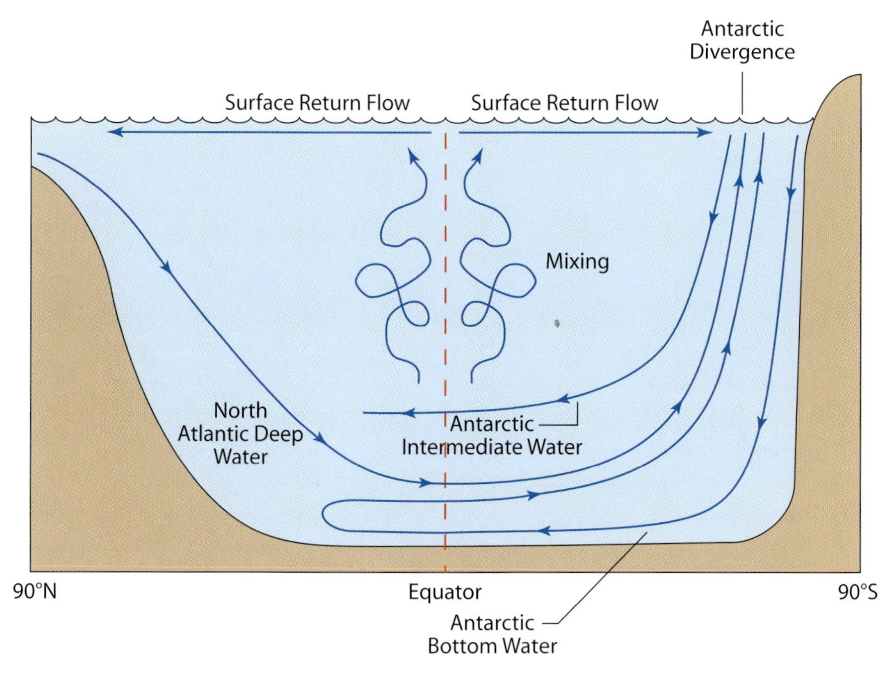

Figure 9.9. The Antarctic divergence draws both North Atlantic Deep Water and Antarctic Bottom Water back to the surface. Water cools and sinks again (as Antarctic Intermediate Water), completing the cycle of thermohaline circulation.

situation typifies tropical oceans year round, as well as temperate seas in summer. When and where the two-layered ocean is present, surface waters are limited to new productivity of only about 7.5 grams of carbon per square meter each year.

Cooling of the temperate and polar seas in fall and winter, in conjunction with the turbulence created by storms, mixes these areas on a seasonal basis. A spring bloom can then follow, but in temperate seas, summer chokes off this bloom as the two-layered ocean is reestablished. In the Arctic, light limits production. Yet despite these limitations, temperate and ice-free arctic seas have a productivity 7 to 20 times that of major ocean gyres.

Upwelling circumvents the dilemma of the two-layered ocean in some tropical, subtropical, and temperate waters, but only where the thermocline is shallow. Thus, coastal upwelling effectively transports nutrients to surface waters along eastern boundaries of the ocean, but not along western boundaries. Similarly, equatorial upwelling effectively brings nutrients to the surface in the eastern part of the ocean, but not in the west. In areas of effective upwelling, annual export production can be 15 to 30 times that of ocean gyres.

The simple physical and biological oceanography we have explored in the last seven chapters explains a surprising number of real-ocean phenomena. Compare our deductions with the data shown in figure 9.10, a map of estimated new productivity for the entire world ocean. The major ocean gyres appear clearly as areas of low productivity (dark purple and maroon), and the effects of equatorial upwelling are readily apparent as tongues of light blue extending westward from South America and Africa. Areas of coastal upwelling are highly productive (yellow and red) along eastern boundaries of the oceans,

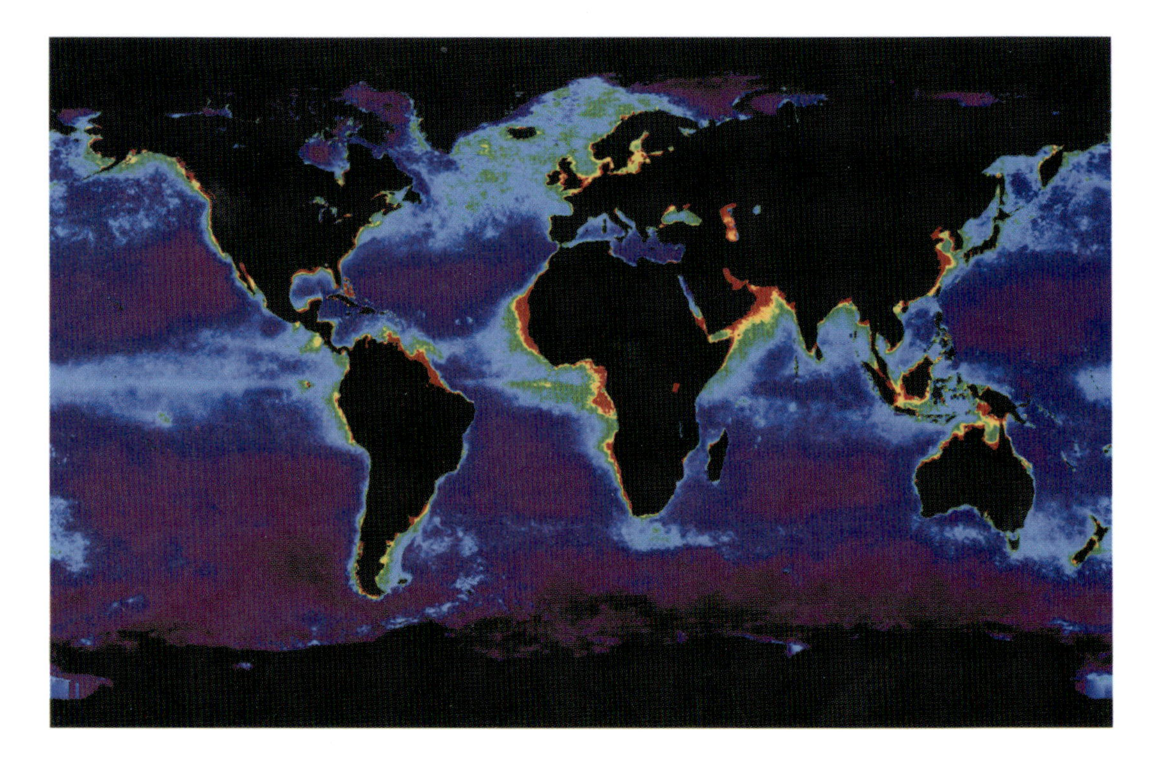

Figure 9.10. The global pattern of new production. Reprinted with permission from P. G. Falkowski et al. (1998), Biogeochemical controls and feedbacks on ocean primary productivity, *Science* 281: 200–206. Copyright 2008 AAAS.

and generally less productive along western boundaries. Shallow areas such as the East China and Yellow Seas, the Arafura Sea and the Gulf of Carpenteria, and the broad continental shelves adjacent to Brazil and southern Argentina are highly productive. The productive Arctic, which we have explained, strikingly contrasts with the unproductive Antarctic, which we have not. Take a moment to feast your eyes on this figure. This is the world ocean, 361 million square kilometers of water surface, and you now can explain how much of it works.

The Curse of Mercator

Lest you get carried away, I should note that the map of figure 9.10 contains a bias. As with many attempts to portray the spherical earth on a flat sheet of paper, the map used in figure 9.10 exaggerates the size of objects near the poles. For example, on this map, Antarctica appears to be roughly the same size as

Asia and Europe combined, and Greenland the same size as South America. In fact, Antarctica is only a quarter as large as Europe and Asia, and Greenland a tenth the size of South America.

This bias overemphasizes processes near the poles and thereby underemphasizes processes near the equator. Thus, the unproductive ocean gyres, as evident as they are in this distorted projection, appear smaller than they actually should, and the lack of productivity in the Southern Ocean—although real and important—appears disproportionately large.

El Niño

We can't leave this overview of the world ocean without discussing one way in which the biologically essential pattern of upwelling can break down.

Recall that the warm pool in the western tropical Pacific and the basinwide tilt in the thermocline are maintained by the continuous friction of the trade and zonal winds (figure 9.11A). Imagine now that the trade and zonal winds decrease slightly in strength. With less tug from these winds, the warm surface water accumulated in the western portion of the tropical ocean will slosh back to where it originated: back to the east. In other words, if equatorial winds die down, warm surface water flows eastward, and the basinwide tilt of the thermocline begins to level out (figure 9.11B).

This effect then cascades through the system. As the wave of water from the west warms surface water in the eastern part of the equatorial sea, the temperature difference between the eastern and western parts of the ocean decreases. The decreased temperature difference in turn mitigates the intensity of the Walker Cell, which then lessens the zonal winds. The decrease in zonal winds allows more warm water to flow eastward, and the effect feeds back on itself. Recall also that the trade and zonal winds power the equatorial divergence. So,

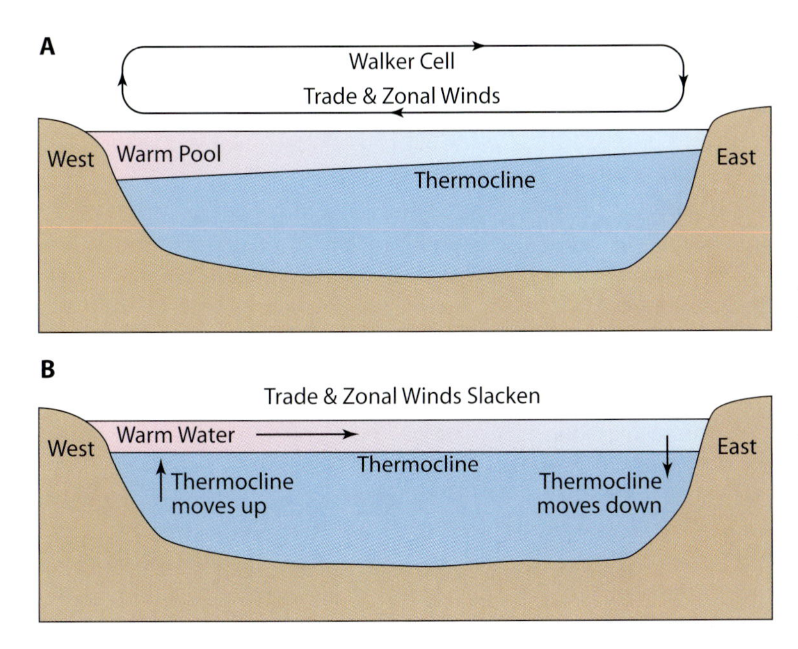

Figure 9.11. Mechanics of El Niño. (A) Strong zonal and trade winds maintain the basin-wide tilt of the thermocline. (B) If winds slacken, the thermocline moves up in the west and down in the east as warm water moves eastward.

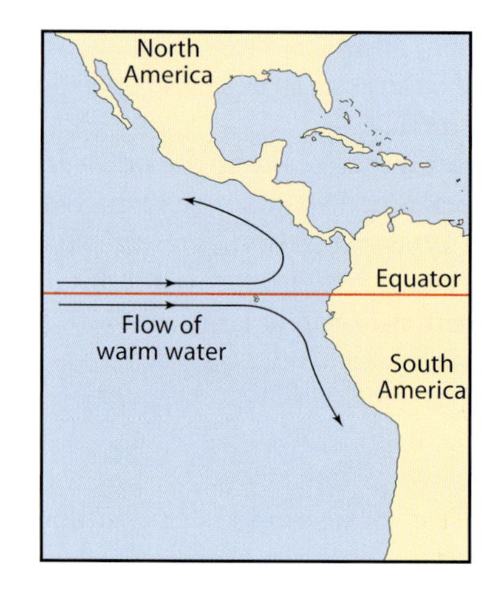

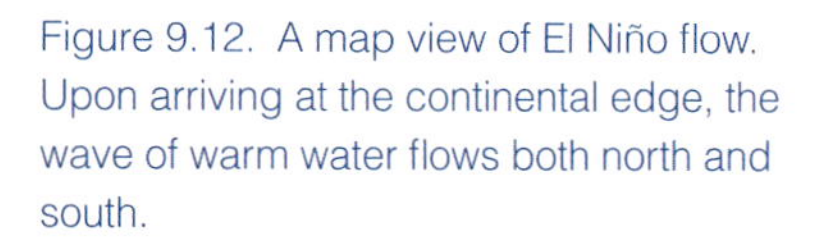
Figure 9.12. A map view of El Niño flow. Upon arriving at the continental edge, the wave of warm water flows both north and south.

as the trade and zonal winds die down, the intensity of the equatorial divergence decreases.

And the effects don't stop there. As warm surface water sloshes along the equator and reaches the eastern boundary of the ocean, it is forced to flow both north and south (figure 9.12). This poleward-flowing warm water pushes down the thermocline along much of the tropical and temperate eastern boundary of the ocean, preventing local coastal upwelling from transporting nutrients to the surface. Without effective equatorial or coastal upwelling, nutrients in the surface water are rapidly depleted, and the food chain collapses for lack of fertilizer.

Times are tough in the western part of the ocean as well. As warm water flows east, the thermocline rises in the west, but it never becomes shallow enough for upwelling to become biologically effective. Thus, until the trade winds pick up, the two-layered ocean dominates, and life in tropical and subtropical seas is difficult.

This cascading scenario is known as *El Niño*, and it plays out every three to seven years in the Pacific Ocean.

The biological effects can be drastic. For example, the rise in sea temperature in the El Niño of 1982–83 killed 95% of the corals in the Galápagos, and 75% to 90% of the corals along the coast of Costa Rica. Catches of coastal fisheries severely decrease after an El Niño as fish, starved for their planktonic food, fail to reproduce, or die. For instance, the catch of anchovies off Ecuador and Peru decreased from 12 million metric tons before the El Niño of 1972–73 to less than 3 million metric tons after. Seabirds, at the top of the food chain, often lack sufficient energy to reproduce in an El Niño year, and they die in droves. Before the El Niño of 1972–73, there were 27.5 million seabirds estimated to live along the coast of Ecuador and Peru; only 1.8 million survived. During severe El Niño events, the shores of the Galápagos Islands are littered with the carcasses of seals, sea lions, dolphins, and whales that have died from starvation.

And there are a myriad of ancillary effects of El Niño. The wave of warm water traveling poleward along the coast can carry fish to novel locales. In the severe El Niño of 1983, for instance, fishermen caught barracuda (a strictly tropical fish) in Monterey Bay, just south of San Francisco. And my colleagues at Hopkins Marine Station in Pacific Grove, California, were able to catch albacore tuna from small boats quite close to shore; in normal years, these tuna stay well out to sea.

The warm-water wave of El Niño, a result of decreased winds in the tropics, also allows local sea level to rise in the eastern Pacific. During the event of 1982–83, for example, sea level at Monterey was nearly 20 cm above normal. Plants and animals that typically would be exposed at low tide found themselves constantly submerged.

El Niño also has numerous effects on the terrestrial environment. These are felt most strongly in coastal Ecuador and Peru, where the arrival of warm oceanic water is accompanied by unusually humid air, resulting in torrential rains. During the El Niños of 1972–73 and 1982–83, flooding in coastal towns overwhelmed sewer systems, leading to epidemics of typhus, dysentery, and hepatitis. In Indonesia during the El Niño of 1997–98, just the opposite happened. As the trade winds slackened, the influx of humid oceanic air to the western Pacific decreased, leading to a severe drought. Smoke from the consequent forest fires made the air nearly unbreathable on Java, Sumatra, and Timor, and contributed to the crash of an airliner on Kalimantan (Borneo).

What triggers an El Niño? Interesting question. If you talk to oceanographers, they may well start where we have, with the trade winds dying off, and show how the change in ocean temperature results. In contrast, if you talk to meteorologists, they may start with a change in ocean temperature and show how that leads to a tapering off of the winds. Oceanographers and meteorologists even have different names for the phenomenon. The term "El Niño" (literally, "the boy," a reference in Spanish to the Christ child) originates with the timing of the event: in Ecuador and Peru, the arrival of warm water at the coast has historically occurred in December. This warming was the first evidence of the oceanographic side of the phenomenon, and the colloquial name for it, El Niño, was introduced into the scientific literature in 1891. Eventually, the term was applied not only to the increase in water temperature, but also to the changes in currents and sea level that accompanied the temperature shift.

In contrast, meteorologists commonly refer to the "Southern Oscillation" in reference to a shifting pattern of atmospheric pressure between Darwin, Australia, and Tahiti. Typically, the surface atmospheric pressure in Darwin is low, while that in Tahiti is high, the pattern one would expect when the Walker Cell is active. During an El Niño event, however, the difference in pressure between Darwin and Tahiti decreases, and this shift in pressure accompanies (causes?) the change in the wind pattern in the tropical Pacific. The phenomenon was first noticed in 1924 by Sir Gilbert Walker (the same Walker who proposed the Walker Cell), and he dubbed it the "Southern Oscillation."

The two viewpoints—El Niño, Southern Oscillation—stayed separate until, by coincidence, the phenomena coincided with the first International Geophysical Year in 1957–58. The additional scrutiny of the occasion revealed the ties between the viewpoints, and as an ecumenical gesture, the two fields now refer to the phenomenon as "El Niño/Southern Oscillation," or ENSO, for short.

In fact, an ENSO event can result from a large number of factors both atmospheric and oceanic, all of which affect each other. Predicting ENSO events is becoming a science, rather than an art, primarily because computer simulations of the atmosphere and ocean are becoming adept at incorporating all the myriad details that affect the process.

As noted above, ENSO events occur sporadically, with intervals commonly ranging from three to seven years. The alternate state of the ocean, in which zonal and trade winds blow with above-average speeds, augmenting the size and intensity of the Pacific warm pool, is commonly known as *La Niña*, Spanish for "the girl." It is perhaps inevitable in the alternation between El Niño and La Niña that there are occasionally years that straddle the fence. These years of average conditions are informally known as La Nada, "the nothing."

Multi-Decadal Oscillations

The Pacific Decadal Oscillation

The oceanic and climatic oscillations of El Niño are only one example of the major shifts through which the ocean cycles. In another example, oceanographers have recently documented a long-term shift in the temperature structure of the Pacific Ocean north of the equator. At times, the sea-surface temperature in the northwestern quadrant of the North Pacific (near Japan) is above the long-term average, while the temperature in the southeastern quadrant of the North Pacific (near Mexico) is below the long-term average. This pattern is the *positive* or *cold* phase of the oscillation—the numerical difference in temperature between northwest and southeast is above zero—also known as *La Vieja*, the old woman. Periodically, this pattern reverses, such that the northwest is cold and the southeast is warm. The reversed pattern is the *negative* or *warm* phase, also known as *El Viejo*, the old man. The cycles in this pattern of temperature are known as the *Pacific Decadal Oscillation*, or for short, the PDO.

The name, although catchy, is somewhat misleading in that it implies that the phenomenon fluctuates with a period of about a decade. Instead, current data suggest that the PDO oscillates with a combination of periods, of which 20 years and 70 years are particularly evident. Thus, Pacific Multi-Decadal Oscillation might be a better term.

The Pacific Decadal Oscillation bears some resemblance to El Niño in that it involves major shifts in the surface temperature of the Pacific, but the resemblance stops there. First, the two phenomena occur on different temporal scales. El Niño cycles last three to seven years, whereas PDO cycles last upwards of 20 years. In fact, each phase of the PDO lasts for so long that the switch from one to the other is often referred to as a "regime shift." Second, the areas affected by the two phenomena are different. The primary effects of El Niño are felt in the tropics, with secondary effects that extend into temperate waters. In contrast, the major effects of the PDO occur in temperate seas, with only secondary effects in the tropics.

Although El Niños and the PDO occur at different scales and have their primary effects in different places, the frequency of ENSO events can apparently be modulated by the Pacific Decadal Oscillation. When the PDO is in its warm phase, in which the waters of the eastern Pacific are warmer than average, ENSO events are more common than usual; when the PDO is in its cold phase, with cool waters in the eastern Pacific, ENSO events are less common than usual. For example, in the thirty-year period from 1947 to 1976, when the

PDO was in its cold phase, only about six years (20%) were spent in El Niño conditions. In contrast, in the twenty-one years from 1977 to 1998, when the PDO was in its warm phase, roughly 17 years (80%) were spent in El Niño conditions.

El Niño and the Pacific Decadal Oscillation also differ in the level at which we currently understand them. As I noted above, computer models are available that predict ENSO events with laudable accuracy, an indication that the basic science of the process is understood. In contrast, oceanographers are only beginning to speculate about the underlying cause of the PDO. In large part, this lack of understanding is due to the long period of the phenomenon. Widespread information about ocean surface temperature is available only for the last hundred years or so, barely enough to encompass one full cycle of the PDO's longer dominant period. With so little information to guide them, it will be a challenging task for oceanographers to figure out the detailed mechanism of the PDO.

As with its mechanism, the biological effects of the Pacific Decadal Oscillation are only beginning to be understood. The few data available relate primarily to the effect of the phenomenon on fisheries, and these will be discussed in chapter 11.

The North Atlantic Oscillation and the Indian Ocean Dipole

Large-scale, long-term fluctuations are not confined to the Pacific Ocean. The North Atlantic is home to an analogous fluctuation, the *North Atlantic Oscillation*. Like the Southern Oscillation, the North Atlantic Oscillation (or NAO for short) is indexed by the difference in atmospheric pressure between two points, in this case between Iceland (at latitude 65° North, where the pressure is typically low) and the Azores (at 38° North, where the pressure is typically high). Unlike the Southern Oscillation, which shifts with a period of 3 to 7 years, major shifts in the NAO have a period closer to 20 years, more similar to that of the Pacific Decadal Oscillation. The North Atlantic Oscillation has been implicated in shifts in the location of the Gulf Stream, and it seems likely that as more evidence becomes available, biological implications of the NAO will become apparent.

The Indian Ocean is also home to a long-period oscillation in surface temperature, known as the *Indian Ocean Dipole*. In this case, the oscillation has a period of three to twelve years, with an average of seven years, and it appears to be independent of both the ENSO cycle and the yearly monsoon. The Indian Ocean Dipole has only recently been recognized, and it is not yet well studied.

As oceanographers accumulate more data, it is likely that other long-term fluctuations will be found. For example, recent measurements of deep-sea currents and the results of computer models suggest there may be cyclic fluctuations in the thermohaline circulation. It will be interesting to keep watch on the field of oceanography over the next few years as researchers debate the existence and significance of long-term ocean cycles.

Summary

In this chapter we have touched upon a variety of mechanisms that, from time to time and place to place, allow the ocean to circumvent its two-layered nature. In winter, the thermocline disappears in temperate seas, allowing nutrients to be mixed up into the ocean's surface layer, setting the stage for algal blooms in spring. In arctic seas, the thermocline never forms, and phytoplankton productivity tracks the seasonal availability of light. These bloom-and-bust cycles complement the more continuous productivity enabled by shallow depths and the upwelling along the equator and in conjunction with eastern boundary currents. Together, the productivity of the shallow, temperate and arctic seas and upwelling areas provide most of the new productivity of the entire ocean (figure 9.10).

These predictable patterns of productivity are modulated by long-term, very-large-scale fluctuations in oceanic conditions. The mechanisms and consequences of cycles such as the Pacific Decadal and North Atlantic Oscillations are just beginning to be glimpsed.

In the last nine chapters we have come a long way toward understanding how the ocean works. Our knowledge of the two-layered ocean—and the physics that can circumvent it—reached a satisfying climax in figure 9.10, in which we were able to account for much of the pattern of ocean productivity. You should take a moment to revel in this accomplishment.

Further Reading

Caviedes, C. (2001). *El Niño in History: Storming through the Ages*. University of Florida Press, Gainesville.

Chavez, F. P., et al. (2003). From anchovies to sardines and back: Multidecadal change in the Pacific Ocean. *Science* 299: 217–221.

Falkowski, P. G., et al. (1998). Biogeochemical controls and feedbacks on ocean primary productivity. *Science* 281: 200–206.

Mann, K., and J. R. Lazier (2006). *Dynamics of Marine Ecosystems* (3rd edition). Blackwell Publishing, Malden, MA.

Philander, S. G. (1990). *El Niño, La Niña, and the Southern Oscillation*. Academic Press, New York.

Sarmiento, J. L., and N. Gruber (2006). *Ocean Biogeochemical Dynamics*. Princeton University Press, Princeton, NJ.

Complexity: Carbon, Iron, and the Atmosphere

The last chapter ended with the satisfying conclusion that we can explain much of the global pattern of primary production in the ocean. It is important not to get carried away, though. The grand simplicity of figure 9.10 tends to obscure complexities in the real ocean that cannot be explained by our understanding to this point, and these complexities can have important consequences for all life on earth. In this chapter, we will grapple with one of these complexities, the role played by the biological carbon pump in regulating atmospheric concentration of carbon dioxide. The knowledge we gain will help us understand how the ocean contributes to the regulation of our planet's temperature.

The HNLC Paradox

In our current, simplified view of the ocean, the surface layer is continuously depleted of nutrients by the biological pump. Only in shallow seas, or areas where the thermocline is shallow and upwelling is active, are sufficient nutrients brought to the surface to allow for continuous new production, and this production quickly gobbles up nutrients as they are delivered. According to this scenario, any area of the surface ocean that has an abundance of light and nutrients should be quickly pumped back into compliance with the two-layered paradigm.

It is with some surprise, then, that we find several major exceptions to this simple picture. The entire Southern Ocean does not fit this pattern, and there are large areas of the subarctic and eastern equatorial Pacific that are similarly anomalous. These areas have high concentrations of nitrate, but low concentrations of primary producers. For decades, these high-nitrate, low-chlorophyll (HNLC) areas have been an open and contentious question in biological oceanography. If nitrate is there, why haven't phytoplankton pumped it out? If nitrate is there, why is productivity low?

The ability to answer these questions is of more than academic interest. If the HNLC areas of the ocean were to become more productive, they could cause the ocean to draw carbon dioxide from the atmosphere at a greater rate than it currently manages. As we will see, an increase in the net rate of CO_2 absorption by the ocean could have a strong effect on the temperature of earth as a whole.

There are several processes that contribute to this important interaction between the ocean and global climate—the greenhouse effect, the carbon pump, and the availability of micronutrients—and we must deal with them separately before we can combine them into a coherent whole. Let's begin with the greenhouse effect.

The Greenhouse Effect

In chapter 6, we briefly delved into the thermal equilibrium that exists between the earth and the sun. As sunlight impinges on the globe, some of it is absorbed and converted to heat, raising the planet's temperature. As temperature rises, earth radiates more heat into space in the form of infrared light. Over the millennia, the earth has heated up to a temperature at which the rate it absorbs heat is almost exactly equal to the rate at which heat is radiated back into space. Currently, this temperature is approximately 15°C, comfortably conducive to life.

But this simple picture glosses over some important details. First, the benign temperature cited here, 15°C, is somewhat misleading. This is the temperature of the planet's surface. But it isn't earth's surface that directly exchanges heat with space, rather, it is the atmosphere. In the upper reaches of the atmosphere, where the bulk of earth's heat is radiated out into space, the temperature is a frigid −18°C. This temperature—not 15°C—is the true equilibrium temperature for earth. To put this another way, in the absence of the atmosphere, the temperature of earth's surface would be 33°C colder than it is now, cold enough to freeze all water on the planet and too cold to support life as we know it.

How does the atmosphere maintain the surface at a temperature 33°C above that of thermal equilibrium? In effect, the atmosphere acts as an insulating blanket. The surface of the planet (at the bottom of the blanket) must heat up to 15°C in order for the top of the blanket (the upper portion of the atmosphere) to be at −18°C, the temperature required for thermal equilibrium.

Why does the atmosphere act as a blanket? Here, the greenhouse effect comes in. As we have seen (chapter 6), the sun's surface has a temperature of about 5500°C, and as a result, it emits light primarily in the visible range of wavelengths.[1] These wavelengths pass easily through the mixture of gases in

[1] The most abundant wavelength of light (in meters) given off by a body at absolute temperature, K, is $0.002897/K$, where absolute temperature (measured in kelvins) is °C+273.15. This relationship is known as Wien's Displacement Law. For example, the absolute temperature of the sun's surface is approximately 5800 kelvins, so the peak wavelength of its light is 500 nanometers, near the middle of the range of visible light. The absolute temperature of earth's surface is 288 kelvins, producing infrared light with a peak wavelength of 10,000 nanometers.

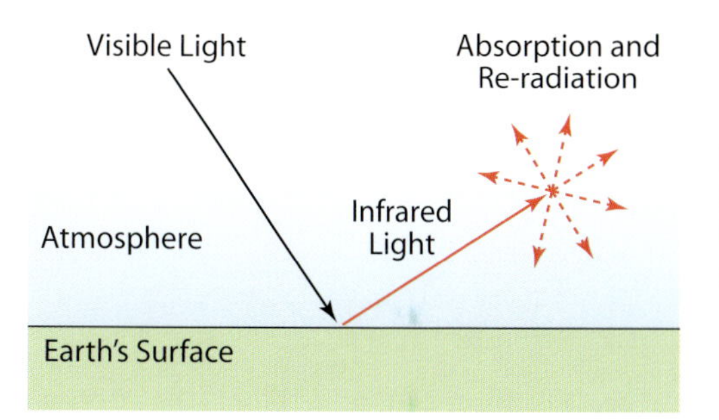

Figure 10.1. The greenhouse effect. Absorption and re-radiation of infrared light causes the atmosphere to act as an insulating blanket.

our atmosphere and are absorbed when they strike land or water. In contrast, because the surface of earth is much cooler than the sun, the light it emits is in the infrared range. On the way from earth's surface to outer space, these long wavelengths can be absorbed by various gases—the greenhouse gases—in the atmosphere.

This absorption heats the greenhouse gases, which then radiate infrared light in turn. But roughly half of this reradiated light is radiated not toward space, but rather, back toward the earth's surface, where it is reabsorbed (figure 10.1). Because much of the infrared light emitted by earth is, in essence, trapped by the atmosphere, the surface must heat up so that the fraction that *does* escape is equal to the rate of heat input from the sun. Thus, it is the propensity of greenhouse gases to absorb infrared light that allows the atmosphere to act as a blanket.[2]

Only certain gases have this effect. Nitrogen and oxygen, which together form 99% of the atmosphere, are virtually transparent to infrared light and therefore have negligible effect on earth's equilibrium temperature. Instead, the most important greenhouse gases are water vapor, carbon dioxide, methane, and nitrous oxide. Currently, water vapor accounts for nearly 65% of greenhouse heating, carbon dioxide about 35%, and methane and nitrous oxide the rest.[3]

Since the beginning of the Industrial Revolution in about 1800, humankind has been prodigiously burning carbon (in the form of trees, coal, oil, and natural gas) and making cement,[4] both of which release CO_2. As a result, the concentration of CO_2 in the atmosphere—which was about 280 parts per million before the Industrial Revolution—has risen almost 30%, to approximately 360

[2] The term "greenhouse effect" is actually a misnomer: greenhouses do indeed keep the plants inside them warm, but they do so primarily by inhibiting convective loss of heat to the surrounding air rather than by absorbing and reradiating infrared light. Proof comes from a greenhouse constructed with special quartz glass transparent to both visible light and the infrared light radiated by its interior: it stayed nearly as warm as a conventional greenhouse. Similarly, greenhouses constructed with inexpensive polyethylene panels work well, even though polyethylene is transparent to infrared light.

[3] A few other gases, notably the manmade chlorofluorocarbons, are potent greenhouse gases, but their concentrations are so low that their overall effect is minor compared to the gases listed here. Chlorofluorocarbons *do* have an important effect on the ozone concentration high in the atmosphere, but that is another matter.

[4] The first step in making cement is to heat limestone (calcium carbonate, $CaCO_3$), converting it to lime (calcium oxide, CaO) and driving off CO_2. Now, limestone is a sedimentary rock, formed

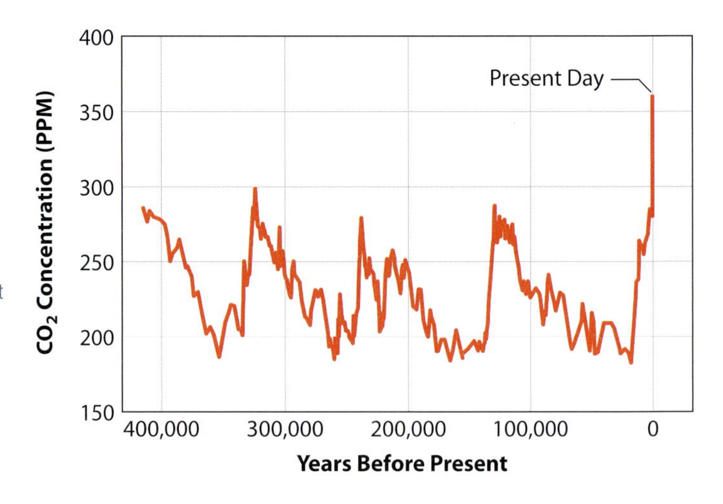

Figure 10.2. Current carbon dioxide concentration in the atmosphere is higher than it has been in at least the last 420,000 years. Data from J. R. Petit et al. (1999), Climate and atmospheric history of the past 420,000 years from the Vostok ice core, Antarctica, *Nature* 399: 429–436.

parts per million as of 2006. Current CO_2 concentrations exceed any others recorded in the last 420,000 years, and they are still increasing (figure 10.2). The release of nitrous oxide and methane has also increased either directly or indirectly from human activities, but about 60% of the anthropogenic (that is, human-caused) increase in greenhouse effect is due to carbon dioxide, and we focus here on that gas.

The elevated concentration of carbon dioxide increases the insulating properties of earth's atmospheric blanket, and as a result the temperature at the planet's surface must rise to maintain the upper ranges of the atmosphere at the temperature required to stay in equilibrium with solar input. The average temperature of the planet has risen about 0.6°C in the past 200 years, and if the rate of CO_2 production continues to grow at the current rate, the temperature will increase by another 1.5 to 5.8°C in the next century.

A temperature increase of this magnitude is worrisome. As global temperature rises, the average temperature of the oceans will also rise, and as a consequence, the volume of water in the sea will expand (see chapter 3). Currently, sea level is rising 10 to 20 cm per century due to this effect. Worse, the rise in surface temperature has the potential to melt or destabilize large portions of the ice sheets in Greenland and West Antarctica. If just half the volume of this ice were added to the ocean, sea level would rise six to seven meters, drowning many of the world's coastal cities.[5] Reduced production of sea ice in the polar

---

from the calcareous remains of marine organisms (e.g., coccoliths and shells). Thus, carbon dioxide emitted in the manufacture of cement is of biological origin, and in this respect it is similar to CO_2 released when burning fossil fuels such as oil, coal, and natural gas.

[5] It should be noted that melting the Greenland or West Antarctic ice sheets would take a long time. Consider these facts: The Greenland ice sheet has a volume of approximately 2.620×10^6 cubic kilometers, with a mass of 2.4×10^{18} kilograms. Let's assume that the average temperature of this ice is −20°C. It takes 1960 joules to raise the temperature of one kilogram of ice 1°C, and 3.34×10^5 joules to melt that kilogram once it has reached 0°C. Thus, approximately 3.73×10^5 joules are required to melt each kilogram of Greenland ice, for a total of 8.96×10^{23} joules. Through the course of a year, an average of roughly 100 joules of solar energy impinges on each square meter of Greenland each second, about 2.2×10^{14} joules per second for the entire island. If we somehow painted Greenland black so that all this solar heat input could be absorbed and used

oceans could slow (or even abolish) the thermohaline circulation, reducing ventilation of the deep ocean. Rising air temperatures could intensify the density stratification of the two-layered ocean, which could in turn reduce ocean primary production. The reduction of primary productivity would then be transmitted down the trophic stream, with detrimental effects on all marine herbivores and predators. Similar dire effects would occur on land.

In recent years, human society has become aware of the potentially detrimental effects of its CO_2 production, and has begun to search for ways to minimize or reverse the trend. It is at this point that the physics, chemistry, and biology of the ocean take center stage in what is unfolding to be an intriguing, and often alarming, interaction between science and society.

There is roughly sixty times as much carbon dissolved in the sea as there is currently present in the atmosphere. As a result, minor changes in the ocean's carbon content can have drastic effects on the CO_2 concentration in the air. For example, the ocean would need to absorb only 0.5% more carbon than it already contains to reduce atmospheric carbon to its pre-industrial levels. On the other hand, if the ocean lost just 0.5% of its current carbon reserve to the atmosphere, the existing problem would double. Clearly, the concentration of carbon dioxide in the atmosphere is in delicate balance with that in the ocean, and if we are to predict and control the detrimental effects of CO_2 production, we must understand this balance. That understanding begins with the biological carbon pump.

Carbon Reservoirs, Carbon Fluxes

In this section we follow the path of carbon as it moves between the atmosphere and the ocean. In the course of this cycle, carbon is incorporated into many different compounds. In the atmosphere, carbon is primarily in the form of carbon dioxide, with a little bit of methane thrown in. In seawater, it is primarily in the form of bicarbonate, with minor contributions from CO_2, carbonate, and organic molecules. Keeping track of this chemistry is complicated, and for our present purposes unnecessary. Instead, we simplify our discussion by focusing on carbon itself, without regard to the specific compound in which it appears. We will return to the chemical complexities at the end of the chapter.

First, let us take inventory of earth's supply of carbon. In doing so, we will deal with large quantities, and we need an appropriately large unit of measurement. For convenience, we measure carbon in *gigatonnes*. A gigatonne is a billion metric tons, where, as we noted earlier, a metric ton is 1000 kilograms.

Until humans came along, the primary source of carbon in the atmosphere was outgassing from volcanoes and other geological sources. Currently, volcanism is thought to add roughly 0.04 gigatonnes of carbon to the atmosphere

---

to melt ice, it would still take 8.96×10^{23} joules $\div$ 2.2×10^{14} joules per second $= 4.1 \times 10^{9}$ seconds to melt the ice sheet. This is roughly 130 years. Even drastically increased air temperatures would be much less effective at heating Greenland than the scenario depicted here. Thus, it will take several centuries to melt either the Greenland or the West Antarctic ice sheet. Of more immediate concern is the rate at which these ice sheets are extruding ice into the sea. As the sheets melt, meltwater may lubricate the base of each sheet, accelerating its seaward flow.

each year, although the amount may have varied in the past. Over geological time, the total release of carbon has added up to about 70 million gigatonnes. Where is all that carbon now?

Approximately 99.9% of it has been taken out of the atmosphere by living organisms and deposited into sediments. Most of this sedimentation occurred in the sea. Thus, if it were not for the ocean's biological pump and its transport of carbon to the seafloor, the concentration of carbon dioxide in earth's atmosphere could be a thousand times what it is now, with truly drastic consequences for the greenhouse effect. As with photosynthetic control of atmospheric oxygen concentration and bacterial control of nitrogen concentration, we again see the role of biology in determining the composition of earth's atmosphere, and in this case, its importance in avoiding a planet-baking greenhouse effect.

Although it is worth pondering this apocalyptic perspective, it must be viewed in proper context. Because of its immense size, the sedimentary reservoir of carbon is essentially inert: buried in rocks and the ocean floor, the vast bulk of earth's carbon reservoir cannot freely interact with the atmosphere. As we briefly discussed in chapter 2, the primary route by which sedimentary carbon is returned to the atmosphere is via seafloor spreading, which over the course of millions of years results in subduction and subsequent outgassing. So yes, *if* all sedimentary carbon were returned to the atmosphere, there would be drastic consequences. But there is no natural mechanism available that can cause this to happen quickly. Instead, century-to-century control of atmospheric carbon dioxide levels depends on several smaller reservoirs that *do* actively interact with the air.

The first of these reservoirs is the ocean interior, the ocean below the thermocline. Currently, there are approximately 36,000 gigatonnes of carbon dissolved in the deep layer of the sea. This reservoir can indeed exchange carbon with the atmosphere, but only over relatively long periods of time. Thermohaline circulation ventilates the ocean interior with a turnover time of 500 to 1000 years, depending on the ocean basin, and this turnover time sets the pace at which the ocean interior interacts with the atmosphere. For example, an atom of carbon taken out of the atmosphere today and delivered to the deep ocean interior by the biological pump will spend 500 to 1000 years (on average) before it reappears in surface waters. Or to put it another way, the carbon welled up from the ocean interior today arrives at the surface in concentrations that reflect the balance of the ocean and atmosphere roughly 500 to 1000 years ago. Because of the slow pace of exchange between the ocean interior and the atmosphere, the carbon content of the ocean interior changes very slowly, and in the short term we can treat the ocean interior in the same way we treat the sediments: for our purposes, the ocean interior is a reservoir of fixed size that does not affect atmospheric concentrations of carbon.

Instead, short-term effects—changes that can be completed in years to decades—are controlled by exchanges among three reservoirs that are an order of magnitude smaller than the ocean interior. Currently, there are about 750 gigatonnes of carbon in the atmosphere itself, primarily in the form of carbon dioxide, as noted above. There are about 920 gigatonnes of carbon dissolved in the surface layer of the ocean, and another 2250 gigatonnes in terrestrial plants and soils.

These three reservoirs can readily exchange carbon amongst themselves, and until the Industrial Revolution, these exchanges were in balance. In other

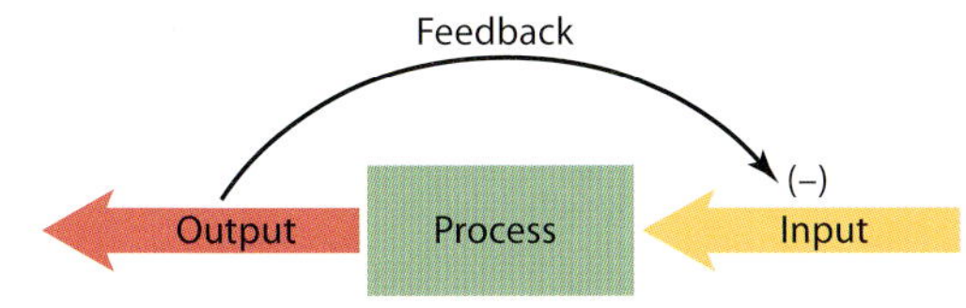

Figure 10.3. Negative feedback provides a mechanism by which a process can maintain a set output.

words, the amount of carbon dioxide released into the atmosphere from terrestrial and marine sources was very nearly equaled by the rate at which carbon dioxide was absorbed. As a result, over at least the last 420,000 years, the concentration of CO_2 in the atmosphere has not varied much, fluctuating in a range from approximately 180 to 300 parts per million. Low concentrations occurred during ice ages, high concentrations when the earth was warm (figure 10.2).

Then along came humans. Tapping into the hitherto inert geological reservoir of carbon by making cement and burning oil and gas, humankind began to tinker with the balance of exchange among atmospheric, terrestrial, and marine reservoirs. Currently, human society adds about 5–6 gigatonnes of carbon to the atmosphere each year, more than a hundred times the natural rate from volcanism, and this excess drives up atmospheric concentration. How will our planet respond?

Before we dive into this topic, we must digress briefly to discuss the phenomenon of *feedback*. Consider the heating system in your house. How does it maintain a comfortable temperature on a cold day? When you turn the system on, the furnace pumps out hot air, and room temperature rises. This increase in air temperature is sensed by a thermostat, and when room temperature reaches a comfortable set point, the thermostat turns the furnace off. The effect of rising temperature on the thermostat is a classic example of *negative feedback*: the output of the system (heat in this case) feeds back to the system's input (the furnace) in a negative fashion (it turns the furnace off). This process is shown schematically in figure 10.3. Negative feedback provides a mechanism by which a system can maintain a constant state, and thus, despite the potential connotations of its name, negative feedback can be a good thing.

Now back to the problem of atmospheric carbon dioxide. It would be comforting to think that negative feedback mechanisms are available whereby the earth could respond to the current increased rate of CO_2 production. For example, if rising concentrations of CO_2 somehow increase the rate at which the "extra" carbon is absorbed, this negative feedback could keep atmospheric concentration of CO_2 at livable levels. Unfortunately, these sorts of feedback mechanisms are few and far between. We consider three.

Photosynthetic Feedback. Photosynthesis by terrestrial plants is perhaps the best example of a potential beneficial feedback mechanism in earth's carbon cycle. Consider the case of photosynthesis in air. Although the concentration of CO_2 in the atmosphere is at an historical high, it is still quite low on an absolute scale: only 360 parts per million. As a result, the rate of primary production in terrestrial plants is at times limited by the availability of carbon dioxide. Rising levels of carbon dioxide may ease this limitation, allowing terrestrial plants to grow at an increased rate, thereby removing some of the extra carbon from the atmosphere: a negative feedback. There are complications that raise serious questions about the long-term viability of this mechanism, but it may at least serve an important short-term role in mediating the effects of CO_2 production.

Unfortunately, photosynthesis cannot work as a feedback mechanism in the ocean. As noted in chapter 4, the concentration of dissolved carbon in the sea—primarily in the form of bicarbonate—is much higher than it is in the atmosphere. As a consequence, the availability of CO_2 is never a limiting factor for marine primary producers. Because CO_2 is not limiting, any increase in atmospheric concentration that makes CO_2 more available to phytoplankton cannot increase their rate of primary production. Only if some other limiting factor is removed can productivity increase, and only if productivity increases can the oceans help reduce the concentration of CO_2 in the air.

Feedback from Upwelling. Are there other limiting factors that would allow ocean primary productivity to increase in response to rising levels of atmospheric carbon dioxide? As we have seen, in the typical two-layered ocean, the availability of nitrate limits the rate of new production. Can the regulation of nitrate availability serve as a negative feedback mechanism? Unfortunately, the answer appears to be "no." The most obvious possibility would be a link between rising carbon dioxide levels and the rate of nitrate delivery by upwelling. For example, if rising global temperatures somehow increased the intensity of upwelling,[6] the rate of nitrate delivery to the surface ocean would be enhanced, increasing the rate of primary production. But this increased production would not draw carbon dioxide from the atmosphere.

To see why, we recall that fixed nitrogen is delivered to the ocean interior by sinking biological material: the biological pump (chapter 6). According to the Redfield formula (equation 4.5), carbon and nitrogen are present in this particulate organic matter in a ratio of approximately 106 carbon atoms to 16 nitrogen atoms. As the sinking organic matter decomposes, its nitrogen eventually accumulates in the form of nitrate and its carbon in the form of bicarbonate, *but the ratio of their concentrations is unchanged*. As a consequence, when water wells up from the ocean interior, it delivers carbon and nitrate in the same 106:16 ratio required by phytoplankton. Thus, any new production allowed by the delivery of nitrate can use bicarbonate derived from the ocean interior rather than CO_2 drawn in from the air, and there would be no appreciable impact on atmospheric carbon dioxide concentration. In short, it is not apparent how regulation of nitrate delivery in upwelling can serve as an effective negative feedback mechanism.

HNLC Feedback. But then there are those anomalous HNLC areas. There is already plenty of nitrate in those surface waters, so primary production could potentially proceed without delivery of carbon from the ocean interior. What

[6] In fact, current models suggest that rising temperature will *decrease* the rate of upwelling. The largest degree of global warming will occur near the poles, tending to decrease the difference in temperature between the poles and the tropics. This reduced temperature gradient will lead to reduced circulation in the Hadley cells, reducing the intensity of the trade winds and thereby imposing ocean conditions similar to a prolonged El Niño.

limits primary production in these areas, and could it be used in a feedback mechanism?

Limits in the HNLC Areas

Over the years, a wide variety of potential limiting factors have been proposed for one or more of the high-nitrate, low-chlorophyll areas of the ocean. Candidates have included:

Cold. It is possible that the low water temperature of the HNLC areas of the subarctic Pacific and the Southern Ocean limit the rate at which phytoplankton can grow. If phytoplankton grow too slowly, the rate at which they take up nutrients could be less than the rate at which nutrients are delivered, even if delivery is itself slow. Although plausible for waters near the poles, this potential explanation suffers from a lack of generality: it clearly cannot explain the large HNLC areas in equatorial seas. Furthermore, there are cold seas that do not have high nutrient concentrations, so cold by itself is insufficient to explain even the high-latitude HNLC areas.

Light. In the Southern Ocean, water is often uniformly cold throughout the water column, and the resulting lack of density stratification allows nutrient-rich water to be delivered to the surface. Thus, in this case it is not a question of why surface waters have a high concentration of nitrate. Instead, the question is why the concentration of phytoplankton is not concomitantly high. But the same mixing that delivers nutrients to the surface can also mix phytoplankton to the dark depths. If, as they are carried about by this mixing, phytoplankton spend too little time in the light near the surface, their overall rate of production may be too low to maintain a high standing crop capable of using up the nutrients. As for the previous explanation, this one also suffers from a lack of generality: it cannot explain the tropical HNLC areas where the thermocline is shallow and mixing is less than energetic.

No Bloom-Forming Species. Different species of phytoplankton have different maximal growth rates. Some species—diatoms and the haptophyte *Phaeocystis* in particular—can grow rapidly when supplied with abundant light and nutrients, forming massive blooms that subsequently sink, thereby pumping nutrients and carbon out of the surface layer. Some other species can grow only slowly even under optimal conditions. If the HNLC areas are home solely to these slow-growing species, the biological pumps tied to their production may be too inefficient to keep up with nutrient supply. This explanation has the advantage of generality: it could apply to any HNLC area. It has a decided disadvantage, however. If it is solely a lack of bloom-forming species that keeps an area high in nutrients and low in chlorophyll, any bloom-forming individuals that arrive on the scene should be primed for a rapid takeover. In systems such as the subarctic and equatorial Pacific, which are crossed by major wind-driven currents, it is difficult to imagine that bloom-forming species have continuously been excluded.

Grazing. If grazers exert sufficient downstream control over primary producers, the rate of production may fall below the level needed to keep up with nutrient delivery. This explanation could apply to any of the HNLC areas, and it remains a leading contender to explain their existence.

Limitation by Nutrients Other Than Nitrate. Lastly, it is possible that primary production in the HNLC areas is indeed limited by availability of nutrients, except that the limiting nutrient is something other than nitrate.

Although clearly possible in theory, for many years it was difficult to imagine how this last limitation might be imposed in practice because no one could point at a nutrient that was limiting. For instance, there are times when and places where the availability of silicon limits the production of diatoms—in coastal upwelling areas, for example—but silicon seems not to be a limiting factor in the HNLC areas.

Then, in the 1980s, advances in analytical techniques allowed oceanographers for the first time to measure the concentration of iron in seawater. It soon became apparent that the concentration of iron in HNLC areas was lower than that in other parts of the sea, and it was proposed that primary production in the HNLC portion of the ocean in general was limited by a lack of iron.[7]

This proposal was of more than academic interest because it raised the possibility of human control. Unlike nitrate, which, according to the Redfield ratio, phytoplankton require in approximately a 0.151:1 ratio with CO_2, iron is required in vanishingly small concentrations, roughly 0.000047:1, $Fe:CO_2$. When oceanographers sat down and did the math, the amount of iron required to fertilize the HNLC areas of the ocean appeared to be within practical reach. Fertilize the ocean! Boost new primary productivity and soak up all that CO_2! The possibility for human control of the atmosphere began to seem real.

The allure of iron went even further. Information from ice cores suggested that wind-borne delivery of iron to the oceans has varied substantially through time. In particular, the supply of iron was high at the beginning of each of the last several ice ages. This record led John Martin, an oceanographer at the Moss Landing Marine Laboratory in California, to propose that fertilization of the Southern Ocean by iron was the *cause* of the ice ages. He once famously quipped that if someone gave him half a tanker of iron to spread on the ocean, he would give the world an ice age.

The advent of the *iron hypothesis*, as Martin's proposal became known, sowed the seeds for an intriguing (and ongoing) interaction between science and society, to which we will return. But first, we need to review the facts.

Iron in the Ocean

As we have noted, iron is necessary for a wide range of biological processes. For example, iron is essential for photosynthesis: it is needed to form ferrodoxin and members of the electron-transport chain, molecules that ferry electrons from place to place in the chloroplast. In respiration, iron is required for molecules of a separate electron-transport chain in mitochondria. Iron is needed by cyanobacteria for the enzymes involved in nitrogen fixation, and a lack of iron inhibits diatoms' ability to take up silicic acid.

[7]The role of iron in limiting marine primary production was originally proposed in 1931 by Haakon Gran, but analytical techniques available in the 1930s did not allow his idea to be tested. By the time improved analyses became available in the 1980s—improvements we discuss later in this chapter—Gran's proposal had been largely forgotten, and he has not been given the credit he is due.

In each case, iron is used as part of the biochemical machinery of cells, but, unlike nitrogen, it is not used as a major component in the formation of new cellular material. As a result, iron is required in much lower concentrations than for the nitrogenous nutrients, nitrate and ammonia. Concentrations of iron as low as 0.03 micrograms per liter of seawater are sufficient to allow phytoplankton to grow. For this reason, until the 1980s, it was commonly assumed that iron would not be a limiting factor in the ocean.

This assumption was apparently validated by the initial direct measurements of iron concentration in seawater, which suggested that iron was quite abundant. Subsequent research, however, revealed that these early measurements were erroneous. Iron is such a common component of research labs that it had contaminated water samples in previous studies, leading to measured concentrations that were too high. Think for a moment of all the different places iron comes into contact with laboratory equipment. Water is delivered in iron pipes. Iron is used in the equipment that makes the glass that goes into test tubes and bottles. The racks in incubators, the hose connectors in the seawater system, the needles in syringes all contain iron. The list is endless. Until oceanographers found a way to contend with these potential sources of contamination, the true concentration of iron in the ocean remained an open question.

Eventually, in the late 1980s, appropriate techniques were devised. To give you a feel for the care involved in these measurements, here is the recipe required for taking an uncontaminated sample of ocean water for culturing phytoplankton. For each sample, a new polycarbonate bottle is used: as noted above, glass won't do because it contains traces of iron. Each bottle is soaked in detergent for a week to remove any dirt, oil, or residue from the manufacturing process. Each bottle is then rinsed three times with deionized water, water filtered to remove any extraneous ions, then filled with deionized water and allowed to sit for 2 to 3 days. It is then rinsed again with deionized water, followed by a rinse with a higher grade of deionized water. While all this rinsing and soaking goes on, the investigator is busy distilling hydrochloric acid needed for the next step. (Hydrochloric acid from the manufacturer must be carefully distilled four times in special quartz glassware to remove any trace of iron.) Once the bottles have been soaked and again rinsed in deionized water, they are filled with 10% hydrochloric acid and soaked for 4 days, after which they are rinsed yet again, filled with high-grade deionized water, and soaked for 1 to 3 days. At last the bottles are deemed clean, but before they can be used, they must be dried. This happens in a special container where the bottles, held upside down, are exposed to a flow of air specially filtered so that no iron-containing dust particles can enter. After drying, each bottle is placed in a plastic bag, which is then placed in another plastic bag, which is in turn placed in a third plastic bag, and the bottle is finally ready to be taken onboard ship for experiments at sea.

But the fun is not over yet. It is a multistep process to actually take a sample of seawater and perform an experiment. Before a sample is taken, each clean bottle is rinsed three times with seawater, a procedure that takes place in a special dust-free clean room with the scientist wearing rubber gloves. The bottle is then filled with the actual sample, which is then again triply bagged. At that point the actual experiment can finally commence.

Through the laborious use of this and similar techniques, oceanographers were at last able to discern the broad pattern of iron concentration in the ocean. In places where primary productivity is high—in coastal upwelling regions, for example—iron is readily available. Even in areas where primary productivity is low—subtropical gyres in the North Atlantic and Pacific, for instance—iron concentrations are sufficiently high so as not to limit production. But in the HNLC areas, iron is present in extremely low concentrations, often less than 0.01 micrograms per liter, well below the concentration needed for growth by phytoplankton and for nitrogen fixation by cyanobacteria.

The low concentration of iron in HNLC ocean water should surprise you. Iron is the most abundant element on earth as a whole—it forms 80% of the mass of earth's core—and it is the fourth most abundant element in earth's crust. As we have seen, scientists have to take extreme measures to avoid contaminating their samples with iron. If iron is so pervasive, why is it so scarce in seawater?

The answer comes in two parts. First, iron is very nearly insoluble in seawater. For example, the *Titanic* sank in 1912, but explorers have visited the wreck in recent years and it is still there. If iron were readily soluble in seawater, the steel plates of the *Titanic*'s hull would have dissolved long ago. Thus, even when bits of solid iron are readily available to a body of seawater, the resulting concentrations are low.

The solubility of iron is related to its state of reduction or oxidation. In its reduced form, iron is readily soluble in seawater. Three billion years ago when free oxygen was absent from earth's atmosphere and ocean, iron was consequently in a reduced state, and the concentration of iron in seawater was presumably high. It was in this iron-rich seawater that the basic biochemistry of life evolved. The photosynthetic machinery that resulted subsequently led to the present high concentration of oxygen in both the atmosphere and the ocean. Exposed to this abundant oxygen, iron is currently present primarily in an oxidized state. This oxidized iron is nearly insoluble, accounting for the anemic ocean of today. Yet again we see the profound effect life has had on the chemical composition of the environment.

The second reason iron is in short supply in the ocean involves the sources by which iron is delivered to the sea. There are two major ones. As iron weathers out of rocks on the continents, it dissolves in river water, which empties into the coastal ocean. As a result, coastal waters are typical replete with iron. However, the rate at which iron is delivered by rivers is slow relative to the rate at which it is sunk by the biological pump. As a result, input from rivers cannot maintain a high concentration in the ocean as a whole.

The more important source of iron for the oceans is airborne dust. As winds blow over deserts and other dusty areas on land, iron-rich particles are suspended in air, and they can be blown out to sea. For example, waters in the western half of the subtropical North Pacific are continuously fertilized by dust blown from the Mongolian deserts. The tropical Atlantic receives a rain of dust blown from the Sahara by the trade winds. Surface waters in these areas have high iron concentrations. In contrast, those areas of the ocean not downwind of a desert, notably the North Pacific, the eastern half of the tropical Pacific, and the Southern Ocean, are depleted of iron.

Fertilization Experiments

Once the general pattern of iron concentration was mapped, the stage was set for experimentation. The basic concept is quite simple: if iron limits primary production in HNLC areas, adding iron to HNLC water should increase its productivity. Testing this hypothesis proceeded in two steps.

The first step was small. Clean bottles were taken to sea, as described above, and filled with HNLC water. Small amounts of iron were added to some bottles and not to others, which served as controls. Both types of bottle were then exposed to sunlight on the ship's deck for several days, and the concentration of phytoplankton was subsequently measured. The results were clear: phytoplankton grew much faster in the fertilized bottles than in the controls, indicating that iron was indeed the factor limiting primary production.

As straightforward as these results are, they are less than satisfying. The simple act of taking seawater out of the ocean and confining it in bottles could have affected the results. In particular, bottle experiments such as these are renowned for changing the behavior of zooplankton, potentially skewing the ecology in the bottles away from that found in nature. Furthermore, bottle experiments cannot measure how much of the new production attributed to iron would sink out of surface waters. To be sure that iron was really the limiting factor in HNLC areas, and to measure its potential for drawing down atmospheric CO_2, tests were needed in the ocean itself.

This led to a grand set of large-scale experiments. The basic technique is this: Iron and a small amount of sulfur hexafluoride (an inert molecule used as a tracer) are dissolved in acid and mixed into a large patch of the surface ocean (tens to hundreds of square kilometers). The acid in this fertilizer is quickly neutralized upon contact with seawater, but the iron concentration of the water is elevated to a level sufficient for phytoplankton growth.

The trick, then, is to keep track of this fertilized water in order to measure the effects of iron fertilization. Iron itself can't be used to identify the experimental patch. After all, the whole idea is that the iron is taken up by organisms, thereby potentially removing it from the water, and as we have seen, it is a major production to accurately measure the concentration of iron in seawater. In contrast, sulfur hexafluoride is not a normal component of seawater, it is not taken up by organisms, and it is easily detectable in extremely small concentrations (a few parts per billion). Thus, by measuring the concentration of sulfur hexafluoride, oceanographers can tell whether a given sample of water comes from the fertilized patch.

After the experimental patch is created, ships steam back and forth across it, measuring factors that characterize the ocean's response to iron: the rate of primary production, the concentration of carbon dioxide and nitrate in the water, the abundance of zooplankton, the rate of particle sedimentation, and others.

The initial large-scale experiment (IronEx I) was conducted in 1993 in the equatorial Pacific, just south of the Galápagos Islands. The results were ambiguous. The chlorophyll concentration in the fertilized patch rose by a factor of four, but there was no measurable reduction in the concentration of either nitrate or carbon. In this initial experiment, iron was applied to the ocean only once, and for reasons that are not entirely understood, the concentration of iron in the experimental patch decreased rapidly before it could be effectively absorbed by phytoplankton. It is thought that this ineffective fertilization contributed to the lack of clear-cut results.

A second experiment (IronEx II) was conducted in 1995. This time the fertilizer was applied in several small doses, and the results were spectacular. Primary productivity rose by a factor of 30 to 40 relative to unfertilized water, and the concentrations of nitrate and carbon were drastically reduced. The fertilized patch was visibly green, and to those onboard the research ship, the sea smelled like new-mown hay. The researchers estimated that their addition of 480 kilograms of iron to the patch had resulted in the net absorption of 350,000 kilograms of atmospheric carbon dioxide.

The success of the IronEx experiments in equatorial waters led to further experiments in the Southern Ocean (SOIREE, 1999; EisenEx, 2000), the southern Pacific (SOFeX, 2002), and the northern Pacific (SEEDS, 2001; SERIES, 2002). In each of these experiments, the addition of iron to the HNLC ocean produced a substantial increase in primary productivity and reduced concentrations of both nitrate and CO_2.

While these responses were expected, another effect was not. The iron hypothesis supposes that the removal of iron limitation should increase not only primary productivity, but also export of new production to the ocean interior. The SOIREE experiment was designed specifically to measure this export, but no increase was found. The researchers proposed that this lack of export was due to a peculiarity of the diatoms in the experimental bloom. As a response to release from iron limitation, diatoms increase their buoyancy, thereby reducing the rate at which they sink. Perhaps the measurements in SOIREE were not carried on long enough to catch the export pulse when it finally reached the ocean interior. SOFeX measured a minor increase in particulate transport to the deep sea, but again less than expected.

In summary, field experiments have verified the identity of iron as the limiting factor in HNLC areas of the ocean. Fertilization of these areas can indeed draw in carbon dioxide from the atmosphere. But the last leg of the iron hypothesis—that this new production is transported to the ocean interior rather than returned to the atmosphere—remains to be demonstrated.

Note that these large-scale field experiments allow us to discard several of the original hypotheses for the occurrence of HNLC areas. SOIREE, EisenEx, and SOFeX all were conducted where water is cold and deeply mixed. Nonetheless, blooms formed, demonstrating that cold, light, and lack of bloom-forming species are not the barriers responsible for lower primary productivity in HNLC areas.

The Ecumenical Interpretation

These experiments leave us with an improved, although neither fully understood nor universally accepted, explanation for the HNLC areas of the ocean. It combines two of the explanations proposed at the beginning of the discussion (grazing and iron), and therefore has come to be know as the *ecumenical interpretation.*

In the absence of sufficient airborne input of iron, large phytoplankton (such as diatoms) are scarce, and primary production in HNLC areas is due primarily to cyanobacteria such as *Prochlorococcus*. Given their small size, these organisms are capable of scavenging the low concentration of iron available. However, as noted in chapter 4, they are incapable of using nitrate as a source of nitrogen, subsisting instead on recycled ammonia. Because *Prochlorococcus* does not take up nitrate, nitrate can accumulate in the surface water as it is gradually mixed up from below. The size of the standing crop of phytoplankton is kept in check by small grazers, flagellates and ciliates, with the result that the surface waters have a low chlorophyll concentration. Thus, high-nitrate, low-chlorophyll water results.

If iron is introduced into this system, two things change. First, it is now possible for the small population of bloom-forming phytoplankton (diatoms in particular) to grow. In the course of a few days, population size rapidly increases, overwhelming the ability of flagellates and ciliates to keep the phytoplankton in check. This bloom draws CO_2 from the atmosphere and nitrate from the water, and presumably converts these ingredients into particulate matter that is exported to the ocean interior.

There are then three factors that could end the bloom. First, the burgeoning phytoplankton population could deplete the available nitrate, at which point new growth would be drastically reduced. Alternatively, if the bloom is formed of diatoms, it could deplete the local stores of silicic acid. Again, new production would be curtailed. And finally, the grazer population could catch up and bring the bloom under control. Because it is difficult for flagellates and ciliates to consume large phytoplankton, this control is likely exerted by copepods, krill, and other large grazers, populations that may require days to weeks to respond to the bloom (see chapter 5).

Regardless of how the bloom is brought under control, within a few weeks of its initiation a new equilibrium is established. If the delivery of iron continues, new production is limited by the usual suspects: by delivery of nitrate to the surface layer or by grazing. If delivery of iron is halted, the bloom-forming species die back, and the system reverts to its initial HNLC state: production dominated by *Prochlorococcus*, a population which is in turn held in check by small grazers.

The Policy Debate

Verification of iron as the factor limiting primary production in HNLC areas was a milestone in biological oceanography, but one that immediately thrust the field into the arena of public policy. If Martin's iron hypothesis is correct—and it looks like it might be—only a small amount of iron would be required to fertilize the ocean. Might this fertilization be a viable mechanism for counteracting the anthropogenic influx of CO_2?

To politicians charged with balancing the needs and desires of human society, this possibility is enticing. On one hand, society needs to reduce carbon emissions or it risks doing grave damage to itself and to all life on earth. On the other hand, society wants to keep driving gasoline-powered cars, using electricity from coal-fired generators, and building roads and buildings from cement. To a politician, it must be tempting to think that we can have both our needs and our desires. We'll just fertilize the ocean, sink all that carbon, and go merrily about our business.

To entrepreneurs, ocean fertilization has the smell of newly minted cash. As soon as large field experiments were successfully carried out, patents were filed on methods to fertilize the sea, and companies were formed to put these patents into action.

These responses put oceanographers in an uncomfortable position. Yes, they were having great fun creating big green patches of ocean in unlikely parts of the globe, and as a result, they had learned a lot about how the ocean works. But they were keenly aware of how much they did not know, and how dangerous it would be to implement any fertilization scheme without further research. Oceanographers were well aware that standard operating procedure for governments and entrepreneurs is to say, "Ready, fire, aim," and it was disconcerting to have to cry, "Wait!" With each new experiment, the oceanographic community felt compelled to accompany their results with a series of warnings:

It probably won't work. The notion that ocean fertilization can effectively offset anthropogenic CO_2 emissions requires that the carbon taken up in primary production be exported to the deep ocean. The large-scale ocean experiments raise questions as to whether the required export would occur, and therefore raise real concern as to whether ocean fertilization can effectively absorb CO_2. Foresightful oceanographers also pointed out that even if fertilization worked in the short term, it was bound to fail in the long run. Yes, it takes the ocean interior 500 to 1000 years to turn over, but turn over it eventually will. Much of the carbon we sequester in the ocean interior now will be returned to the atmosphere later. So, at best, ocean fertilization is only a short-term fix.

Even if it works, if we start, we can't stop. A single application of iron to the HNLC ocean can at best absorb a fixed amount of carbon dioxide. Because humankind seems intent on continually adding CO_2 to the atmosphere, negating these emissions would require continual fertilization of the ocean.

If it works, it will cause untold damage. Every ocean fertilization experiment has shown that adding iron to the HNLC ocean drastically changes the ecology of these surface waters. Where cyanobacteria dominated, now there are blooming populations of diatoms and haptophytes. Where the primary producers had primarily taken up ammonia, now they use nitrate. The mix of grazers is different. And no one knows what the long-term effects of these changes might be. If fertilization works, one effect of clear concern is the delivery of a drastically increased amount of organic material below the thermocline. As this material decomposes, oxygen will be consumed, increasing the size and severity of the oxygen minimum zone in the deep ocean. Because science doesn't even have a full inventory of the organisms that currently live in this vast habitat, there is no accurate way to predict the effect of decreased oxygen levels.

To date, these warnings seem to have their intended effect: as of 2007 no large-scale fertilization of HNLC areas has been attempted. However, the story is not over. As global warming worsens and society begins to pay the price, the quest for an easy, short-term solution will intensify. It will be interesting to follow the interaction among oceanographers, politicians, and entrepreneurs in coming years.

Iron and Ice Ages

What of John Martin's hypothesis that fertilization of the oceans is the cause of ice ages? Here, the jury is still out.

Current thought is that ice ages are related to periodic changes in the tilt of earth's axis and the shape of its orbit around the sun. These changes affect both the intensity of sunlight arriving at earth during a given season and the amplitude of seasonal change in intensity at different latitudes. These cycles have periods of approximately 23,000, 41,000, and 100,000 years, and they are called *Milancovitch cycles* in honor of Milutin Milancovitch (or Milancović), the Serbian mathematician who carefully decoded their orbital physics.

Measurements from ice cores taken in Greenland and Antarctica show that earth's temperature and other measures of climate have indeed varied at periods of approximately 23,000, 41,000, and 100,000 years, strong evidence that Milancovitch cycles play an important role in the timing of ice ages. It is less clear, however, exactly how the Milancovitch fluctuations control temperature. The variations in solar heat input are themselves quite small, too small to directly account for the temperature swing between glacial and interglacial conditions. Instead, these small changes are assumed to be amplified by other processes on earth that then form the proximate cause for drastic temperature fluctuations.

The iron hypothesis is one of these potential amplifying mechanisms. In this scenario, shifts in the earth's tilt and orbit reduce the input of solar heating to the polar regions, causing the polar ice caps to grow slightly. Consequent changes in patterns of wind and rain result in increased delivery of dust to the Southern Ocean. The iron contained in this dust fertilizes the sea, allowing the biological pump to draw carbon dioxide from the atmosphere. The reduced concentration of greenhouse gas in turn decreases the temperature of earth's surface, allowing the polar ice caps to grow even more, and the cooling process feeds back on itself. Voilà, an ice age!

Although this scenario is plausible, it is only one of several proposed mechanisms, all of which may contribute to the overall effect. The science of ice ages is currently in that tantalizing state where enough is known for creative thought, but not enough is understood to reliably discern the truth. As with the interaction between oceanographers and politicians, it will be intriguing to follow the science of ice ages as the story unfolds in the next few years.

Further Complications: Carbonate Concentration

Aside from their effect on temperature, elevated concentrations of carbon dioxide in the atmosphere are already having measurable effect on marine life. To understand one prime example—the potential dissolution of coral skeletons and calcareous shells—we need to delve into the chemistry of aqueous carbon dioxide.

In chapter 4, we noted that when carbon dioxide dissolves in water, much of it immediately forms carbonic acid, H_2CO_3, which then dissociates into a bicarbonate ion (HCO_3^{2-}) and a hydrogen ion (H_3^+):

$$CO_2 + H_2O \rightleftharpoons H_2CO_3 \rightleftharpoons H^+ + HCO_3^- \qquad \text{(Eq. 10.1)}$$

We now take this reaction one step further. Bicarbonate can itself dissociate, releasing another hydrogen ion and forming a carbonate ion (CO_3^{2-}):

$$HCO_3^- \rightleftharpoons H^+ + CO_3^{2-} \qquad \text{(Eq. 10.2)}$$

A carbonate ion formed in this fashion can then be combined with a calcium ion to form calcium carbonate, the basic building block for the shells of foraminifera and mollusks and the skeletons of corals:

$$Ca^{2+} + CO_3^{2-} \rightleftharpoons CaCO_3 \qquad \text{(Eq. 10.3)}$$

Under current oceanic conditions, the formation of calcium carbonate is no particular problem. There are plenty of calcium ions available in seawater, a result of the weathering of rocks through the ages, and carbonate ions are present in saturating concentrations in most of the ocean. This benign situation will likely change, however, as the concentration of CO_2 in the atmosphere rises. To see why, we must first explore the balance of chemical equations.

Each of the reactions shown above is reversible, as depicted by the two-headed arrows. For example, just as a bicarbonate ion can dissociate into a hydrogen ion and a carbonate ion (equation 10.2), a hydrogen ion and a carbonate ion can combine to produce a bicarbonate ion. Left to themselves, each of these reactions settles into an equilibrium. In the case of the reaction shown in equation 10.2, the equilibrium is such that currently most of the carbon in the system is in the form of bicarbonate with little in the form of carbonate; that is, the equation is biased to the left.

What happens if we add bicarbonate ions to the system of equation 10.2? The concentration of bicarbonate on the left-hand side of the equation initially increases, but some small fraction dissociates to form hydrogen and carbonate ions. In other words, in this reaction, any increase in bicarbonate concentration leads to a "flow" to the right in the equation, causing an increase in hydrogen ion and carbonate concentrations. In this case, the flow would be small because it is against the bias of the equation. Conversely, if we add hydrogen ions to the right-hand side of the system, some of them combine with carbonate ions to form bicarbonate. In fact, because the equilibrium is strongly biased toward bicarbonate, if we add hydrogen ions to the right-hand side of the equation, most of them combine with carbonate and flow to the left. The same effect would apply if we added carbonate ions to the system; most of them would drag hydrogen ions with them as they formed bicarbonate and flowed to the left in the equation.

This example demonstrates Le Chatelier's principle, a guiding concept of chemistry. In a reversible reaction, addition of any reactant or product causes the reaction to flow in a manner that reduces the extent of the change. The amount of flow depends on the equilibrium ratio of the various concentrations involved in the reaction, that is, whether the equilibrium is biased to the left or to the right in the chemical equation.

We can now apply this principle to the combination of reactions shown above. As additional atmospheric CO_2 dissolves in seawater, it adds to the concentration of bicarbonate and hydrogen ions (equation 10.1). These increases then affect the equilibrium between bicarbonate and carbonate (equation 10.2). Deducing the effect is trickier in this case because we add molecules to both sides of the equation. The bicarbonate formed by the dissolution of CO_2 adds to the left-hand side of equation 10.2 (which creates a tendency for the equation to flow to the right), while the hydrogen ions add to the right-hand side of equation 10.2 (which creates a tendency for the equation to flow to the left). But, because the equilibrium of equation 10.2 is biased strongly to the left, the addition of hydrogen ions outweighs the addition of bicarbonate: there is net flow to the left, and the concentration of carbonate ions is reduced.

In summary, as new carbon dioxide dissolves in seawater, the concentration of carbonate ions *decreases*.

So, as human society adds CO_2 to the atmosphere, the changed atmosphere has the effect of decreasing the concentration of carbonate in the surface ocean, which in turn makes it more difficult for corals to form their skeletons and animals such as foraminifera and mollusks to form their shells. If the current trend in carbon dioxide emissions continues, oceanographers speculate that within a few decades these animals will no longer be able to form calcium carbonate. Even before then, reduction in carbonate concentration (coupled with high pressure of the ocean depths) will begin to dissolve calcareous oozes and sediments.

As distressing as it is to contemplate a world without corals and seashells, a decreased rate of calcium carbonate formation also has a positive note. To see why, we again apply Le Chatelier's principle. When calcium carbonate is formed (equation 10.3), carbonate ions are taken out of solution. This change affects the equilibrium in equation 10.2. The removal of carbonate ions causes some bicarbonate to dissociate and flow from left to right in this reaction. This shift tends to minimize the change in concentration of carbonate, but it adds to the concentration of hydrogen ions in seawater. The increased concentration of hydrogen ions then affects the reaction shown in equation 10.1. Adding hydrogen ions to the right-hand side of this equation causes the reaction to move to the left, with the result that the concentration of carbon dioxide is increased. In summary, the current production of calcium carbonate skeletons and shells serves to increase the concentration of carbon dioxide in surface waters. If, in the future, the formation of calcium carbonate is curtailed—and especially as calcareous sediments dissolve—the opposite will happen: the concentration of carbon dioxide in surface waters will decrease fractionally, allowing the ocean to absorb more CO_2 from the atmosphere than it otherwise would.

If all of this seems a bit complicated to you, you are in good company. The chemistry of carbon in seawater is complex, and the full details are beyond the scope of this book. Good introductions to the full system can be found in Pilson (1998) and Sarmiento and Gruber (2006), and a brief explanation of the mathematics involved in calculating the concentrations of bicarbonate and carbonate ions is given in the appendix to this chapter. For present purposes, it may be best simply to remember the salient results of the chemistry:

(1) Increasing the concentration of CO_2 in seawater, an unavoidable consequence of increasing the concentration of CO_2 in the atmosphere, decreases the concentration of carbonate, potentially making it difficult to form calcareous skeletons and shells.

(2) On the other hand, dissolution of calcareous sediments or any decreased rate of skeleton formation may partially offset the increase in CO_2 concentration.

In this discussion, we have concentrated on the carbon compounds involved in equations 10.1, 10.2, and 10.3. However, in both the scientific and popular literature, these equations are often instead described in terms of the hydrogen ions involved. We have just noted that when carbon dioxide dissolves in seawater, the concentration of carbonate ions decreases, but at the same time, the concentration of hydrogen ions increases. This increase in H^+ reduces the pH of seawater; that is, it becomes more acidic. The pH of surface seawater, which currently averages 8.05, has decreased by about 0.1 since the beginning of the Industrial Revolution and the pH of the ocean will continue to decrease as more CO_2 dissolves in the sea. It is important to remember, however, that it is not the pH of seawater (oceanic acidification) that is of most immediate concern; rather, it is the accompanying effect on carbonate concentration.

Summary

This chapter serves notice that the simple picture we have constructed of the two-layered ocean is just the first step toward a practical knowledge of how the ocean works. For each generality we have drawn, there are exceptions, and some of these exceptions—such as the HNLC areas—have the potential to be vitally important. Nonetheless, the principles we have learned in our exploration of the two-layered ocean remain valid and serve as a foundation on which to build a more detailed understanding of the sea.

Further Reading

Chisholm, S. W., and F.M.M. Morel (eds.) (1991). What controls phytoplankton production in nutrient-rich areas of the open sea? *Limnology and Oceanography* 36: 1507–1965.

Doney, S. C. (2006). Ocean acidification. *Scientific American* 294(3): 58–65.

Mann, K., and J.R. Lazier (2006). *Dynamics of Marine Ecosystems* (3rd edition). Blackwell Publishing, Malden, MA.

Pilson, M.E.Q. (1998). *An Introduction to the Chemistry of the Sea*. Prentice-Hall, Upper Saddle River, NJ.

Royal Society (2005). *Ocean Acidification Due to Increasing Atmospheric Carbon Dioxide*. The Royal Society, London.

Sarmiento, J. L., and N. Gruber (2006). *Ocean Biogeochemical Dynamics*. Princeton University Press, Princeton, NJ.

Appendix

Chemical Calculations for Marine Carbon

Concentrations of dissolved carbon dioxide (CO_2), bicarbonate ions (HCO_3^-), and carbonate ions (CO_3^{2-}) in seawater are governed by three chemical reactions. First, there is the dissolving of gaseous CO_2 in water to form carbonic acid:

$$CO_2 + H_2O \rightleftharpoons H_2CO_3 \qquad \text{(Eq. 10.4)}$$

Next is the dissociation of carbonic acid to form a hydrogen ion and a bicarbonate ion:

$$H_2CO_3 \rightleftharpoons H^+ + HCO_3^- \qquad \text{(Eq. 10.5)}$$

And finally, there is the dissociation of the bicarbonate ion to form another hydrogen ion and a carbonate ion:

$$HCO_3^- \rightleftharpoons H^+ + CO_3^{2-} \qquad \text{(Eq. 10.6)}$$

In the ocean, these three processes come to an equilibrium, and we desire to know the concentrations of each molecule when that equilibrium is achieved.

For each chemical reaction, equilibrium occurs when the concentrations of the products are present in a certain ratio to the concentrations of the reactants. Consider, for example, the equilibrium between gaseous carbon dioxide and carbonic acid (equation 10.4). Equilibrium is achieved when

$$K = \frac{[H_2CO_3]}{[CO_2][H_2O]} \qquad \text{(Eq. 10.7)}$$

Table 10.A1

Empirically measured equilibrium constants for carbon dioxide in seawater.
In the following equations, T is the absolute temperature (that is, °C + 273.15) and S is the salinity (in practical salinity units, see chapter 3). One atmosphere is 1.01×10^5 pascals. Equations from Sarmiento and Gruber (2006), *Ocean Biogeochemical Dynamics* (Princeton University Press, Princeton, NJ), p. 325.

Solubility constant of carbon dioxide (moles per kilogram per atmosphere)

$$\ln(K_0) = -60.2409 + 93.4517\left(\frac{100}{T}\right) + 23.3585\ln\left(\frac{T}{100}\right)$$
$$+S\left[0.023517 - 0.023656\left(\frac{T}{100}\right) + 0.0047036\left(\frac{T}{100}\right)^2\right]$$

Dissociation constants of carbonic acid and bicarbonate (moles per kilogram)

$$-\log_{10}(K_1) = -62.008 + \frac{3670.7}{T} + 9.7944\ln(T) - 0.0118S + 0.000116S^2$$
$$-\log_{10}(K_2) = 4.777 + \frac{1394.7}{T} - 0.0184S + 0.000118S^2$$

Here, the square brackets denote concentration. For instance $[H_2CO_3]$ is the concentration, in moles per kilogram of solution,[8] of carbonic acid. The value of K, the *equilibrium constant*, is affected by temperature and salinity, and its magnitude under different conditions is measured experimentally.

Actually, carbon dioxide and water in this relationship form a special case. First, the concentration of water is so high (nearly 56 moles per kilogram) that it does not change appreciably when carbon dioxide is dissolved. As a consequence, the concentration of water is effectively constant and can be absorbed into K, producing a new constant, K_0. In addition, as a practical matter, it is easier to measure the concentration of gaseous CO_2 by the pressure it exerts, measured in atmospheres, rather than the number of moles per kilogram. This pressure (technically, the *partial pressure* of CO_2) is denoted pCO_2. Taking these ideas into consideration, we can rewrite equation 10.7:

$$K_0 = \frac{[H_2CO_3]}{pCO_2} \quad \text{(Eq. 10.8)}$$

The new equilibrium constant, K_0, has units of moles per kilogram per atmosphere. The value of K_0 at a given temperature and salinity has been empirically determined and can be calculated from the relationship cited in table 10.A1.

In similar fashion, an equilibrium equation can be written for the dissociation of carbonic acid into a hydrogen ion and bicarbonate:

[8] As noted earlier, one mole is 6.02×10^{23} molecules. The mass of a substance (in grams) equal to its molecular weight contains 1 mole of that substance's molecules. For example, the molecular weight of CO_2 is 44 daltons, and 44 grams of CO_2 contains 6.02×10^{23} CO_2 molecules. 6.02×10^{23} is Avogadro's number.

$$K_1 = \frac{[H^+][HCO_3^-]}{[H_2CO_3]} \qquad \text{(Eq. 10.9)}$$

In this case, all concentrations are measured in moles per kilogram, and the equilibrium constant K_1 therefore has units of moles per kilogram.

The equilibrium equation for the dissociation of bicarbonate into hydrogen ions and carbonate ions is:

$$K_2 = \frac{[H^+][CO_3^{2-}]}{[HCO_3^-]} \qquad \text{(Eq. 10.10)}$$

Like K_1, K_2 has units of moles per kilogram of solution. As with K_0, values of K_1 and K_2 depend both on temperature and salinity, and can be calculated from relationships shown in table 10.A1.

At this point, we have three known values (K_0, K_1, K_2), and five unknown values (pCO_2, $[H_2CO_3]$, $[H^+]$, $[HCO_3^-]$, and $[CO_3^{2-}]$). In addition, we have three equations relating these values, which tells us that we can solve algebraically for at most three of our unknowns. Thus, to proceed, we must specify values for two of our five unknown concentrations. We could choose any two, but for present purposes it will be best if we choose to specify the pressure of carbon dioxide and the concentration of hydrogen ions. As we will see, this will allow us to calculate the concentrations of bicarbonate and carbonate ions as a function of values that humankind is affecting: the atmospheric concentration of carbon dioxide and the pH of seawater.

We begin by noting that, by definition, $\mathrm{pH} \equiv -\log_{10}([H^+])$. Thus,

$$[H^+] = 10^{-\mathrm{pH}} \qquad \text{(Eq. 10.11)}$$

Inserting this equality into equation 10.10, we see that

$$K_2 = \frac{10^{-\mathrm{pH}}[CO_3^{2-}]}{[HCO_3^-]} = \frac{[CO_3^{2-}]}{10^{\mathrm{pH}}[HCO_3^-]} \qquad \text{(Eq. 10.12)}$$

Rearranging this equation allows us to solve for $[HCO_3^-]$ in terms of pH and $[CO_3^{2-}]$:

$$[HCO_3^-] = \frac{[CO_3^{2-}]}{10^{\mathrm{pH}}K_2} \qquad \text{(Eq. 10.13)}$$

The value for $[HCO_3^-]$ we have just calculated, along with our new term for $[H^+]$ (equation 10.11), can then be substituted in equation 10.9:

$$K_1 = \frac{10^{-\mathrm{pH}}\dfrac{[CO_3^{2-}]}{10^{\mathrm{pH}}K_2}}{[H_2CO_3]} = \frac{[CO_3^{2-}]}{10^{2\mathrm{pH}}K_2[H_2CO_3]} \qquad \text{(Eq. 10.14)}$$

Next, we rearrange to solve for $[H_2CO_3]$:

$$[H_2CO_3] = \frac{[CO_3^{2-}]}{10^{2pH} K_1 K_2} \qquad \text{(Eq. 10.15)}$$

Continuing our chain of substitutions, we insert this value for the concentration of carbonic acid into equation 10.8:

$$K_0 = \frac{\dfrac{[CO_3^{2-}]}{10^{2pH} K_1 K_2}}{pCO_2} \qquad \text{(Eq. 10.16)}$$

Simplifying and rearranging this equation, we finally arrive at a relationship of practical consequence:

$$[CO_3^{2-}] = K_0 K_1 K_2 pCO_2 10^{2pH} \qquad \text{(Eq. 10.17)}$$

That is, if we know the three equilibrium constants and can specify the pH of the ocean and the pressure of carbon dioxide in the atmosphere, we can calculate the concentration of carbonate ions. As you might expect, the higher the pressure of carbon dioxide, the higher the concentration of carbonate ions. Note, however, that lowering the pH—that is, making the ocean more acidic—*reduces* the carbonate concentration. This is the nonintuitive effect noted in the chapter. Note also that pH appears as an exponent in equation 10.17, whereas the pressure of carbon dioxide enters the equation as a simple multiplier. As a result, carbonate concentration is much more sensitive to a change in pH than it is to a change in CO_2 pressure.

Now that we know $[CO_3^{2-}]$, we can return to equation 10.13:

$$[HCO_3^-] = \frac{[CO_3^-]}{10^{pH} K_2} \qquad \text{(Eq. 10.13, redux)}$$

Substituting our newly found formula for $[CO_3^{2-}]$ (equation 10.17) allows us to express bicarbonate concentration as a function of ocean pH and carbon dioxide partial pressure:

$$[HCO_3^-] = \frac{K_0 K_1 K_2 pCO_2 10^{2pH}}{10^{pH} K_2} = K_0 K_1 pCO_2 10^{pH} \qquad \text{(Eq. 10.18)}$$

For a given pressure of carbon dioxide, making the ocean more acidic—lowering the pH—decreases the concentration of bicarbonate ions. Note that bicarbonate concentration is proportional to 10^{pH}, whereas carbonate concentration is proportional to 10^{2pH} (equation 10.17). Thus, bicarbonate concentration is much less sensitive to a change in pH than is carbonate concentration.

For a given ocean pH, increasing the concentration of carbon dioxide in the atmosphere increases the concentration of bicarbonate, as it does for carbonate. Again, because pH appears as an exponent, any change in it has a larger effect than does a change in CO_2 concentration.

In summary, empirical measurements of the equilibrium constants for the reactions of carbon dioxide in seawater allow us to calculate concentrations of carbonate and bicarbonate ions in the sea. These concentrations are functions of both ocean acidity and the partial pressure of atmospheric CO_2.

The answers we have calculated here are fine as far as they go, but you may have noticed a problem. In solving for the concentration of carbonate and bicarbonate ions, we have assumed that we can specify pH and pressure of carbon dioxide independently. In reality, it is not that simple: the primary factor currently affecting the pH of the ocean is the rise in atmospheric carbon dioxide concentration. In other words, pCO_2 and pH interact. For instance, increasing the concentration of atmospheric carbon dioxide (which, by itself, would increase the carbonate concentration in the ocean) also decreases the pH of the ocean (which, by itself, would *decrease* carbonate concentration). The net effect depends on the precise relationship between pCO_2 and pH in seawater, a relationship that involves the overall capability of the ocean to buffer itself against changes in acidity. This buffering capacity in turn involves other ions we have not included here. We will not delve into these complications. If you are interested in a more precise calculation of carbonate and bicarbonate concentrations, please consult the texts by Pilson (1998) and Sarmiento and Gruber (2006) listed in *Further Reading*.

Fisheries

In the last chapter, we examined how human society's production of carbon dioxide is raising the temperature of our planet and delved into the role that the ocean can potentially play in ameliorating this problem. We now turn to another interaction between humankind and the sea, one that is more direct but no less fraught with complications. Let's explore the science and management of fisheries.

A Brief History of Fishing

Before we begin this exploration, however, we need to define the scope of our inquiry. For our purposes, the term "fishery" applies to the harvesting of any edible living product from the sea. For example, there are (or have been) fisheries for fish, whales, squid, crabs, seals, otters, lobsters, mussels, clams, sea urchins, and seaweed. Here, we will focus on the fisheries of greatest historical and economic impact: the harvest of fish and whales.

Human fisheries have existed since the Stone Age. Bones from a wide variety of fish as well as the shells of clams, cockles, oysters, mussels, and marine snails are common elements in ancient middens around the world. Hooks and harpoons fashioned from bone and horn, and nets made of flax or hemp, have been found in several Neolithic archeological sites in Europe, and Saharan cave drawings from 5500 years ago show a man angling for fish from a canoe. In this respect, the technology of fishing has changed very little. Like Stone Age fishermen, we still use hook and line, nets, and harpoons to harvest food from the sea.

For centuries, advances in fishing technique were incremental. Greek literature describes lobster traps woven from rushes. With the advent of metallurgy, hooks were made first of copper, then bronze, and then iron. Nets were originally deployed from shore in the form of small cast nets, lift nets, and handheld seines. With development of new technology for boats—more effective sails and oars—nets became larger and more sophisticated. For example, the beam trawl allowed adept fishermen to drag their fishing apparatus across the seabed, scaring fish up into a bag-shaped net. As fishermen acquired new technology, fishing was conducted farther from shore.

As fisheries expanded, societies evolved legal structures to deal with the inevitable friction among fishermen. Who owned what? And who had the right to fish where? Unlike land, where ownership of each plot of earth was often well defined, the ocean, and the fish hidden in it, were viewed by ancient societies as a common resource. To the Romans, for example, fish were part of the *res nullius*, things that belonged to no one. Only when a fisherman actually took possession of his prey was it considered his property; up to that point it was fair game. The open nature of the sea (in a legal as well as a spatial sense) helped to foster the expansion of fisheries, and this tradition continued until quite recently.

Growing familiar with the sea, fishermen were emboldened to go hunting in ways that would give you or me pause today. For example, in both northern Europe and the Arctic, men ventured out to kill whales, an interaction in which the animal being hunted was far larger than the open boats of the hunters. In this combat, whales had the advantage of size and speed, but this was offset by their need to come to the surface for air. Through skill and persistence, fishermen in small rowboats, kayaks, or umiaks were able to harpoon the beasts. A skillful thrust might sink the harpoon into a vital organ—the lungs or heart—and kill the whale outright. But more commonly the harpoon and its attached line served mainly to tether the leviathan to the hunters' boat or to floats made from animal skins. The bleeding whale's efforts to dive against the tether eventually caused it to tire sufficiently so that it could be approached in relative safety and killed.

The catch from these early fisheries—both fish and whales—was brought to shore for processing. Fish were preserved, first by drying them in the sun or over a smoking fire, and then by packing them in salt. These dried meats were high in protein and resistant to spoiling, and a staple food for the common folk throughout Europe and Asia. Whale meat was dried and salted as for fish, but in addition, the blubber was rendered into oil for lamps.

From the beginning of recorded history in about 3500 BC, the global catch from ocean fisheries grew slowly, more or less keeping pace with the growth of human population. Fifty-five hundred years ago, when the man in the cave drawing dangled his hook and line, there were only about 10 million people on earth. In 1 AD, during the heyday of Rome, the human population was approximately 230 million, and that number held steady for a thousand years as plagues and wars constrained population growth. By 1500 AD, however, global population had nearly doubled to 450 million, and in response fisheries began to expand their scope. For example, in the late 1400s, around the time when Columbus sailed to the New World, large fleets of fishing boats sailed into the North Sea from Scandinavia, the Netherlands, Scotland, and England to catch herring. Basques, followed by the English, went even farther afield, sailing across the Atlantic to catch cod on the Grand Banks east of what is now Newfoundland.

Human population growth accelerated around 1800 AD, fueled by the Industrial Revolution, and by the mid-1800s, when the global population was around 1.3 billion, the revolution reached the fishing industry. Steam engines allowed ships to move with less regard for the wind, vastly improving fishermen's capability to set large nets and drag large trawls, and steam winches allowed them to haul in the resulting catch. The purse seine was invented in 1826 and put into common practice beginning in 1850. This innovation allows fishermen to surround whole schools of fish, such as sardines and herring, with a net.

The bottom of the net is then cinched off, forming the purse, and the entire school is winched onboard. In 1892, the otter trawl was invented. By replacing the old-fashioned beam as a means to open the mouth of the net, the "doors" of an otter trawl allowed fishermen to drag larger nets across the bottom. Steam power also made it easier to travel to distant waters, and the invention of mechanical refrigeration plants allowed fish to be brought back fresh. On land, railroads opened new markets for fresh fish in areas far removed from the sea.

Although early human fishing had impacted a few species—the Aleuts of the Pacific Northwest decimated the local sea otters beginning 2500 years ago, and European colonists nearly extirpated sea turtles in the Caribbean—the industrialization of the late 1800s gave humankind its first serious opportunity to impact the global populations of marine organisms. Fishermen leapt at the chance. Global human impact was felt first by the great whales. Demand for whale oil for illumination and baleen for corsets increased in the 1800s, and shore-based whalers began to reduce northern stocks of slow-moving whales (right, bowhead, humpback, and sperm whales) even before whaling was industrialized. But in 1873, the advent of the steam-driven whaleboat armed with harpoon guns and explosive harpoons gave fishermen a new and decisive advantage over whales, and allowed them to catch larger, faster prey. By 1900 whale populations in both the North Atlantic and North Pacific were in serious decline. In response, the industry moved its operations to the Southern Ocean, where virgin populations of humpback, blue, fin, sei, and minke whales were depleted in turn. This process was abetted by the invention of the factory ship in 1925. With these ships in the fleet, whalers no longer needed to return to shore to process their catch. Instead, whales were hauled up a ramp on the stern of the boat, butchered, and processed at sea. Against a background of dwindling whale populations worldwide, the International Whaling Commission (IWC) was formed in 1949. Its stated mission was to preserve whales as a viable fishery, but its powers of enforcement were virtually nonexistent. Stocks continued to decline until, in the 1980s, public outcry forced IWC member nations to call for a moratorium on whaling. A few whale populations are recovering, the California gray whale for instance, but even now a few countries—notably Norway, Japan, and most recently Iceland—still hunt whales.

Following in the footsteps of the whaling industry, pressure from industrialized fishing began to affect other fisheries early in the twentieth century. As early as 1893, there was concern in England over the depletion of haddock and plaice: more and more boats were fishing these stocks, but fewer and fewer fish were being caught. Great Britain and other European nations considered regulating the fisheries by mandating an increase in the mesh size of nets to allow small fish to escape. But before these regulations could be implemented, worries over declining stocks were overshadowed, first by World War I (1914–18) and then by World War II (1939–45). In a grand, unintentional experiment in fisheries sciences, these global wars severely reduced the number of boats fishing in the North Atlantic. The fish stocks recovered.

After World War II, the world's fisheries underwent a second industrialization, driven in large part by the continuously accelerating pace of human population growth. Global population, which in 1900 had been 1.6 billion, grew nearly 60% by mid-century to 2.5 billion. This larger population demanded more fish, not only for direct consumption, but also for fish meal and oil to feed to pigs and chickens as food supplements. In response to this demand,

Japan, the Soviet Union, and Spain built large fleets of factory boats—similar to those pioneered by the whaling industry—to harvest as much of the world's fish as possible. These "distant-water" fleets could travel to any part of the ocean, catching, cleaning, freezing, and processing fish as they went, returning to port only to unload and head out again. Building on the technology developed in World War II, acoustic "fish finders" and spotter planes allowed the fleets to find fish with increased precision, and in the 1970s and '80s, satellite thermal images of the ocean helped fishermen position themselves at abrupt shifts in temperature where fish were likely to be found. As a result of this second industrialization, the global catch of fish increased steadily from about 20 million metric tons in 1950 to more than 80 million metric tons in the 1980s.[1] Approximately 200 million people were employed worldwide in some aspect of the fishing industry, and up to 19% of the protein consumed by humans came from fish.

By the mid-1970s, the pace of global fishing, especially fishing by distant-water fleets, became a concern to nations around the world and led to a fundamental change in how society viewed the oceans. "Cod wars" between Great Britain and Iceland erupted when Iceland, in an attempt to control its local fishery, extended its territorial waters 4, then 12, then 200 miles out from its coast. Sparked by this audacity, country after country began to question the assumption of *res nullius* as they saw local fisheries resources disappear into the holds of Japanese, Russian, or Spanish factory ships. As a means to control the drain of these resources and to exert sovereignty over what they considered "their part of the ocean," most maritime countries of the world extended their territorial sovereignty by instituting exclusive economic zones (EEZs) extending 200 miles out from shore. The new territorial claims took effect on January 1, 1977.

Since that time, three changes of note have happened in the world's fisheries. First, a variety of attempts have been made by various countries to regulate fisheries within the newly claimed exclusive economic zones. We will examine some of these attempts, and will see why these regulations have had mixed success. Second, for a few years the global catch of fish held steady at 75–85 million metric tons per year. This plateau was maintained in part by new regulations instituted after 1977, but to a greater degree it reflected continued declines in some fisheries offsetting the expansion of others. There are now indications that the global catch is declining, a topic to which we will return later in this chapter. And finally, human population has continued to grow. From the 2.5 billion people on earth in 1950 have sprung the 6.4 billion people alive in 2007, and population is projected to increase to at least 9 billion before it levels off. This increased population places increased demands on earth's fisheries, and creates an ever-strengthening interaction between human society and the ocean.

Renewable Resources vs. Mines

To understand the implications of this interaction, we need to understand the principles of renewable resources.

[1] Initial estimates of maximal global catch were as high as 100 million metric tons, but these estimates have since been revised downward. It appears that for years the Chinese inflated their reported catch to match political expectations.

Imagine that you have just discovered gold in your backyard. Sitting under the lawn is a vein of high-grade ore just waiting to be tapped and sold. What do you do? Well, the gold isn't doing you any good sitting in the ground; to realize a profit, some effort is required on your part. The more effort you put in—the more you dig, the more processing equipment you buy, the more electricity you feed the equipment—the more gold you recover per year. In other words, the rate of return from your fixed reservoir of gold is proportional to the effort you expend. Of course, the faster you dig the gold out, the sooner it is gone. These are the pertinent facts of a *mine*, a resource of fixed, finite size.

Contrast this scenario with a resource that grows through time. The example we take, as you might expect, is that of a population of animals that reproduces. What kind of return can you get for the effort you expend to utilize this kind of renewable resource? As we will see, it is likely that the rate of return is *not* proportional to the effort expended. In fact, in many circumstances the more effort you put in, the less benefit you receive. To explain this important effect, we return to a subject introduced in chapter 5, the limited growth of populations.

The Logistic Curve Revisited

Let's review the basic concepts. In the presence of abundant resources and the absence of grazing or predation, a population of a given species grows exponentially. For example, if one female fish produces five female offspring each year and then dies, the population of female fish increases from 1 to 5 at the end of the first year, 5 to 25 at the end of the second year, 25 to 125 at the end of the third year, and so forth.

However, this pattern of growth cannot last long. After only twenty years of this exponential growth, there would be nearly 100 million million fish in the population. This for a hypothetical fish that produces just five young each year. In contrast, a large cod can spawn 9 million eggs a year, leading Alexandre Dumas to explain that

> It has been calculated that if no accident prevented the hatching of the eggs and each egg reached maturity, it would take only three years to fill the sea so that you could walk across the Atlantic dryshod on the backs of cod. (*Le Grande Dictionnaire de Cuisine*, 1873)

Such codly excursions are impossible because, at some point, the growing population begins to feel the pinch of diminishing resources: they begin to be limited by the rate at which food is provided or by the rate at which waste products can be removed. In any real system, exponential growth is halted as the population reaches the carrying capacity of its environment.

As we saw in chapter 5, this pattern of limited growth can be modeled using the logistic equation, a relationship graphed in figure 11.1. When the population is small, growth is exponential, but as the population size begins to deplete the available resources, the rate of growth decreases, and through time the population asymptotically approaches its limit. The logistic model of population growth is governed by two parameters: (1) the intrinsic rate of growth, set by how many offspring a female can produce each year, and (2) the carrying

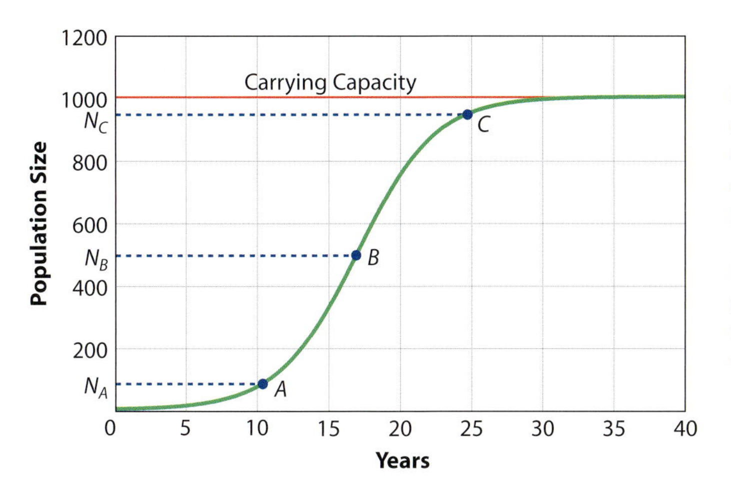

Figure 11.1. The logistic curve of population growth. Growth is relatively slow at points *A* and *C* (corresponding to population sizes N_A and N_C). In contrast, growth is rapid at point *B*, corresponding to a population size (N_B) midway between 0 and carrying capacity.

capacity, the maximum number of fish that can sustainably be supported by the environment. (For a more thorough explanation of both the logistic equation and these parameters, see the appendix to chapter 5.)

It will be useful to take one more step in our exploration of logistic growth. Consider point *A* on figure 11.1. At this time, the population has size N_A, and the rate of growth is equal to the slope of the graph at point *A*. In this particular case, growth is slow because there are few individuals in the population. At a later time, point *B* near the middle of the logistic "hill," the population is larger (N_B), and its growth rate is faster. At point C, the population is larger still (N_C), but because this size is near carrying capacity, the slope of the curve is gentle and the growth rate is low.

We can use these points, and others measured in similar fashion, to remove time from consideration and define directly the relationship between population size and growth rate (figure 11.2). Growth rate is a parabolic function of population size. At small size, growth is slow because there are few individuals to reproduce. At intermediate population size, growth rate is maximal; there are many individuals reproducing and resources are still abundant. Above this intermediate size, however, the rate of growth declines as resources become limiting, until at carrying capacity the rate of growth is 0. At carrying capacity, a new individual can be born only when an old individual dies.

For the logistic curve, the peak growth rate occurs when the population is half its maximal size. For the growth curves of actual populations, which may differ in detail from this logistic model, the population size at maximal growth rate may be slightly different, but it is always at some intermediate size.

Harvesting the Resource

In chapter 5, we examined this pattern of growth from the perspective of a prudent consumer. To a consumer, the growth in the prey population from one year to the next represents "surplus production" that can be harvested without repercussion. Thus, figure 11.2, which depicts how population size determines growth rate, as shown by the left-hand ordinate, can equally well show how

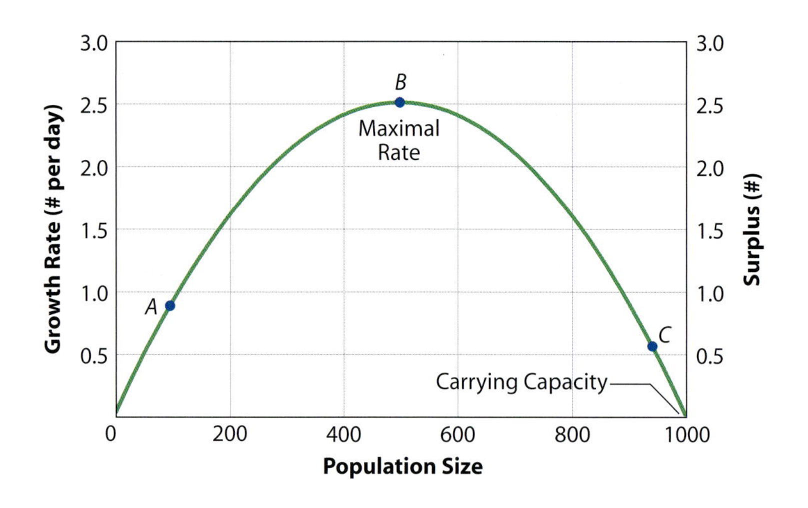

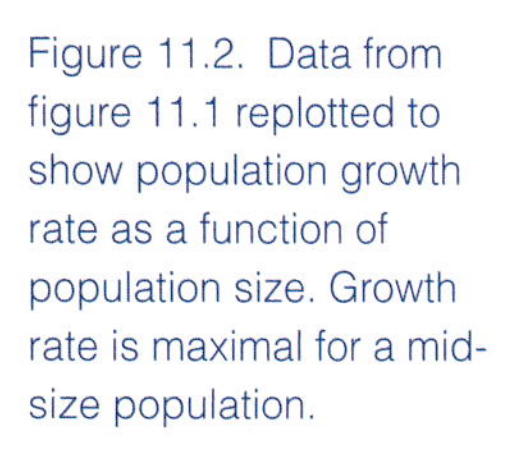
Figure 11.2. Data from figure 11.1 replotted to show population growth rate as a function of population size. Growth rate is maximal for a mid-size population.

population size determines surplus production, shown by the right-hand ordinate. If in a given period the consumer takes from the prey population only the growth in population (the surplus), the prey population returns to its initial size and can then begin the process anew. Thus, if harvested prudently, the prey population is a renewable resource, one that can be tapped again and again. However, the total sustainable amount that can be taken each year depends on the growth rate of the prey population. If the consumer maintains its prey at either a high or low population size, surplus production is low (see figure 11.2), and the return is low. Alternatively, if the consumer maintains its prey at an intermediate population size, surplus production is high and sustainable return is maximal.

What if the consumer is greedy and takes more each year than its prey produces? In that case, the prey population is no longer a renewable resource; instead it is more akin to a mine, a resource that can be depleted. As long as the consumer takes more than the prey population produces, the size of the population declines, and eventually it will disappear.

In this chapter, the prey of concern are the edible animals of the sea, and we—human society—are the consumers. The basic message is clear. The ocean presents us with a tremendous opportunity—a renewable resource of tasty, healthful food—and all we have to do is manage the fishery prudently. However, if each year we take more than the resource produces, we will eventually mine ourselves out of business. When viewed this way, it all seems so simple.

Estimating Sustainable Yield

But of course there are complications. These come in two basic varieties. First, to avoid harvesting too much from a population, we must be able to predict how the population grows—that is, how much surplus will be produced in a given fishing season—and such predictions can be extremely difficult. Second, society must use its knowledge of fisheries science to act in a prudent fashion, and that, too, has proven problematic. First, let's examine the difficulties inherent in reliably predicting sustainable yield.

Perhaps the single largest problem for fisheries scientists is accurately measuring carrying capacity and intrinsic rate of increase, the parameters that tell us where on the logistic curve a population lies. Let's start with carrying capacity. In theory, estimating this parameter is simple. If you assume that in the absence of fishing the population is at its carrying capacity—a reasonable approximation—all you need to do is count the number of fish in the population before you start to fish. But what in theory is simple, in practice is virtually impossible. One cannot simply look at the ocean and see how many fish are present. An English fisheries scientist summed up the problem this way: "Counting fish is as easy as counting trees, except they are invisible and they move." As a result, estimating the "virgin stock," the unexploited population size, is extremely difficult and prone to large errors.

Measurements of intrinsic rate of growth present different difficulties. Again, in theory, the process should be simple: take a small number of the fish in question, give them plenty of food, and see how many young they produce in a year. In practice, that won't work. Reproductive output depends not only on the amount of food, but also the type of food, the temperature of the water, and additional factors that are poorly understood. Some of the effects of environmental variation can be forecast—reduction in primary productivity caused by an El Niño event, for instance, can severely impact reproduction in herbivorous fish that feed on phytoplankton—but other factors are currently unpredictable. For many fish stocks, the number of fish produced in a year has no correlation with primary production; there are simply "good years" and "bad years" that seem to happen at random, making it very difficult to estimate the appropriate intrinsic rate of increase to use in a logistic model. At present, the best one can hope for is to measure a realistic range of intrinsic rates of increase.

Because of the difficulties inherent in measuring carrying capacity and intrinsic rate of growth, simple logistic models—as useful as they are as an intellectual tool for thinking about population growth—are difficult to apply to real-world fisheries. The cod fishery on Canada's Grand Banks provides an excellent example.

As noted in chapters 2 and 9, the Grand Banks is an area of shallow ocean east of Newfoundland. The shallow bottom, intense storms, mixing between the Labrador Current and the Gulf Stream, and the seasonal disappearance of the thermocline all ensure that the waters of the Grand Banks are rich with phytoplankton, and cod, among many other species, have historically used the resulting food web to grow in abundance. From very early times (perhaps even before Columbus) Basque fishermen caught cod on the Grand Banks, and the presence of cod was a driving force behind both British and French colonization of eastern Canada.

Cod were fished sustainably for 400 years, until at least the 1850s. In retrospect, there were hints that the stocks were beginning to show stress by the early 1900s, but this stress became readily apparent only after extraordinarily heavy pressure was applied by the arrival of distant-water fleets in the 1960s (figure 11.3). In response, in 1977 the Canadian government extended its EEZ to include most of the Grand Banks and began to exclude foreign ships from the fishery.

Canadian policy makers then turned to the government's fisheries scientists for advice, and a detailed model of the cod population was constructed, similar

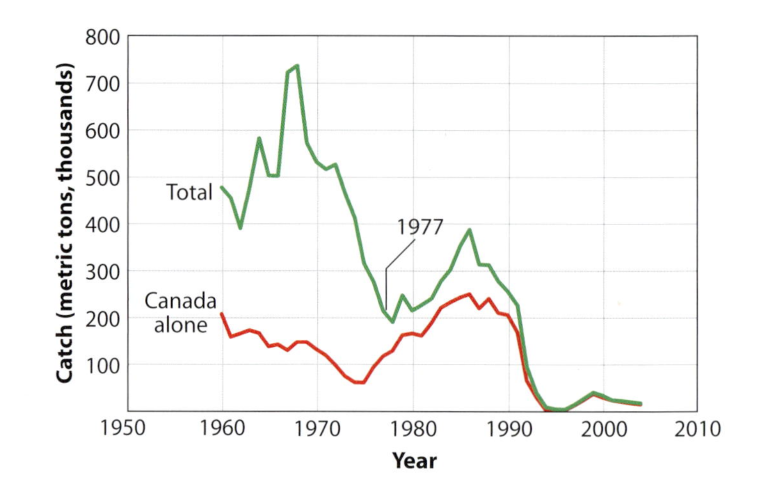

Figure 11.3. The history of the Canadian cod fishery from 1960 to 2005.

in principle to the logistic model we have discussed. Various estimates of intrinsic growth rate and carrying capacity were obtained and debated, and official values were selected. Based on the model using these parameters, regulations were then imposed, limiting the catch of cod to amounts that were thought to be sustainable. Canadian fishermen followed the regulations, and the fish stocks initially increased. But beginning in about 1987, despite all scientific advice and governmental regulation, the population collapsed and has not recovered. In retrospect, it is clear that estimates of population size and reproductive capability were substantially in error.[2]

Given our current inability to predict population growth, the prudent course of action for fisheries managers is to incorporate a "safety factor" into the harvest. Assume for the moment that a fish population is currently at the size required for optimal growth. If industry plays it safe and takes fewer fish than the yearly surplus, the worst that can happen is a growth in population. This growth moves the population "uphill" on the logistic curve, so that growth the next year is not at an optimal rate. However, the increased population is essentially money in the bank, an investment that can be cashed in if desired by carefully increasing harvest in the future.

In contrast, if industry does not incorporate a sufficient safety factor, it could take too many fish in a bad reproductive year. In this case, population size moves downhill on the logistic curve, again reducing the rate of growth, but also potentially depleting the population beyond the point of recovery, as happened with Canadian cod. The more natural variation in population growth, the larger the safety factor needed to maintain a renewable resource, and the more difficult it is to maintain the resource at its optimal size.

These sorts of problems currently make fisheries science a challenging field. But progress is being made both in our ability to measure the marine environment (and thereby to measure and predict the natural factors controlling reproduction and carrying capacity) and in our ability to realistically model the

[2] Some of this error was due to optimistic selection of parameters by government scientists, abetted by the reluctance of policy makers to accept bad news. For a disturbing exposé of the interaction between science, government, and fishermen, see Michael Harris's *Lament for an Ocean* in the list of further reading.

effects of natural variability and harvesting. As time goes on, our understanding of the science of fisheries will undoubtedly progress, allowing scientists to provide more robust and reliable information to the fishing industry about how best to manage their marine resources.

Population Size Structure

Unfortunately, lack of reliable data and accurate predictions is not the largest problem facing the world's fisheries. Instead, the root problem is one of human nature. To see why, let's explore two pertinent and interacting aspects of the fishing industry: the effect of fishing on the size structure of populations, and the role of fiscal optimism.

We begin with size structure and the factors that control it. Figure 11.4 shows a hypothetical (but representative) survivorship curve for fish in the absence of harvesting by humans: a plot of who dies when. On the abscissa is the age of the fish, which can also be taken as a measure of the size of individuals: the older the fish, the larger the fish. On the ordinate is the probability that a fish will die in the next year.

Life is tough for young (small) fish. They are readily eaten by predators, and their chances of surviving a year in the sea are minimal; that is, their chances of dying are high. The older and larger they become, however, the lower the probability they will die in a given year, and fish of reproductive size have a high probability of surviving. At some age, of course, fish die of old age, but for many species this age can be quite high. There are some species—orange roughy for example—that do not even become reproductive until they are 25 to 30 years old, and they can live and reproduce for over a century.

Given this pattern of survivorship, what is the size structure of the resulting population? Each year, young fish are produced in profusion, but most of them die. As a consequence, by the end of the year there are few small fish left in the population. However, the survivors grow, and because their chance of subsequent death is low, these older, larger fish accumulate from one year to the next. The result is a distribution skewed to large size (figure 11.5).

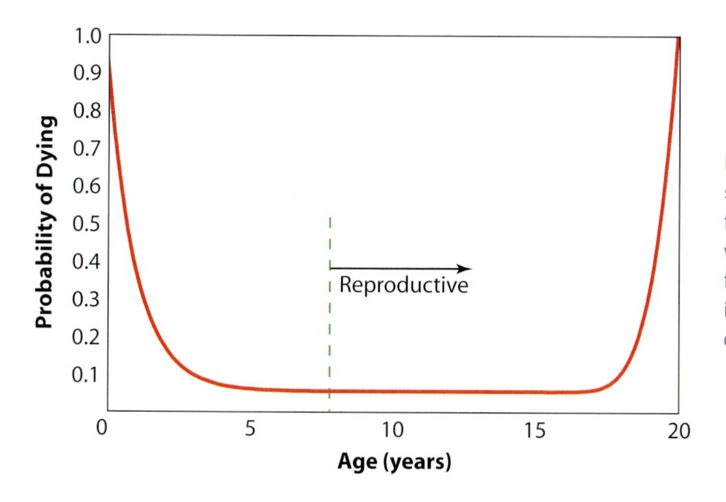

Figure 11.4. A hypothetical survivorship curve for marine fish. The probability that a fish will die in a given year is high for very young and very old individuals, but low for much of a fish's life.

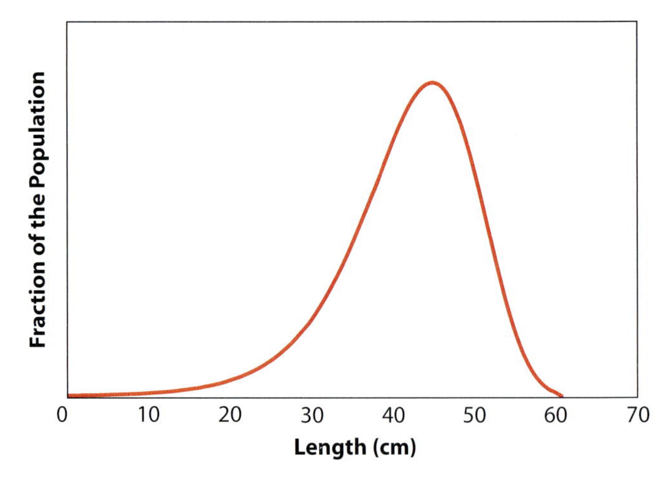

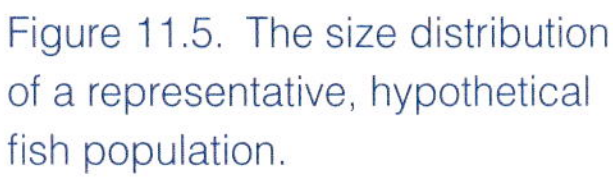
Figure 11.5. The size distribution of a representative, hypothetical fish population.

Because of their large size, these fish can produce many young. For example, a female red snapper 42 centimeters long (with a mass of 1.1 kg) can produce about 4400 eggs. In contrast, a female 61 centimeters long (weighing 12.5 kg) produces 9,300,000 eggs, a 212-fold increase. Thus, because the size distribution of a virgin population is skewed to large size, it is highly fecund.

Unfortunately, this size distribution also sets the population up as easy pickings for the first fisherman who comes along. Fishermen often capture fish using nets. Fish smaller than the mesh size of a net can escape, but fish larger than the mesh size are retained and can be dragged to the surface. Thus, nets are size-selective filters, intrinsically designed to catch large fish, and for a population that consists mostly of large fish, nets are an extraordinarily effective means of harvest.

Life Cycle of a Fishery

At this point human nature comes into play. Imagine you are the lucky first fisherman who encounters an untapped population of fish. Dragging your net through the water, you easily catch an abundance of the large fish that live in the area, and when you take these fish to market, they fetch a good price.

What happens next? If your single boat can turn this kind of profit, imagine what you could do with several boats. Using your initial successful catch as a proof of concept and your boat as collateral, you go to the bank and take out a loan to buy more boats and gear and to hire other fishermen. Back at the fishing ground the next year, the fishing is again good: there are still plenty of large fish left, catch per effort is high, and the money flows in.

The inevitable response to this initial good fortune is *fiscal optimism*, an exceedingly dangerous concept. Word gets around that the fishing is good, and other entrepreneurs take out loans, buy boats, and flock to the area. Investment is made in shore-side facilities to support the fishery—canneries, freezing plants, trucking routes—and money is paid out to publicize the catch and build a market.

It is then that the process begins to go sour. As the number of boats increases, so does the number of fish caught. If the fishery is not carefully regulated, within a few years more fish are caught than is sustainable, and the industry begins to mine the population.

How would fishermen know that they are overfishing? The first indication is given by the size of fish in the catch. As we have seen, the large fish of the population are initially captured. If too many large fish are netted, younger fish cannot grow fast enough to replace them, and the average size gets smaller. Thus, the first clue that the population is not replacing itself is a reduction in the size of fish caught.

What is the industry's response? The prudent response would be to decrease fishing pressure and allow the stock to recover, but that is unlikely because decreasing the number of fish caught would decrease profits. So, the industry instead begins to use nets with a smaller mesh size, which allows them to catch the same mass of fish as caught in years past. To the fishermen, the change matters little because they are paid by the pound, not by the size of fish they catch. But it is disastrous for the fish population. Catching the same mass of smaller fish means that a larger number of fish are caught, and population size decreases faster than it did before.

As the size of the fish decreases, two things happen. First, fewer large fish are left to reproduce, so average reproductive output decreases for the puny females left in the population. In other words, as the size of fish decreases, the ability of the population to recover is jeopardized. Second, as both population and individual size decrease, the effort required to catch a given mass of fish rises. Thus, through time, the catch per effort goes down. This is the second sign that the fish population is being harvested in a nonrenewable fashion. At this point, the fish resource is being mined, but unlike a gold mine in which the rate of return is proportional to effort, for this "fish mine," the rate of return is *inversely* proportional to effort.

Again, the prudent response is to decrease the pressure on the fish population, thereby giving it a chance to recover. However, because of the initial fiscal optimism that created the fishery, reducing fishing effort is virtually impossible. Having taken out loans to pay for boats and gear, fishermen have to maintain a steady income to pay the interest on those loans. The only way to keep a steady income is to catch the same mass of fish each year, but as we have seen, maintaining mass requires catching a larger number of fish because fishing with nets decreases the average size of individuals. Thus, the inevitable result of fiscal optimism is a process of positive feedback that rapidly depletes the fish stock. The total catch (expressed as pounds of fish brought to market) may stay constant for a few years as fishermen increase their efforts, but eventually they are catching fish smaller than reproductive size. With few, if any, fish left to reproduce, the population crashes, and the fishery dies.

One might think that reduced catch per effort would drive a fishery out of business before it had completely depleted its prey population. Indeed, if market forces alone governed the fishery's demise, bankruptcy might well be the result. But when a fishery dies, fishermen are put out of work, and this loss of jobs is difficult for society to accept. Instead, societies are prone to ease the immediate pain of fishermen and those dependent on the industry by subsidizing the fishery, allowing the industry to realize a profit when it would otherwise be losing money. The influx of subsidies thus allows the fishing effort to persist, and fish stocks to be even more severely depleted. Currently, nearly a quarter of the money flowing into global fisheries comes from government subsidies.

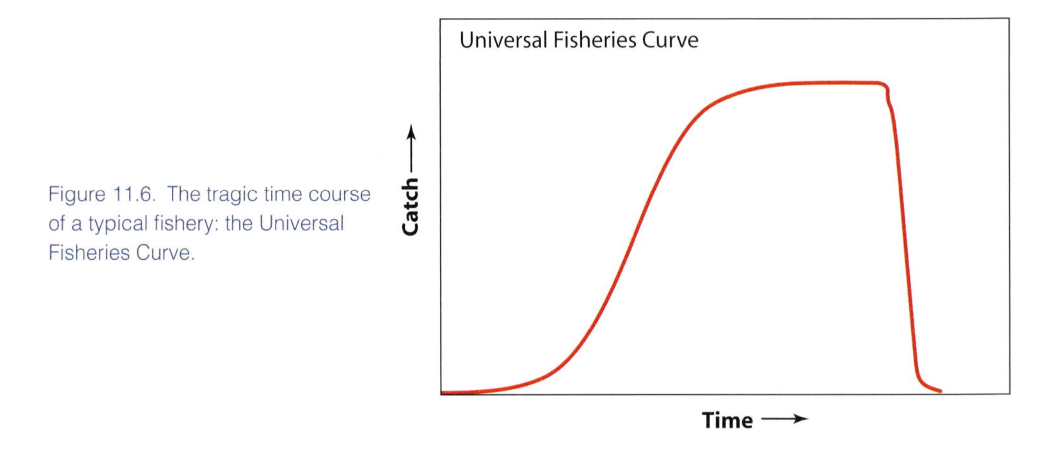

Figure 11.6. The tragic time course of a typical fishery: the Universal Fisheries Curve.

This tale of woe is depicted in figure 11.6. To recap, in the first few years after a fishery opens, the total catch rises rapidly as fiscal optimism takes effect. A few years follow in which the total catch holds constant, but fish size and catch per effort plummet. And lastly there is the inevitable crash.

Surely, you might think, fishermen would have learned from their history and devised some way to avoid this ruinous progression. But no. This process has been repeated so many times in so many places that figure 11.6 could aptly be termed *The Universal Fisheries Curve*. From menhaden, herring, haddock, hake, plaice, and cod in the Atlantic to sardines, rockfish, bocaccia, anchovy, thresher, blue, and angel sharks, Chilean sea bass, yellowfin tuna, skipjack tuna, and Dover sole in the Pacific, virtually every fishery that has existed has followed this disastrous course.

The Tragedy of Fishing

There are several tragedies in this story. First, there is a tragedy for fishermen. Once a fisherman takes out a loan to buy a boat or gear, he or she has little choice other than to keep fishing regardless of the state of fish stocks. The burden is made heavier in cases where the fishing trade is passed down from generation to generation. In many families, fishing became a way of life at a time when the human population was small enough not to put undue pressure on prey populations. Now, when unsustainable pressure is put on fish stocks, these families know no other way to make a living. The result is a doomed generation of fishermen caught in circumstances beyond their control.

Second, there is a tragedy for society, a classic example of what William Lloyd (1794–1852) termed "the tragedy of the commons." The problem in this case is one of ownership. In a typical case, no single fisherman owns a defined part of a fishery. Instead, the fishery belongs to society as a whole: it is *res nullius*, a common resource. As a result, there is unequal disposition of rewards and punishment. An individual fisherman who continues to fish while knowing the population is declining still receives all the rewards for catching those fish. The negative result of his or her actions—a smaller population to be fished in the future—is shared by the fishery (and society) as a whole. Thus, when benefits are weighed against costs on an individual basis, it is easy to see why a fisherman could—indeed, logically should—choose to keep fishing. The tragedy, of course, is that it is easy for *every* fisherman to make this logical choice, and as a result, fisheries collapse.

It might seem that the tragedy of the fishing commons is fueled by selfishness on the part of fishermen. Can't individual fishermen see that their insistence on personal gain is destroying the general good? However, when faced with family and the bank demanding money, the decision to keep fishing is not one of altruism versus greed. Instead, the choice for fishermen is often one of survival versus bankruptcy, and the decision is simple: they keep fishing. The fiscal and regulatory system of the fishing industry, as it currently stands, is the true culprit here, a matter to which we will return shortly.

And lastly, the story of human fisheries is a potential tragedy for the oceans. By harvesting fish at the present rate, we are performing a massive ecological experiment with potentially drastic results. For example, in recent decades the fishing industry has decreased the populations of pollock and mackerel in the North Pacific. The consequences have been far-ranging and surprising. First, the population of Steller sea lions (which eat pollock and mackerel) has decreased by 79% in the western Aleutian Islands, 73% in the eastern Aleutians, and 31% in the central Gulf of Alaska. Second, because there are fewer sea lions to eat, killer whales are now turning their attention to sea otters, an endangered species. In contrast, pollock and mackerel eat planktonic crustaceans. So when the fish populations declined, the population of crustaceans increased, and populations of the birds that eat the crustaceans, such as the least aucklet, are on the rise. Needless to say, none of these effects was predicted or considered when fishermen began catching pollock and mackerel.

Similar effects have happened elsewhere. In the Caribbean, corals live in a delicate balance with benthic algae. In an undisturbed, "natural" reef, the algae are kept in check by herbivores: fish, sea turtles, and sea urchins. On many islands, fishing in the twentieth century severely depleted the populations of turtles and herbivorous fish. As a result, when disease killed most of the urchins in the early 1980s, there were no effective herbivores left, and algae rapidly overgrew the corals. Live corals, which in 1977 accounted for 52% of the surface area of Caribbean reefs, were reduced to 3% in the early 1990s, and algal cover rose from 4% to 92% over the same period.

The opposite effect occurred in the Gulf of Maine. When cod were plentiful, they ate small sea urchins and thereby kept the urchin population in check. With few urchins to eat them, kelps grew abundantly. With the demise of the cod populations in the 1980s, urchins ran rampant, and the former kelps forests became barren.[3]

Harvesting of fish by humans has other effects as well, most of which are negative. For example, many fishing techniques capture animals other than those intended. For every pound of shrimp brought to market, five pounds of other species are killed and discarded. Until the 1980s, the nets used to catch tunas also caught dolphins; 133,000 dolphins were killed in 1986 alone. Bottom trawls, used to catch scallops for instance, gouge swathes through the substratum, disturbing or destroying benthic plants and animals. For many years, drift nets were used to catch a variety of species in open-ocean habitats. In this

[3] A fishery then arose to harvest the urchins. It followed the Universal Fisheries Curve and crashed after a few years.

method, a mile-long "curtain" of netting was suspended from buoys and allowed to drift for a period, catching anything large enough to be entangled in the mesh. Some of this catch was valuable, but many unintended species were also caught and killed: birds, dolphins, and sea turtles to name just a few. If nets came loose from their moorings or were lost, the resulting "ghost gear" could drift for years, killing fish the entire time. Drift nets have now been banned by a U.N. treaty.

Many open questions remain as to the ecological effects of fishing. For example, some researchers have found evidence that humans are "fishing down the food web." That is, as time goes on and stocks of fish at high trophic levels become depleted, fishing effort expands to species lower on the trophic pyramid. In one sense, this effect is positive. By taking fish at lower trophic levels, humans are making more efficient use of primary production. For instance, intense fishing efforts in the Mediterranean Sea have traditionally focused on low-trophic-level species, and these fisheries have been sustainable for centuries and appear stable today. Alternatively, the reduction in trophic level of current catch could simply be another indicator of overfishing. If overfishing continues at high trophic levels while expanding down the food web, the ecological effects may be substantial. As we have seen, predators can exert strong controls on prey at lower trophic levels. Removal of predators may therefore have complex cascading effects through the entire trophic process.

Fishing can also reduce the diversity of species present in an area, and this reduction in diversity is associated with negative effects on a wide variety of "ecosystem services." For instance, areas with reduced diversity show a reduction in "nursery habitat," the habitat required for fish to reproduce, and an increase in harmful algal blooms and coastal flooding.

A Global Perspective

When I was growing up in the 1950s and '60s, it was common knowledge that the oceans were earth's "great, untapped food resource." Yes, human population was growing, but feeding all those new mouths wasn't going to be a problem because we could always catch more fish. As noted above, over the past 50 years human population has more than doubled, but the bounty from the sea never materialized. The catch of fish has leveled off and may be decreasing. In other words, the great, untapped resource was an illusion, and in recent years, oceanographers have drawn together several lines of evidence to explain why.

We begin with the mass of fish caught each year for each of the world's fisheries, data collected by the U.N. Food and Agriculture Organization. Knowing how many tons are harvested of a given species of fish, we can calculate how many tons of carbon they contain. Knowledge of the food web utilized by that species—that is, how many trophic levels separate fish from primary producers—then allows us to calculate the amount of new primary production required to yield that tonnage of carbon in the fish. This procedure is then repeated for each species in the overall world fishery.[4] The bottom line is that

[4] This approach is outlined in D. Pauly and V. Christensen (1995), Primary production required to sustain global fisheries, *Nature* 374: 255–257. Pauly and Christensen provide data for all the world's fisheries, including those in freshwater. Only the data for marine species are used here.

approximately 85 million metric tons of marine fish are currently caught each year, requiring the fixation of 2.7 billion metric tons of carbon.

This value can be put into global perspective. The current estimate for the total yearly new production of the ocean is 16 billion metric tons of carbon. In other words, humankind's current fishing efforts require 17% of the ocean's total sustainable productivity.

"What's the problem?" you might ask. Heck, we could take five times as much and still stay within the bounds of sustainability. But humans are not the only users of the sea's new production. By taking nearly a fifth of the ocean's sustainable production, we decrease by nearly a fifth the sustainable production available to other predators operating on the same trophic level. For example, when we act as top predators by catching cod, hake, or salmon, we take food away from tunas and other billfish, dolphins, toothed whales, sharks, and even some squid. When we act as intermediate predators (by catching anchovies, for example) there is less sustainable production to be eaten by the squid, birds, and marine mammals that use this resource today. Marine birds alone are estimated to catch 70 million metric tons of fish each year, nearly as much as humans take by fishing, and although much of that take is currently in open-ocean areas not heavily fished by us, any increase in fishing pressure would undoubtedly put us in competition with birds.

The value of 17% cited above can itself be misleading. This percentage reflects the fraction of the ocean's overall new production required by current fisheries. However, in certain areas, the percentage is much higher. For example, in the temperate and polar shelf areas of the ocean, fisheries require 35% of the total net production. In these areas, 40–50% of net production is new production (chapter 6). Thus in these shelf areas, current fisheries require that 70–88% of all sustainable production be directed to human use.[5]

The values cited here for the global catch of fisheries are misleading in another respect as well. These are figures for the catch of fish brought to market, but many other fish are caught and killed in the process and simply discarded. This "by-catch," such as that mentioned above for shrimp and tuna fisheries, amounted to an additional 20 million tons per year in the 1980s, about a quarter of the market catch. In recent years, the mass of by-catch has been reduced to approximately 8 million metric tons per year, in part due to advances in fishing gear and regulations by some countries prohibiting the discard of by-catch. However, the recent reduction in by-catch has less of an impact on the ocean than one might expect: much of the reduction is simply a matter of semantics. Markets have developed for fish traditionally considered to be by-catch, allowing much of the mass of by-catch to be shifted into the category of regular catch. It is comforting to know that by-catch is now being put to good use, but the inadvertent capture of fish is still a drain on the stocks involved.

It is difficult to form general conclusions about the ecological effect of fisheries' by-catch. The carcasses of the dead animals are returned to the sea, so by-catch does not result in a net loss of nutrients or carbon and may benefit

[5] If 35% of total net production goes to fisheries and 50% of net production is new production, fisheries account for 0.35/0.50 = 70% of new production. If only 40% of net production is new production, fisheries account for 0.35/0.40 = 88% of new production.

carrion-feeding species. On the other hand, the capture and killing of these animals acts as a substantial source of mortality that otherwise would not be present.

Current Status of the World's Fisheries

As of 2005, the U.N. Food and Agriculture Organization estimated that 75% of the world's fisheries were fully or overexploited. This figure bolsters the conclusion reached above, that current fishing efforts are bumping up against the limits of sustainable primary production. As we have seen, the worldwide catch of fish, which rose steadily from 1950 until the late 1980s, has since held constant at 75–85 million metric tons of fish per year, and there are some indications that it is actually declining.

This plateau and possible decline is due to several factors. First, the creation of exclusive economic zones and the implementation of conservation efforts in many fisheries have helped rein in the fishing industry. This effect is likely not as profound as one might suppose. Just because countries have jurisdiction over their territorial waters does not automatically mean that those waters are managed to the benefit of fish populations. As we will see below, regulations are often not imposed until after fish stocks are depleted. Furthermore, the 1982 U.N. Convention on the Law of the Sea dictated that any country that does not fully exploit the fisheries in its waters must open its EEZ to fishing by other countries. As a result, many countries now "rent" their waters to distant-water fleets and the fishing continues.

The lack of growth in global fisheries catch is also due to the sequential depletion of fish stocks as the Universal Fisheries Curve has been applied to one species after another. For instance, it has been estimated that 33% of known U.S. fish stocks are overexploited or depleted. Illustrative examples come from California, a state renowned for its environmental awareness. Despite regulatory efforts, a review of the state's fisheries in 2001 gave the grim results shown in figure 11.7. It is worth taking a moment to peruse these graphs in detail.

The first thing to notice is the similarity in the curves: all but one exhibit the boom-and-bust nature of the Universal Fisheries Curve. The second thing to notice is the range of species: from invertebrates (abalone, urchins, squid, and shrimp) to vertebrates (the fish), benthic (abalone, urchins, and Dover sole) to pelagic, offshore (yellowfin tuna) to inshore. Lastly, I have noted on each graph when restrictions were first placed on the fishery in an attempt to stabilize the catch. When no date is noted, no restrictions are in place. Without exception, restrictions, if implemented at all, were implemented in response to the imminent collapse of the fishery, too late to maintain the fishery at anything like its maximal sustainable yield.

The sole deviant graph in this group is that for the market squid (figure 11.7H). Despite drastic fluctuations thought to be due to El Niños, this fishery has grown dramatically in recent years in response to burgeoning markets in Europe, Japan, and China. But one has to wonder where this market is headed. What will this curve look like in 2020? To help you make your guess, I should point out that currently there are two primary restrictions placed on the squid fishery in California: (1) the fleet can catch no more than 107,000 metric tons

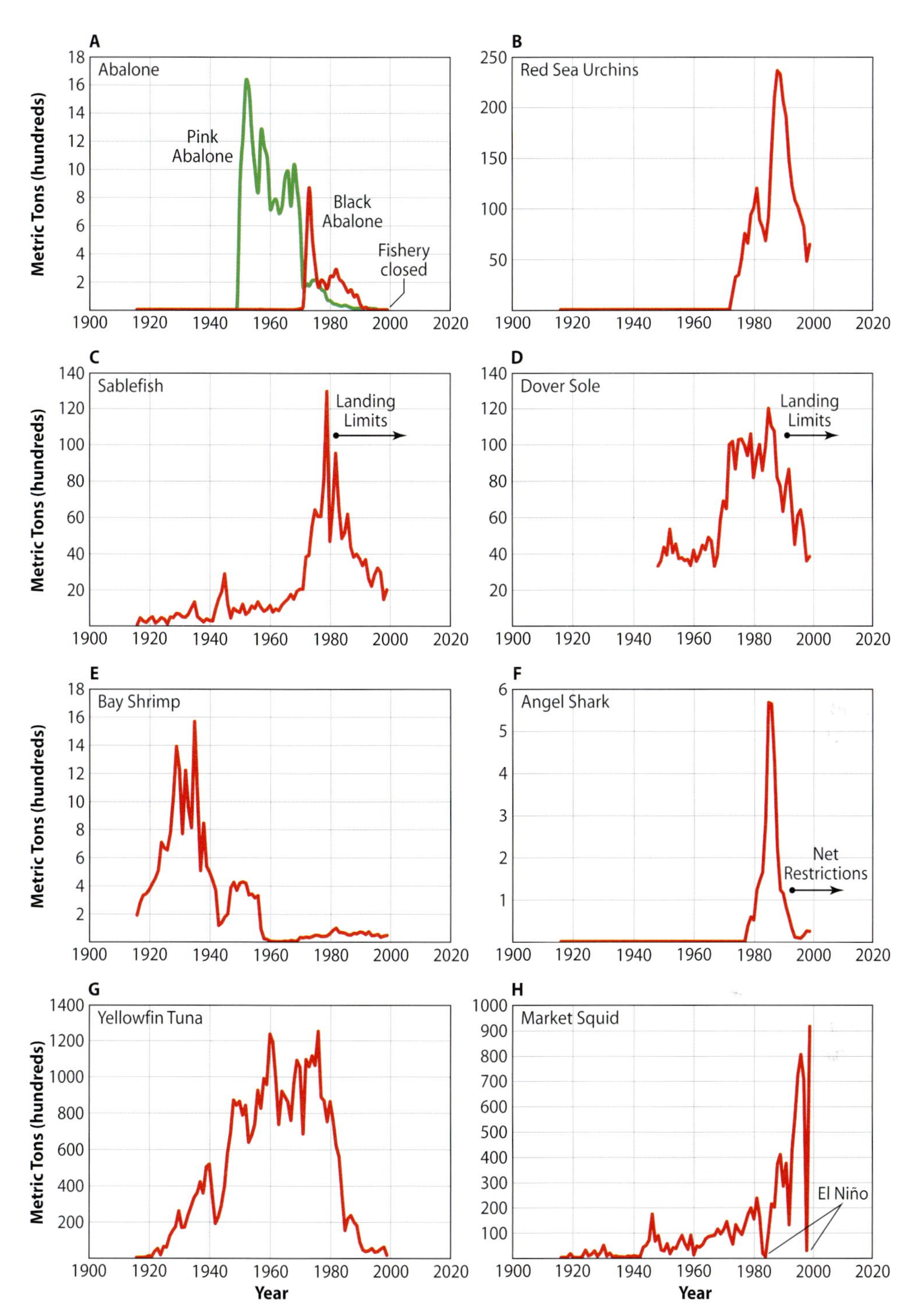

Figure 11.7. Representative examples of California fisheries. Data from *California's Living Marine Resources: A Status Report* (2001), California Department of Fish and Game.

per year, somewhat in excess of the maximum value ever caught to date, and (2) the lights used to attract squid must be less than a certain wattage.

The fisheries curves shown in figure 11.7 are an admittedly biased group. They were not chosen randomly; rather, they were picked to illustrate the points discussed above. But the bias in my choice is worrisomely small. Of the 78 commercially fished marine species reported in the 2001 review, only 15 have documented histories substantially different from the Universal Fisheries Curve, and 7 of these 15 show trends similar to that of market squid in which it is too early to tell what the long-term outcome will be.

While the status of the world's fisheries is generally grim, there are exceptions. For example, rock lobsters in New Zealand and Western Australia, abalone in Tasmania, salmon in Alaska, and squid in the Falkland/Malvinas Islands are well-managed, stable, productive fisheries. Similarly, lobster and crab fisheries along the Atlantic shores of Canada and New England are booming and appear to be sustainable. However, even this good news has its dark side. Historically, the principle predator of juvenile lobsters and crabs in these areas has been cod. So, the increased catch of lobsters and crabs is at least in part due to the drastic overfishing of cod.

Fixing the Problem

Several mechanisms have been implemented to aid the world's troubled fisheries. The first involves a change of objective. Historically, the primary goal of fisheries management has been to achieve maximal sustainable yield. Given sobering examples like that of the cod fishery on the Grand Banks, some fisheries managers have given up trying to attain maximal sustainable yields and have instead called for implementation of a "precautionary principle," mandating that populations be maintained at sizes much nearer their carrying capacity. The new objective is to maintain sustainable fisheries, even if they are less than maximally productive.

Changes are also suggested for the methods used to approach this goal. Traditionally, fisheries have been managed indirectly by regulating fishing effort: by limiting the number of days open to fishing, the number of boats allowed in the fishery, or the type of gear permitted. These strategies have often not worked as planned. For example, until 1995 the halibut fishery off Alaska was regulated primarily by adjusting the number of days that the fishing grounds were open. As the intensity of fishing effort increased, open season became shorter and shorter until fishing was limited to just two to three days. This period was so short that a boat couldn't catch a full load of fish, come into port to unload, and make it back out to the fishing ground before the season closed. As a result, boats took on unsafe loads of halibut, and many sank.

In response to this and other failures of indirect regulation, methods are now being tried that directly regulate the number of fish caught. This is often accomplished by taking the previously common resource—the fishery—and dividing it among a set number of fishermen. Individual quotas of fish are auctioned or sold, and the owners of quotas are allowed to take up to a specified number of fish, a number that can be changed from year to year as the situation warrants. Often, these quotas are transferable: the owner of a quota can sell it to another fisherman. A system of this sort was implemented for the Alaskan halibut fishery in 1995, and the stock now appears to be stable: the fishing grounds are open 245 days per year, and the number of vessels in distress has decreased by two thirds.

The quota method of regulating fisheries effectively circumvents the tragedy of the commons. Because there are limits to how many fish each fisherman can take, there is no monetary pressure to take more. Instead, there is incentive for fishermen to maintain the stock both individually and collectively by cooperating with other fishermen. The larger the stock of fish, the less effort each fisherman must make to catch his or her quota, and the larger the individual profit. Similarly, if the stock is allowed to grow, the average size of fish increases, and as a result, so does the monetary return for the fixed number of fish each fisherman can catch. Maintenance of a large, stable fish stock also increases the market value of each fishing quota. The healthier the population of fish, the more money fishermen can ask if they decide to sell their "stock" in the fishery.

The largest disadvantage of a quota system is that it manages fisheries one species at a time, neglecting strong connections between fisheries and between fisheries and un-fished species. For example, what is good for the cod fishery in New England may be bad for the lobster fishery, and any decision by these fisheries affects the local kelp and urchin populations. In such cases, different fisheries are in competition with each other, and there is no clear mechanism by which quotas can resolve the conflict. Instead, what is required is a larger scale of regulation, what has been termed an *ecosystem-based approach*.

The ecosystem-based approach to fisheries management is in its infancy. One mechanism currently being implemented to address problems at this level is to set aside areas that are off-limits to fishing. The idea is that these *marine protected areas* establish refuges in which prey populations can flourish, providing both a buffer against depletion of stocks and a source of fish, which in theory move out of the protected area to fishing grounds where they can be legally captured. Whereas quota systems protect one single species, a marine protected area can help to maintain all the species in a given ecological network.

As one might guess, creation of marine protected areas is controversial. Although models suggest that marine protected areas can be effective, there is little consensus as to the optimal size and distribution of areas. In the field, preliminary results are encouraging, but too few areas have been established and monitored for sufficient time to provide definitive data. As you might suspect, fishermen often oppose creation of marine protected areas. To them, the cost is clear: an area in which they have always been able to fish is now off-limits. In contrast, the benefits are uncertain. At best, an increase in fishable stocks outside the protected area will not occur until well into the future.

One other strategy bears mention. It is possible to take pressure off wild fish stocks by farming fish in captivity. Several species—most notably tuna, shrimp, salmon, and mussels—are currently farmed commercially. The ecological benefits of farming are open to question, however. For example, tuna are farmed in large near-shore netted pens, where they are fed fish caught at sea. In effect, this "farming" is analogous to the feed lots in which cattle are fattened for market. Captive tuna eat as much food as they would in the wild, so no pressure is taken off the flow of carbon out of the sea. Farmed salmon and shrimp are not fed live food, but are fed fish meal and fish oil from other species caught at sea. Waste products from salmon and shrimp farms, along with antibiotics used to keep the animals healthy, are often dumped into the ocean, creating potential new problems.

In contrast, farming mussels as well as other suspension-feeding invertebrates appears to be a valuable strategy. The animals are raised attached to ropes that dangle into the water, and they filter their food (primarily phytoplankton) from water that passes by. Because mussels feed low on the trophic pyramid, they make efficient use of primary production.

In summary, a variety of strategies are available to solve the problems faced by global fisheries. What is now required is a mandate from human society to implement these strategies.

Complications

Consider figure 11.8, the catch history for the sardine and anchovy fisheries in California. By now, the shape of these curves looks all too familiar, and you can be excused for assuming that they are further examples of overfishing. Indeed, until recently the sardine fishery in California was presented in textbooks as the poster child for the dark side of human exploitation of the sea.

In fact, overfishing may not be the primary cause of the demise in the sardine and anchovy fisheries. The first clue that overfishing might not be the culprit came when biologists compared population records for anchovies and sardines from various places around the Pacific basin. Even though fishing pressure differed substantially from California to Peru to Japan, the population of anchovies in each place declined at the same time, and each anchovy decline was accompanied by a subsequent increase in the population of sardines. When, about 25 years later, the sardines declined, the anchovies came back. From this evidence, it seemed likely that some environmental factor, rather than fishing pressure, was synchronizing the rise and fall of these fish populations across the entire ocean.

Recall from chapter 9 that the North Pacific is subject to periodic, large-scale changes in its pattern of temperature. In this Pacific Decadal Oscillation

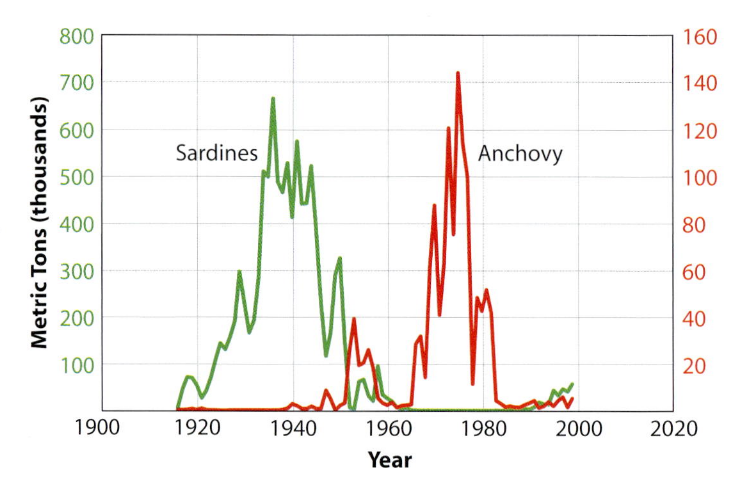

Figure 11.8. The sequential crash of California's sardine and anchovy fisheries is likely due to both overfishing and the Pacific Decadal Oscillation.

(PDO), water in the eastern half of the basin is warmer than average for roughly 25 years (the El Viejo phase), alternating with about 25 years when water is colder than average (the La Vieja phase). During El Viejo, sardines thrive, and during La Vieja, anchovies thrive. Thus, oceanographers now think that it may be something about the Pacific Decadal Oscillation that controls the population size of these fish, contributing to the pattern seen in figure 11.8. (It still seems likely, though, that overfishing exacerbated the declines.)

The mechanism by which the PDO affects sardines and anchovies is unknown. It cannot simply be temperature. For example, during El Viejo, water is warm in the eastern Pacific and cold in the west, but sardines thrive in both places. Similarly, temperatures are skewed in the opposite direction during La Vieja, but anchovies thrive both east and west. More research will be needed to define the mechanistic link between fisheries and large-scale phenomena such as the Pacific Decadal Oscillation.

Large-scale, long-term environmental effects such as this cloud our understanding of fish biology and complicate the management of fisheries.

Conclusion

Managing the ocean's fisheries will be a revealing test of humankind's ability to live sustainably on earth. Although much is left to learn about fish populations, our knowledge of how the ocean works, coupled with enlightened regulation and a modicum of unselfish behavior, is sufficient for us to coexist with healthy, resilient populations of edible marine organisms. However, if society chooses to ignore that knowledge and to selfishly sacrifice long-term good for short-term gain, the fisheries of the world are doomed.

Further Reading

Cohen, J. E. (1995). *How Many People Can the Earth Support?* W.W. Norton & Co., New York.

Cushing, D. M. (1988). *The Provident Sea.* Cambridge University Press, Cambridge.

Ellis, R. (1991). *Men and Whales.* Alfred Knopf, New York.

Field, C. B., et al. (1998). Primary production of the biosphere: integrating terrestrial and oceanic components. *Science* 281: 237–240.

Harris, M. (1998). *Lament for an Ocean: The Collapse of the Atlantic Cod Fishery: A True Crime Story.* McClelland & Stewart, Inc., Toronto.

Hilborn, R., et al. (2003). State of the world's fisheries. *Annual Review of Environment and Resources* 28: 359–399.

Iverson, R. L. (1990). Control of marine fish production. *Limnology and Oceanography* 35: 1593–1604.

Kurlansky, M. (1997). *Cod: A Biography of the Fish that Changed the World.* Walker & Co., New York.

National Res. Council (1999). *Sustaining Marine Fisheries.* National Academy Press, Washington, DC.

Pauly, D., and V. Christensen (1995). Primary production required to sustain global fisheries. *Nature* 374: 255–257.

Smith, T. D. (1994). *Scaling Fisheries: The Science of Measuring the Effects of Fishing 1855–1955*. Cambridge University Press, Cambridge.

Warner, W. W. (1976). *Beautiful Swimmers: Watermen, Crabs, and the Chesapeake Bay*. Penguin Books, New York.

Warner, W. W. (1984). *Distant Water: The Fate of the North Atlantic Fisherman*. Penguin Books, New York.

An Invitation

As we bring this text to a close, let's return for a moment to our starting point. Imagine yourself gazing out at the ocean from shore, or even better, from a ship or airplane. What do you see? If I have done my job, you will see more than you saw before. Instead of a featureless extent of water, you should see (at least in your mind's eye) an immense life support system with serious intrinsic problems. The clear blue water of the tropics, which previously triggered thoughts of surfing vacations and pastel coral reefs, now brings to mind the biological pump that accounts for that water clarity, causing you to ponder the lack of primary production in ocean gyres. Similarly, ocean breezes should feel different, raising thoughts of turbulent mixing, Ekman spirals, and upwelling. If you see a fishing boat go by, you might wonder what trophic level it is sailing out to exploit.

As you contemplate the ocean, try to review the information you have gained in the last 11 chapters. Two things should happen. First, the more you think about the central concepts presented here, the more sense they should make. As with any new knowledge, what you have learned about how the ocean works will take time and rumination to really soak in. Second, you should quickly bump up against the limits of what you know. This is, after all, just an introduction to oceanography. There is much left to learn, and as you absorb the basic principles of oceanography, new questions will occur to you. Here are a few ideas to get you started.

Benthic Organisms

What about those benthic animals we ignored in chapter 5? You may recall that we justified their neglect by noting that they occupy only a small fraction of the marine habitat. Is that a valid criterion? Probably not. The euphotic zone accounts for only about 2.5% of the marine world, but we spent much of this text exploring it in detail. Can't benthic organisms be similarly important despite their relatively small habitat? Certainly, benthic organisms play a large role in the process of sedimentation. Once an organic particle lands on the seafloor, it does not inevitably become a permanent part of the sediment. Burrowing worms and clams may stir it into or out of the mud, or it may be eaten by some scavenging organism, leading eventually to the reintroduction of its carbon and nutrients into the water column when the scavenger dies. These and similar processes help to determine what fraction of carbon exported from the surface is sequestered for the long term in sediments, and what fraction stays in the water column where it can be transported back to the surface. Benthic organisms also play important roles in ocean shallows. Coral reefs, for instance, support some of the most diverse assemblages of plants and animals on earth, acting as oases of life in many otherwise depauperate tropical seas. How do they do that?

Phosphorus

In our discussion of the two-layered ocean, we have focused on nitrogenous nutrients (nitrate in particular) as the typical limiting factor in primary production. Even allowing for complexities such as the regional lack of iron, the availability of fixed nitrogen forms the linchpin of our understanding of carbon and energy flow in the ocean. This view may well be shortsighted. Although fixed nitrogen is currently available to the oceans in limiting amounts, its availability is subject to biological control. Organisms can make more of the stuff! There is a ready supply of elemental nitrogen in the atmosphere, and cyanobacteria can convert it into useable nutrients. It is therefore possible that, at some time in the future, the current balance between nitrogen-fixing and denitrifying bacteria will shift, and fixed nitrogen will be available in abundance.

In that case, phosphorus (typically in the form of phosphate) is poised to become the limiting factor regulating primary production. Unlike the influx of nitrogen, which is under biological control, the only major influx of phosphate to the ocean is from rivers, a source that organisms cannot affect. Thus, the ocean's content of phosphorus is not subject to biological control, and as a result, phosphorus may be the ultimate limiting factor in primary production. How would the ocean be different if phosphate, rather than nitrate, controlled productivity?

Day Length

A day is 24 hours long, a fact so engrained in our existence that it has not been necessary to discuss the matter here. When we noted that phytoplankton photosynthesize during the day and rely on respiration at night, we simply assumed that this cycle is repeated every 24 hours. In fact, the length of earth's day hasn't always been 24 hours: it used to be much shorter.

The culprit is the tides. The rise and fall of the tides one observes locally is actually part of a global wave of water motion powered by a gravitational tug of war between the earth, the moon, and the sun. If earth were not rotating, one crest of this wave (a tidal bulge) would be on the side of the globe facing the moon, and a second crest would be on the side of earth opposite the moon

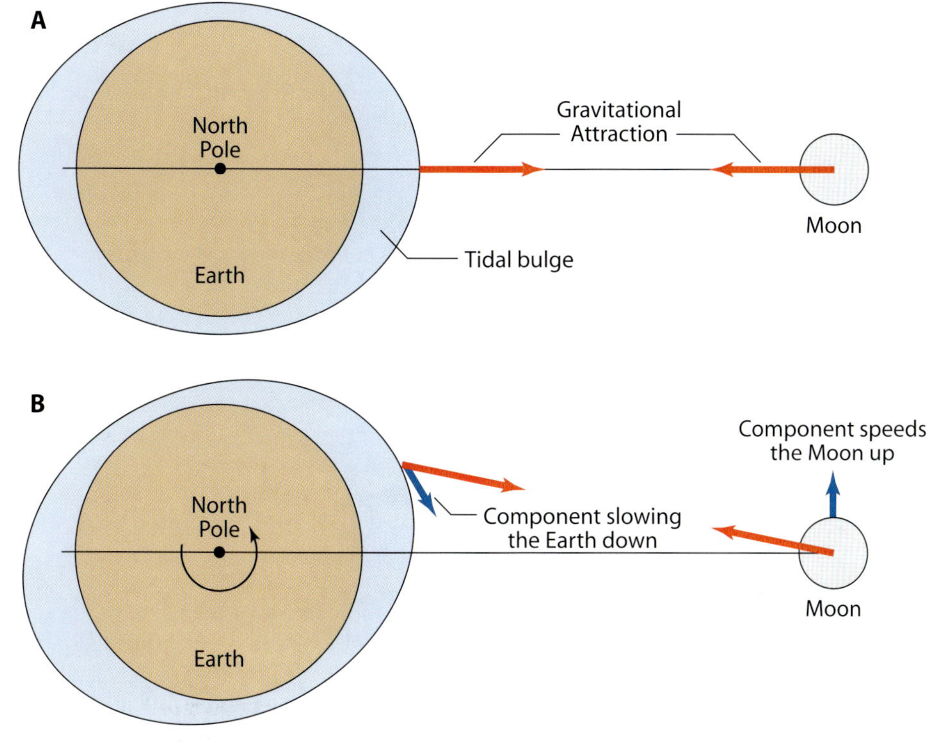

Figure 12.1. Tidal effects on day length. (A) If the earth were not rotating, the primary tidal bulges would lie directly in line with the moon. (B) Friction with the rotating earth causes the tidal bulges to advance, and the moon's gravitational pull on the bulge slows earth's rotation. At the same time, the gravitational pull of the bulge speeds the moon in its orbit.

(figure 12.1A). However, earth does rotate, and friction between the solid planet and ocean water drags the tidal wave slightly in the direction of earth's rotation (figure 12.1B).

This displacement of the tidal bulges is in turn resisted by a gravitational pull between the earth and the moon, with two results. First, a net force is applied to the moon, tending to speed it up in its orbital journey. As the moon speeds up, it retreats from earth, an effect measured by timing the reflection of laser beams off mirrors left on the moon's surface by the Apollo astronauts. The moon moves away from earth at a rate of about 3 centimeters per year.

Second, the pull of the moon on the tidal bulge applies a net force tending to slow earth's rotation. As rotation slows, day length increases. Or, looking back in time, we can deduce that in the past, a day was less than 24 hours.

Biology provides a means to measure this effect. Ancient corals produced calcareous skeletons, laying down a ring of calcium carbonate each day. The thickness of these rings varied annually, allowing paleontologists to count the number of days per year. Furthermore, the age of the fossil corals can be measured using radioactive decay. Combining these bits of information, biologists can measure the length of the day at times far in the past. For example, 400 million years ago there were about 420 days in a year, so each day was only about 21 hours long. It is interesting to contemplate how life would have been different in ancient seas with this shorter day.

Continental Drift

In chapter 2, we briefly explored the topic of continental drift, noting that 200 to 225 million years ago, Pangaea split into two: *Laurasia* in the northern hemisphere and *Gondwanaland* in the southern hemisphere, separated by the shallow *Tethys Sea* along the equator. Subsequent splitting, travel, and amalgamation by the continents rearranged the shape of earth's oceans, and their distribution relative to the poles. How did these changes affect life in the sea? One need not look into deep history to encounter potential effects of continental drift. As recently as 3 million years ago, there was a connection between the Pacific and Atlantic Oceans where the Isthmus of Panama is today. In what ways did the rise of the isthmus, and the subsequent severing of this connection, affect ocean currents and global climate?

Trapping Light Energy without Chlorophyll

As described in chapter 4, photosynthesis requires chlorophyll to capture light from the sun and direct it into the biochemical machinery that stores energy in carbohydrates. Recently, microbiologists have used genetic techniques to locate and describe another molecule, proteorhodopsin, that can convert light energy into a form useful to living cells. Proteorhodopsin is a protein found in the cell membrane of a number of marine bacteria and archaea. It uses light energy to pump hydrogen ions into the cell, and the resulting H^+ concentration provides a means for the cell to produce ATP, the standard biochemical currency of energy. It is currently unclear whether proteorhodopsin might also be used in carbon fixation, but even if its use is confined to the production of ATP, it provides a hitherto unforeseen pathway by which marine organisms tap into the sun's energy. How will we have to revise the trophic stream of energy to include this new information? This is just one example of the many questions that remain open regarding the role of bacteria and archaea in the sea.

Braking Thermohaline Circulation

Rising temperatures threaten to melt sea ice in the Arctic Ocean, and the ice sheet on Greenland already shows signs of unprecedented melting. As these ice reserves melt, they release large amounts of freshwater into the surface layer of the sea, which could have drastic effects. Recall that thermohaline circulation is driven in part by the production of high-density brine in the Arctic. If, due to the melting of ice, surface water is too fresh to attain sufficiently high density when cooled, the production of North Atlantic Deep Water will be curtailed, which in turn will reduce the rate at which warm surface water flows from the tropics into the North Atlantic. Without this input of warm water, Northern Europe will be substantially colder than it is now.

Currently, the potential effect of global warming on thermohaline circulation is being explored primarily in computer simulations, but through time more empirical evidence will be gathered. Will that information support or contradict the dire prediction of colder winters in Europe? It will be interesting to watch the story as it unfolds.

A Final Thought

Actually, you could do more than watch the action from the sidelines. Oceanography is a rapidly expanding field, and there is plenty of room for new ideas, new perspectives, and new people. Use this text as a starting point to learn more about the sea and its life. Follow your curiosity about the ocean. Who knows where it will lead you.

Index